AF334486

The Science of Heterogeneous Polymers

The Science of Heterogeneous Polymers

Structure and Thermophysical Properties

V. P. PRIVALKO
Academy of Sciences of the Ukraine, Kiev, Ukraine

and

V. V. NOVIKOV
Odessa Polytechnical Institute, Odessa, Ukraine

JOHN WILEY & SONS

Chichester · New York · Brisbane · Toronto · Singapore

Other Wiley Editorial Offices

John Wiley & Sons, Inc., 605 Third Avenue,
New York, NY 10158-0012, USA

Jacaranda Wiley Ltd, 33 Park Road, Milton,
Queensland 4064, Australia

John Wiley & Sons (Canada) Ltd, 22 Worcester Road,
Rexdale, Ontario M9W 1L1, Canada

John Wiley & Sons (SEA) Pte Ltd, 37 Jalan Pemimpin #05-04,
Block B, Union Industrial Building, Singapore 2057

Library of Congress Cataloging-in-Publication Data

Privalko, V. P.
   The science of heterogeneous polymers : structure and
thermophysical properties / V. P. Privalko and V. V. Novikov.
      p.   cm.
   Includes bibliographical references and index.
   ISBN 0-471-94167-0 :
   1. Polymers.   2. Inhomogeneous materials.   I. Novikov, V. V.
   II. Title.
QC173.4.P65P75   1995   94-19288
   530.4′ 13-dc20   CIP

British Library Cataloguing in Publication Data
A catalogue record for this book is available from the British Library

ISBN 0 471 94167 0

Typeset in 10/12pt Times by Thomson Press (India) Ltd. New Delhi
Printed and bound in Great Britain by Biddles Ltd, Guildford, Surrey

Preface

While the problem of structure–property relationships for polymers has been the subject of several excellent books, so far the effect of heterogeneity apparently has not been treated in a systematic manner. The notion of heterogeneity depends, of course, on the criterion involved; for example, any single polymer chain may be considered either heterogeneous or homogeneous on the scales much below or much above the size of the chain-repeating units, respectively. At the other extreme, any binary polymer system always will be heterogeneous on the molecular micro scale as soon as the two components are chemically different, although on the supermolecular, macro scale it may be either a homogeneous, single-phase system (for compatible components) or a heterogeneous, two-phase one (when the components are incompatible). It is the latter thermodynamic criterion of heterogeneity that will be used throughout this book to identify the phase state of polymers.

In this context, filled polymers, blends of incompatible polymers and phase-separated block copolymers may be regarded as structurally similar systems since all are characterized by the presence of two different phases separated by an interface, one phase (sometimes both) being continuous (this is usually referred to as the 'matrix phase'), and the other existing as discrete inclusions within the former. In view of such structural similarity, qualification of each of the above systems as belonging to the same family of 'microheterogeneous materials' (MHM) seems justified.

To date, the composition dependence of the different (say, thermoelastic) properties of MHM was traditionally discussed within the framework of one of the whole series of theoretical approaches, all of which assumed that any 'effective' property of MHM (P) depended not only on the 'partial' properties of each individual component (P_1 and P_2) and volume content of a disperse phase (φ), but also on the phase morphology of a particular binary system. Implicit in all these approaches (which subsequently will be referred to as 'pragmatic') was the concept of a sharp, 'mathematical' (i.e. infinitely thin) interface between two components. However, in the course of a continuous accumulation of experimental data the invalidity of that concept became more and more evident. In fact, one had to recognize that for all kinds of MHM the interactions between the components invariably resulted in a smearing-out of the 'mathematical' interface into a 'physical' interfacial region (sometimes of quite respectable thickness). Indirectly, this effect manifested itself, for example, as the dependence of the effective property of a filled polymer on the filler particle size (keeping the filler loading constant). Evidently, any theoretical approach claiming validity (referred to as the 'physical' approach) had to account explicitly for the contribution from such a 'boundary interphase' (BI) with changed properties.

These considerations motivated the partitioning of this book into two parts. Part I is a review of the most important aspects of formation and structural characterization of a BI in different MHM. Part II is devoted to an analysis of the applicability of current 'pragmatic' and 'physical' approaches to a theoretical description of the composition-dependent properties of MHM. The field is so vast that reference has been made only to the papers that the authors considered most representative; it is still hoped, nevertheless, that the major contributions have not been overlooked, so that the reader will be provided with an adequate presentation of both the experimental evidence for the existence of a BI in polymer MHM of whatever composition, and the theoretical tool to account for the contribution of a BI to the thermal transport and thermoelastic parameters of such materials.

The general approaches to the issues treated in this book undoubtedly reflect the strong influence of Professors Yu. S. Lipatov and G. N. Dulnev, with whom the authors had the privilege to collaborate over many years. Much of the original data referred to in this book were provided by Drs N. L. Rymarenko, Yu. D. Besklubenko, S. S. Demchenko, T. Ya. Shimchuk, G. V. Titov, K. D. Petrenko, V. V. Korskanov, G. I. Khmelenko, V. P. Azarenkov, A. P. Arbuzova, A. V. Shmorgun, V. V. Vorona and V. I. Shtompel, as well as by A. A. Usenko, R. L. Shapoval and A. V. Baibak.

The decision to start this project was catalyzed by Professor M. Lewin of the School of Applied Sciences and Technology, The Hebrew University of Jerusalem, whose attention and encouragement are gratefully acknowledged. The final revision of the manuscript was carried out while the senior author (V.P.P.) served as the Lady Davis Visiting Professor at the Department of Chemical Engineering, Technion-Israel Institute of Technology. Thanks are due to Professors M. Narkis and D. Hasson of that Department, as well as to Professor A. Siegmann of the Department of Materials Engineering for their hospitality. The permission of Dr S. Kenig to use the computer facilities of the Israel Plastics and Rubber Center is also acknowledged with gratitude.

Our final credit is to Dr Eleonora Privalko for her unfailing patience and good humor in the onerous task of correcting misprints and overall editing of the manuscript.

Contents

Notations

A	energy density of elastic deformation at the interface
B	parameter of the Vogel–Tamman equation; interaction energy density for binary blends
b_0	chain thickness
C	number of external degrees of freedom per macromolecule
C_c	heat capacity of a crystal
C_g	heat capacity of a glass
C_{ijkl}	effective moduli of elasticity
C_l	heat capacity of a liquid
C_p	isobaric heat capacity
D	diffusion coefficient
d	particle size; thickness of a polymer layer; space dimensionality
E	longitudinal elasticity modulus (Young's modulus)
F_i	initial molar ratio of monomers in the reaction mixture
F_1	molar fraction of component 1 in a copolymer chain
f	free volume fraction; cross-link functionality
f_g	free volume fraction at T_g
$f_i(s)$	force acting on the surface s
f_{ij}	relative content of an ij dyad
G	Gibbs free energy; shear modulus
G_B	bulk crystallization rate
G_c	Gibbs free energy of a crystal
G_{ex}	excess Gibbs free energy
G_l	Gibbs free energy of a liquid
G_R	radial growth rate
$g(\varphi)$	free energy density of a homogeneous binary system
H	enthalpy
H_{ex}	excess enthalpy
H_{ex}^g	excess enthalpy at T_g
$\langle h \rangle$	mean end-to-end distance of a macromolecular coil
$\langle h_0 \rangle$	mean end-to-end distance of an unperturbed macromolecular coil
$\langle h_f \rangle$	mean end-to-end distance of a model freely rotating chain
J_T	contribution of thermal density fluctuations to the intensity of small angle X-ray scattering
J_t	true intensity of small-angle X-ray scattering
j	structure-sensitive property
K	bulk modulus

K_a	molecular packing coefficient in the amorphous state
K_c	molecular packing coefficient in the crystalline state
K_g	molecular packing coefficient in the glassy state
K_n	crystallization rate constant
k	Boltzmann's constant
$\langle L \rangle$	mean thickness of polymer interlayer between filler particles
$\langle L^* \rangle$	'critical' thickness of polymer interlayer between filler particles
l	thickness of a polymer crystal
l_b	length of main chain bond
l_D	characteristic Debye distance of intermolecular interactions
M	molar mass; mobility factor
$\langle M_c \rangle$	mean molecular mass for the onset of entanglements
$\langle M_n \rangle$	number-average molecular mass
$\langle M_w \rangle$	weight-average molecular mass
m	parameter of the nucleation regime
N	number of polymer molecules; concentration of nuclei
N_A	Avogadro number
N_c	coordination number of a Voronoi polyhedron
n	shape parameter of a growing crystal
$\boldsymbol{n}$	vector-normal
n_h	number of holes
n_p	number of polymer molecules
P	pressure; partial property
$\tilde{p}$	reduced pressure
P^*	characteristic pressure
P_g	glass transition pressure
P_i	internal pressure
P_{ij}	partial property of an ij dyad
$\boldsymbol{P}_i$	vector of polarization
p	number of main-chain bonds; degree of polymerization; probability of linking an arbitrarily chosen chain unit to an identical unit in a copolymer chain
Q	small-angle X-ray scattering invariant; distribution index in copolymers
$\boldsymbol{q}$	heat flux
q^i	rate of structural rearrangements
q^+	heating rate
q^-	cooling rate
$\boldsymbol{q}_n(s)$	normal heat flux to the surface s
R	gas constant
$\langle R_g \rangle$	mean gyration radius of a macromolecular coil
r	radius of meniscus curvature; particle size
$\boldsymbol{r}$	vector-radius
S	entropy; scattering vector

S_{comb}	combinatorial entropy
S_{ex}	excess entropy
S_{ijkl}	effective moduli of compliance
S_{R}	surface area of a growing crystal
S/V	surface-to-volume ratio
s	specific surface area of a filler
T	absolute temperature
$\tilde{T}$	reduced temperature
T_{a}	annealing temperature
T^*	characteristic temperature
T_{c}	crystallization temperature
T_{cr}	critical temperature
T_{f}	fictive temperature
T_{g}	glass transition temperature
T_{m}	melting point
T_{m}^0	'equilibrium' melting point; melting point of a pure component
T_0	temperature of zero free volume of a liquid
T_2	temperature of zero entropy of a liquid
t	time; critical conductivity index
t_{a}	annealing time
$t(s)$	temperature on the surface s
U	internal energy
$U_i(s)$	displacement of the surface s
V	volume; molar volume
$\tilde{V}$	reduced volume
V^*	characteristic volume
V_{d}	molar volume of a diluent
V_{M}	molar volume of a binary system
V_{p}	molar volume of a polymer
v	specific volume
v_{c}	specific volume of a crystal
v_{ex}	excess volume of mixing
v_{f}	free volume; specific volume of a filler
v_{g}	specific volume of a glass
v_{h}	hole volume
v_{l}	specific volume of a liquid
v_{p}	specific volume of a polymer
v_{s}	specific volume of a solid
v_{∞}	specific volume of a liquid at the densest packing
v^{O}	occupied volume
w	weight fraction
w^*	'critical' filler content
X	non-linearity parameter; degree of crystallinity
y	fraction of occupied lattice sites

Z	order parameter
z_{ij}	conditional probability of formation of an ij dyad
Z_g	order parameter at T_g
Z_m	parameter of the energy barrier to crystal nucleation
z	coordination number of a fluid lattice
α	thermal expansion coefficient; degree of transformation; coefficient of swelling of a macromolecular coil
α_{Bl}	thermal expansion coefficient of boundary interphase
α_f	thermal expansion coefficient of a filler; thermal expansion coefficient of a free volume
α_g	thermal expansion coefficient of a glass
α_h	thermal expansion coeffieint due to holes
α_{ij}	tensor of thermal expansion coefficients
α_l	thermal expansion coefficient of a liquid
α_m	maximum chain extension factor
α_p	thermal expansion coefficient of a polymer
α_{sep}	degree of phase separation
β	coefficient of isothermal compressibility; non-exponentiality parameter; fluctuation wavenumber; critical index for the density of an infinite cluster
β_h	coefficient of isothermal compressibility due to holes
β_{cr}	critical fluctuation wavenumber
γ	thickening rate parameter for polymer crystals
γ_l	surface tension of a liquid
$\Delta\alpha$	increment of the thermal expansion coefficient at T_g
$\Delta\beta$	increment of compressibility at T_g
ΔC_p	increment of heat capacity at T_g
ΔE	activation energy of segmental transport
ΔH_{conf}	conformational contribution to the heat of solution
ΔH_d	heat of dissolution
ΔH_f	heat of wetting of a filler
ΔH_g	excess enthalpy of a glass
ΔH_l	heat of solution of a liquid
ΔH_m	melting enthalpy of a semi-crystalline polymer
ΔH_m^0	true enthalpy of crystal melting
ΔH_{pf}	heat of polymer/filler interaction
ΔH_r	contribution to heat of solution from regular solution theory
ΔH_s	heat of solution
ΔH_{sp}	energy gain due to specific interactions
ΔH_v	contribution of the excess volume to heat of solution
Δh	activation enthalpy
ΔG_s	excess surface free energy of a growing crystal
ΔG_v	bulk crystallization driving force (difference between Gibbs free energies of amorphous and crystalline phases)

Δl	interparticle gap
ΔP	capillary pressure
Δr	thickness of boundary interphase
ΔS	excess entropy
ΔS_{dis}	disorientation entropy
ΔS_{el}	elasticity entropy difference
ΔS_g	excess entropy of a glass
ΔS_m	entropy of crystal melting
ΔS_p	placement entropy difference
ΔS_v	restricted volume entropy difference
ΔT	degree of undercooling
Δv_m	volume change on crystal melting
$\langle \Delta \rho^2 \rangle$	experimental value of the mean square of electron density fluctuations
$\langle \Delta \rho^2 \rangle_0$	theoretical value of the mean square of electron density fluctuations
δ_p	solubility parameter of a polymer
δ_s	solubility parameter of a solvent
ε^*	characteristic energy
ε_h	hole energy
$\langle \varepsilon_{ij} \rangle$	strain tensor
ζ	critical index for the contour length of an infinite cluster
ζ^*	longitudinal dimension of a critical nucleus
ξ	correlation length
η	shear viscosity
η_g	shear viscosity at T_g
η_g^v	volume viscosity at T_g
θ	reduced undercooling
λ	relative elongation; coefficient of heat conductivity; radiation wavelength
λ_i	conductivity of the ith component
λ_{max}	characteristic fluctuation wavelength
$\lambda_{\parallel}$	conductivity along the strata
$\lambda_{\perp}$	conductivity transverse to the strata
μ_i	chemical potential of the ith phase
v	critical index for a correlation length
v_i	Poisson's coefficient of the ith component
v_{BI}	volume content of boundary interphase
$v(\varphi_0)$	number of elastically active chains in a network
π_{ex}	excess internal pressure
ρ	density; resistivity
ρ^*	characteristic density
χ	parameter of thermodynamic interactions; correlation length
χ_{cr}	'critical' parameter of thermodynamic interactions
χ_{sp}	parameter of thermodynamic interactions at the spinodal
σ	stiffness parameter of an unperturbed macromolecular coil; excess

	free energy at the interface, melt/lateral side of a crystal nucleus; dispersion of the smearing function
σ_e	excess free energy at the interface, melt/end (i.e. fold containing) side of a crystal nucleus
$\langle \sigma_{ij} \rangle$	stress tensor
σ_{ls}	surface free energy at the liquid/solid interface
σ_s	surface free energy of a solid
τ	relaxation time; critical index for elasticity moduli
τ_g	relaxation time at T_g
φ	volume fraction
φ^*	limiting filler fraction
φ_c	percolation threshold
φ_M	filler fraction at the highest packing density
$\langle \nabla t \rangle$	mean temperature gradient
$\langle \nabla \varphi \rangle$	mean concentration gradient
$\{ \cdots \}_i$	operator of averaging by coordinate i

Abbreviations

AAPB	amorphous/amorphous polymer blend
ACP	alternating copolymer
AMA	alkyl methacrylate
AN	acrylonitrile
AP	aromatic polyamide
AR	arylate
AS	alkylstyrene
B	butadiene
B1	butene-1
BCP	block copolymer
BI	boundary interphase
BRE	bulk representative element
C	bis-phenol A carbonate
CAPB	crystallizable/amorphous polymer blend
CCPB	crystallizable/crystallizable polymer blend
Cl	cellulose
CL	caprolactone
CPVC	chlorinated poly(vinyl chloride)
CU	carbonate urethane
DCE	differential cylindrical element
DMS	dimethylsiloxane
DSC	differential scanning calorimetry
E	ethylene
EE	equivalent element
EMA	effective medium approximation
EME	effective modulus of elasticity
EO	ethylene oxide
FHLM	Flory–Huggins lattice model
GP	glass powder
H1	hexene-1
HE	hydroxy ether
HM	hole model
InC	infinite cluster
IPN	interpenetrating polymer network
IsC	isolated cluster
IVC	iodine-substituted vinyl carbazole
LCST	lower critical solution temperature

LFM	lattice fluid model
MHM	microheterogeneous material
MMA	methyl methacrylate
MS	methylstyrene
NMR	nuclear magnetic resonance
O	olefin
OE	oligoester
OMA	octyl methacrylate
PB	polybutene-1
PC	polycarbonate
PCL	polycaprolactone
PCU-c	cross-linked poly(carbonate urethane)
PD	packing density
PE	polyethylene
PEO	poly(ethylene oxide)
PiPMA	poly(isopropyl methacrylate)
PMMA	poly(methyl methacrylate)
PMMA-c	cross-linked poly(methyl methacrylate)
PP	polypropylene
PPO	poly(2,6-dimethyl-1,4-phenylene ether)
PS	polystyrene
PSSO	phenyl-sil-sesquioxane
PU	polyurethane
PVAc	poly(vinyl acetate)
PVC	poly(vinyl chloride)
PVDF	poly(vinylidene fluoride)
PVME	poly(vinyl methyl ether)
P2VP	poly(2-vinyl pyridine)
PVT	pressure–volume–temperature
RCP	random copolymer
RSRG	real-space renormalization group
S	styrene
SAN	styrene-acrylonitrile copolymer
SAXS	small-angle X-ray scattering
SP	silica powder
SPU	segmented polyurethane
SSA	step-by-step averaging
TC	transcrystalline
TEC	thermal expansion coefficient
UCST	upper critical solution temperature
VA	vinyl acetate
VC	vinyl carbazole

Part I

STRUCTURAL CHARACTERIZATION OF HETEROGENEOUS POLYMERS

1 Filled Polymers

1.1 THERMODYNAMICS OF THE MELT STATE

Perhaps the simplest model of a filled polymer in the melt state is that of a suspension of disperse solid particles in a continuous, structureless fluid phase. Even in the absence of any kind of 'specific' interactions at the interfacial boundary, the perturbation of the fluid structure in the immediate vicinity of a solid surface may still be expected. It follows from purely geometrical considerations that either the coordination number of fluid molecules z or the fraction of unoccupied lattice sites φ_h (and hence the reduced density) will deviate more from those for the bulk state, the shorter is the distance to the solid wall [1–3]. The contribution of this effect to the combinatorial entropy of mixing n_h empty sites and n_p polymer molecules each occupying p lattice sites may be estimated with the aid of familiar Flory–Huggins approximation [4, 5]:

$$S_{comb} = -k\{n_h \ln \varphi_h + n_p \ln(1 - \varphi_h) - n_p(p-1)\ln([(z-1)/e]\} \tag{1.1}$$

where k is Boltzmann's constant. Turning now to the experimentally measurable quantity, internal pressure $P_i = T(\partial U/\partial V)_T = (\partial S_{comb}/\partial V)_T$ (where $V = (n_h + pn_p)v_O/N_A$ is the molar volume, v_O and N_A are the volume of a lattice site and Avogadro number, resp.) and noting that $(\partial S_{comb}/\partial V)_T = (\partial S_{comb}/\partial n_h)_T(\partial n_h/\partial V)_T = (N_A/v_O)(\partial S_{comb}/\partial n_h)_T$, we obtain, after rearrangement (in the infinite chain length limit, $p \to \infty$) [6]

$$P_i \cong -(RT/v_O)(1 + \ln \varphi_h) \tag{1.2}$$

Thus, we may predict, qualitatively, that the internal pressure in the melt state of a continuous polymer phase in a filled polymer may either increase in the case of densification (i.e. when φ_h decreases) or decrease in the opposite case.

In light of these arguments, the apparent experimental evidence for the decreased internal pressure of poly(vinyl acetate) filled with titanium dioxide [7] may be attributed to a looser packing of polymer segments near the filler surface. However, the values of P_i estimated by treatment of experimental pressure–volume–temperature (PVT) data for several filled polymers (polyethylene/graphite, Nylon 6/graphite, polypropylene/silica) in the melt state with the aid of a modified Van der Waals equation [8], turned out higher compared with pure polymers, the values of P_i apparently correlating with the filler specific area. The origin of this discrepancy remains obscure; perhaps, the precision of the latter data was not too high because of the rather narrow pressure and temperature intervals of the measurements involved. More extensive data were obtained in the *PVT* studies of polystyrene (PS) and poly(methyl methacrylate) (PMMA) melt blended with untreated glass powder (GP), calcium oxide (CaO) and sodium chloride (NaCl) [9–13]. Mean particle dimensions, d, for all fillers were similar (about $2\,\mu m$), while the surface energies were assumed to increase in the order, $NaCl > GP > CaO$. These data will be analyzed below in more detail.

Inspection of the representative VT plots for PS/Gp at normal pressure (Fig. 1.1) revealed the following characteristic features.

(i) Both specific volume in the melt state (i.e. about the glass transition temperature $T_g = 365\,K$), v_1, and its temperature coefficient (i.e. specific expansivity), dv_1/dT, as expected, decrease with the filler weight fraction, w.

(ii) While the values of v_1 for the 'initial' (i.e. molded below a pressure of about $100\,MPa$) filled samples in the first heating run tend to 'overshoot' those in the second run in the temperature interval from T_g to $423\,K$, at higher temperatures this overshoot is steadily decreasing.

(iii) In the second heating run of samples with high filler loadings (i.e. at $w > 0.2$) we observe a seemingly continuous transition to a smaller slope, dv_1/dT, above about $433\,K$.

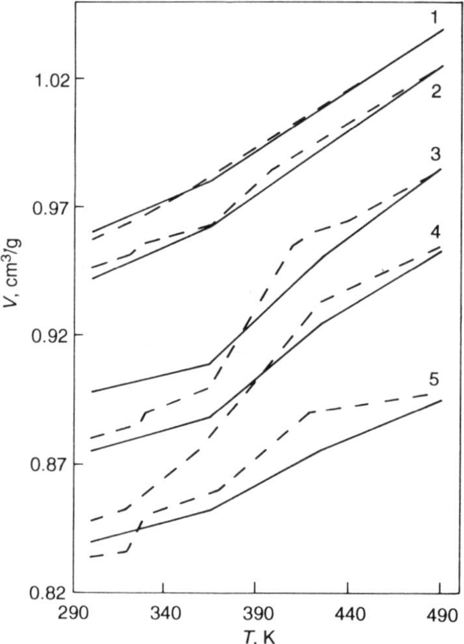

Fig. 1.1. Specific volume vs. temperature plots for PS/GP at normal pressure in the first (broken lines) and second (solid lines) heating runs. Filler content (from up downwards): 0; 0.05; 0.20; 0.30 and 0.50 (the data for the latter were shifted upwards by $0.1\,cm^3/g$). For clarity, similar plots for samples with $w = 0.01$ and 0.05 are not shown

Observation (ii) may be regarded as a sort of 'memory effect' of filled samples to previous thermal and mechanical history which could not be immediately erased by heating above T_g due to slowing down of the relaxation processes. On the other hand, observation (iii) for presumably completely relaxed samples has apparently a thermodynamic origin. Finally, observation (i) requires closer scrutiny.

As can be seen from Fig. 1.2, at $w = 0.01$ the specific volume is below the line of additivity (broken line) while at higher w values positive deviations from additivity are observed. The latter effect in no way may be related to voids in loose, polymer-free aggregates of filler particles since extrapolation of the linear portion of the experimental plot at $w > 0.3$ to $w = 1.0$ yields $v_s = 0.410$ ec/g, which is the specific volume of a monolithic glass.

These data suggest that the polymer component in highly filled samples of PS is in the expanded, loosely packed state. A roughly similar conclusion also proved valid for all other filled systems. Contrary to intuition, however, it turned out that for highly filled samples the 'partial' (i.e. calculated per pure polymer) compressibility of polymer melts, β_1, does not correlate with the corresponding 'partial' specific volume, v_1. It that may be suggested that the likely source of density deficit in those samples is an increase of the 'geometrical' fraction of the available free volume [14], while the 'configurational' (or 'kinetic') free

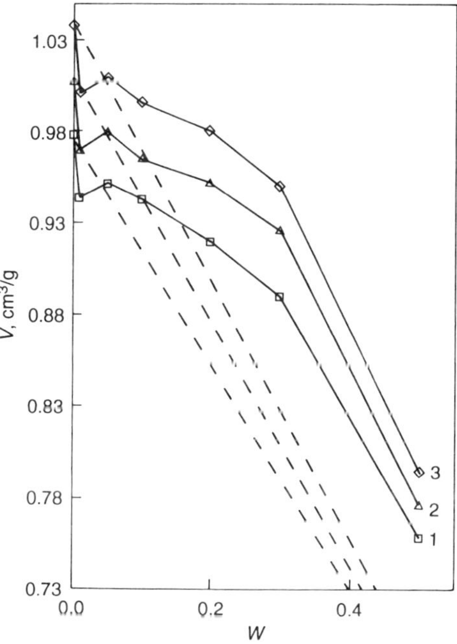

Fig. 1.2. Composition dependence of specific volume of PS/GP at 470 ($\diamond$), 425 ($\triangle$) and 365 K ($\square$), respectively. Linear additivity is shown as broken lines

volume fraction responsible for the thermodynamic properties of polymer melts is affected less.

This idea was tested further by treatment of the experimental PVT data in the melt state domain of PS and PMMA containing different fillers within the framework of the Frenkel–Hirai–Eyring hole model (HM) [15–17]. Implicit in this model is the assumption that the excess (with respect to a hypothetical densely packed state of volume v_∞) thermodynamic properties of a liquid depend on the relative population of holes (i.e. fraction of configurational free volume), each characterized by formation energy, ε_h, and molar volume, v_h. The relevant HM equation of state has the following form [18]:

$$v_l = v_\infty[1 + \sigma^{-1}\exp(-X)] \tag{1.3}$$

where $X = (\varepsilon_h + Pv_h)/kT$ and σ is essentially a fitting constant. Values of the characteristic parameters of the HM presented in Table 1.1 exhibit a rather wide variation depending on the nature of both polymer and filler, as well as on filler content. The significance of each single value is fairly difficult to assess; more reasonable information may be gained, however, from a theoretical analysis of the thermodynamic quantities.

A unique feature of the HM is the predicted S-shape of the equilibrium VT isobars with a rather extended quasi-linear portion around the inflexion point T' which may be calculated as [18]

$$T' = X/2k \tag{1.4}$$

As can be seen from a comparison of Figs 1.3 and 1.1, the observed linear temperature dependence of experimental v_l at normal pressure for pure PS fits exactly the quasi-linear portion of the theoretical isobar calculated by eq. (1.3) with data from Table 1.1. Lower values of ε_h, according to eqs (1.3) and (1.4), should result in a shift of the quasi-linear portions on the theoretical isobars to lower temperatures; hence, in the same temperature interval a gradual transition to the high-temperature portion of the S-shaped curve with smaller slope should be observed. Apparently, this provides a natural explanation (i.e. a decrease in ε_h) for observation (iii) with highly filled samples of PS.

In a similar fashion, we can rationalize the experimental values of thermal expansivities, α_l, compressibilities, β_l, and internal pressures, P_i, using the corresponding theoretical expressions [18]:

$$\alpha_h = (v_\infty/v_l\sigma T)X\exp(-X) \tag{1.5}$$

$$\beta_h = (v_\infty v_h/v_l kT)\exp(-X) \tag{1.6}$$

$$P_h = \varepsilon_h/v_h\sigma \tag{1.7}$$

together with the numerical values of the relevant parameters from Table 1.1. Unfortunately, no systematic correlations of the cited parameters with sample composition could be found; as a general trend, however, one may notice that β_h and P_h for filled PS are higher and lower, respectively, compared with pure

Table 1.1 Parameters of HM model

Parameter	Filler content, w									
	0	20			50			70		
	—	GP	CaO	NaCl	GP	CaO	NaCl	GP	CaO	NaCl
	Filled PS									
$v_\infty \times 10^3$ (cm^3/g)	890	910	850	875	820	850	790	910	990	800
v_h (cm^3/mol)	19.3	17.4	11.8	17.7	19.6	7.1	15.2	23.7	15.2	14.4
ε_h/k (K)	920	690	485	600	415	265	475	675	525	485
σ	0.85	1.01	1.57	1.75	0.95	1.61	1.26	1.61	1.52	1.14
G_{ex}/kT	—	0.2	0.8	2.6	1.0	−2.0	1.3	3.0	1.1	0.5
	Filled PMMA									
$v_\infty \times 10^3$ (cm^3/g)	785	795	760	—	790	—	750	700	905	720
v_h (cm^3/mol)	17.2	18.8	17.9	—	21.6	—	12.8	7.3	10.8	12.1
ε_h/k (K)	630	630	725	—	630	—	575	285	400	525
σ	1.49	1.46	1.10	—	1.46	—	1.24	1.11	2.06	0.92
G_{ex}/kT	—	0.2	−0.8	—	0.6	—	−1.6	−5.6	−1.0	−3.4

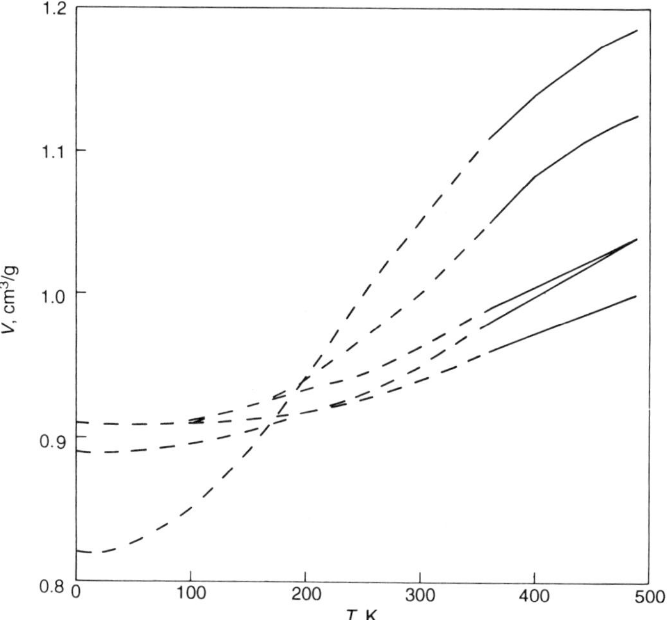

Fig. 1.3. Theoretical (broken lines) and experimental (solid lines) specific volume (per pure polymer) vs. temperature plots at normal pressure for PS/GP. Filler content (from up downwards): 0.5; 0.2; 0.1; 0.5; 0 and 0.01, respectively

polymer, whereas the reverse is true for PMMA, although the partial specific volumes of the polymer component in filled samples of both PS and PMMA were higher than those for each pure polymer. This discrepancy seems to indicate that the implicitly assumed model of a homogeneous, loosely packed polymer medium in filled systems should be revised.

Apparently, a more appropriate model should allow for the non-homogeneous packing density (PD) distribution in a polymer medium between filler particles. According to this model [13] (Fig. 1.4), polymer–filler interactions should lead to the formation of a layer with increased local PD in the immediate vicinity of the solid surface; however, the thickness of this densified surface layer, Δr_s, will be fairly small owing to a rapid decay of the interaction energy with distance [19]. On the other hand, when the distance to the solid is very high (say half the gap between the adjacent filler particles, $L = d[(\varphi_M/\varphi)^{1/3} - 1]$, where φ is the actual volume fraction of filler particles and φ_M is that at the highest random packing density), the local structure of the polymer will remain unperturbed; hence the PD will be that of a pure polymer.

At intermediate distances, two possible patterns of PD evolution may be envisioned. In the case of weak interfacial interactions the structural changes in the surface layer will also be small; hence the PD will decay smoothly to the

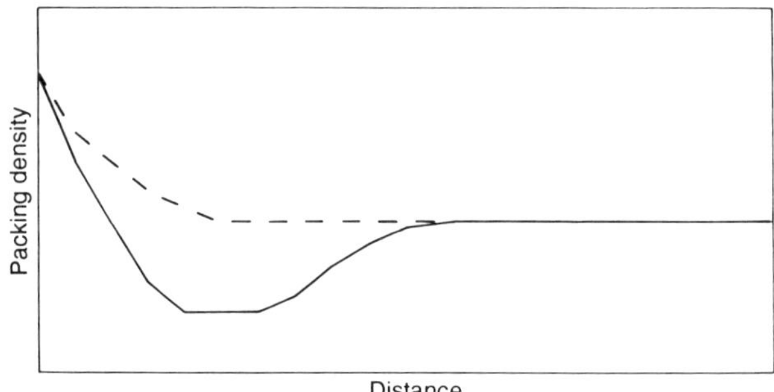

Fig. 1.4. Schematic of the packing density profile for strong (solid line) and weak (broken line) polymer–solid interactions

bulk value (broken line in Fig. 1.4). On the other hand, when interfacial interactions are strong, the structural misfit between the densified surface layer and the unperturbed phase will be so drastic that the PD will pass through a minimum (solid line in Fig. 1.4). It can be easily verified that this model allows for rather wide variations of the local PD across the interparticle gap, L, keeping the mean PD essentially unchanged.

These admittedly oversimplified considerations are, nevertheless, consistent with both indirect experimental PVT data referred to above and direct measurements of the PD in thin polymer films cast on solid substrates with different surface energies [20, 21].

The final basic problem concerns the effect of fillers on the thermodynamic state of polymer melts. The Gibbs free energy of any binary system may be defined as follows:

$$G = G_1 \varphi + G_2(1 - \varphi) + G_{ex} \tag{1.8}$$

where G_i is the free energy of the ith component, φ is the volume fraction of component 1 and G_{ex} is the excess free energy characterizing the deviation from additivity owing to interactions between the components. It is the sign of this latter quantity which allows us to judge the thermodynamic state of a binary system. In the case of filled polymers, for obvious reasons G_{ex} will be simply the difference between the values of G for a polymer component in filled samples and in the pure state.

In the framework of HM, the free energy of a liquid may be expressed as [9–13]

$$G/kT = -v_m/v_h\sigma \tag{1.9}$$

Once again the calculated values of G_{ex} (Table 1.1) exhibit a considerable scatter both in magnitude and sign; nevertheless, we can make the broad conclusion that G_{ex} is predominantly positive for filled PS and predominantly negative for filled PMMA. In terms of the structural model outlined above

(Fig. 1.4), the loosely packed structure of PS in filled samples is thermodynamically less stable compared with a pure polymer; on the other hand, alteration of densely packed and loosely packed microregions seems to impart higher thermodynamic stability to PMMA in filled samples.

It is pertinent to emphasize here that these conclusions were arrived at from a thermodynamic analysis based on HM, which takes into consideration only the 'configurational' (i.e. volume-dependent) properties of polymer melts and thus should be considered as semi-quantitative. A more complete understanding would have been obtained when other (e.g. conformational) contributions to the excess thermodynamic quantities of polymer melts were explicitly taken into consideration [22–24].

1.2 GLASS TRANSITION IN FILLED, NON-CRYSTALLINE POLYMERS

Amorphous polymers are notorious with regard to the difficulties in direct quantitative characterizion of their structure. These difficulties would seem even more enormous if we attempted to verify the model of structural changes in a polymeric component of a filled system (cf. Fig. 1.4). Under such circumstances it becomes imperative to specify a phenomenological parameter which would provide the possibility of quantitative characterization of the structural state of an amorphous polymer.

1.2.1 KINETIC ASPECTS

As already recognized many years ago [25–28], in the course of cooling of an equilibrium melt at a constant rate $q^- = \mathrm{d}T/\mathrm{d}t$, from some starting temperature T' located sufficiently above the glass transition temperature T_g, the rate of structural rearrangements within the melt, $q^i = \mathrm{d}Z/\mathrm{d}t$ [where $Z(T,P)$ is an unspecified structural parameter], is higher compared with the rate of external cooling q^-. This situation satisfies the criterion of internal equilibrium which may be expressed as equalization of the external temperature, T, on the one hand, and the internal (or 'fictive') temperature of the melt, T_f, on the other hand (Fig. 1.5). As the external temperature decreases further, however, q^i begins to lag behind q^-, so that the instantaneous structure of the melt, $Z(T_\mathrm{f}, P)$, will exhibit increasing deviations from the equilibrium one, $Z(T, P)$. In other words, at any external temperature $T'' < T_\mathrm{g}$ the internal structure frozen in the glassy sample, $Z(T_\mathrm{g}, P)$, would correspond to that, $Z(T_\mathrm{f}, P)$, which the equilibrium melt would have had at the fictive temperature, $T_\mathrm{f} > T_\mathrm{g}$.

Pursuing this line of reasoning, it is easy now to realize (Fig. 1.5) that the glassy state structure of the same liquid at the same temperature, $T'' < T_\mathrm{g}$, will be more remote from the equilibrium (i.e. the fictive temperature T_f of the glass

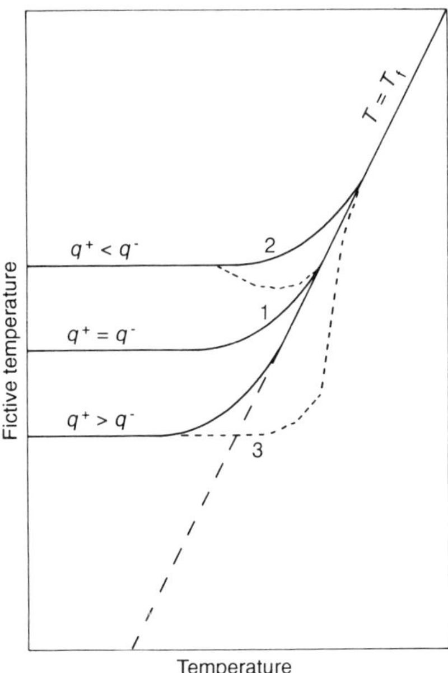

Fig. 1.5. Schematic of the evolution of the fictive temperature vs. temperature in the course of either cooling the glass-forming liquid at three different rates (solid lines) or subsequent heating of the glass at a single rate (dotted lines)

will be higher) the higher is the cooling rate of the equilibrium melt, q^-. On subsequent heating at a rate q^+, three possibilities may be envisioned (Fig. 1.5).

(i) $q^- = q^+$. Unfreezing of the glassy structure (i.e. evolution of T_f with T) proceeds via approximately the same route as in the course of freezing (curve 1).

(ii) $q^- > q^+$. The glassy structure will start unfreezing (i.e. T_f on heating will approach the equilibrium line $T_f = T$) at a somewhat lower temperature compared with T_f on cooling (curve 2).

(iii) $q^- < q^+$. The rate of structural rearrangement in the glass lags behind q^+; hence, the glass structure will be 'overheated' at first (i.e. the transient fictive temperature T_f will 'overshoot' above the equilibrium line $T_f = T$), after which an accelerated approach to the latter line will occur (curve 3).

Experimentally, the information on the pattern of structural evolution attending the melt after a jump-link change of one of thermodynamic variables (T, P), keeping the other constant, may be obtained by monitoring the relaxation of any structure-sensitive property j (say, specific volume, refraction index, enthalpy, etc.) Making use of an empirical observation that the relaxation rate is higher the greater is the deviation of property j in the transient state (j) from that in the new equilibrium state (j_∞), i.e.

$$dj/dt = -(j - j_\infty)/\tau \qquad (1.10)$$

we obtain after integration [29]

$$\delta = (j - j_\infty)/(j_0 - j_\infty) = \exp[-(t - t_0)/\tau] \tag{1.11}$$

where j_0 corresponds to the onset of relaxation at the moment t_0, and τ is the relaxation time which may be conveniently expressed as

$$\tau = \tau_0 \exp(-\Delta h/kT) \tag{1.12}$$

where τ_0 is the pre-exponential constant, and Δh is the relevant activation energy.

It turns out, however, that the curves $\delta = f(t)$ in the glass transition interval are not superimposable both in isothermal and/or isobaric conditions at sufficiently wide variations of the amplitudes of external perturbations [i.e. in a broad range of the difference $(j_0 - j_\infty)$], as well as in iso-δ conditions when the temperature or pressure jumps are also large [29, 30]. The intrinsic non-exponentiality of relaxation process apparent from the former observation may be empirically accounted for [31–33] by replacing the simple exponential equation, eq. (1.11), with a fractional-exponential equation, eq. (1.13)

$$\delta = \exp[-(t - t_0)/\tau]^\beta \tag{1.13}$$

where $0 < \beta < 1$ is a measure of the relaxation spectrum width [i.e. $\beta = 1$ corresponds to a single relaxation time as is implicit in eq. (1.11), whereas the other extreme, $\beta = 0$, corresponds to an 'infinitely broad' spectrum].

The latter observation is the manifestation of a structural memory of a glassy sample to its formation history (cf. Fig. 1.5); i.e. the probability of relaxation depends both on the external (T) as well as the internal (T_f) temperature frozen in below T_g. In this case, eq. (1.12) should be replaced by eq. (1.14) [34]

$$\tau = \tau_0 \exp\{(X\Delta h/kT) + [(1 - X)\Delta h/kT_f]\} \tag{1.14}$$

where $0 < X < 1$ is the measure on non-linearity of relaxation.

Now, the evolution of T_f during a sudden cooling of an equilibrium melt from T' to $T'' < T_g$ (or, equivalently, during a subsequent jump-like heating from T'' to T') may be expressed with the aid of an obvious linear relationship [35]

$$T_f(t) = T' + (T'' - T')(1 - \delta) \tag{1.15}$$

with δ and τ specified by eqs. (1.13) and (1.14), respectively. The validity of the latter equation is limited, however, to a rather small amplitude of temperature change, $\Delta T = T'' - T'$; to overcome this difficulty, the total interval, ΔT, will be assumed to consist of a series of increments (ΔT at time t_1, ΔT_2 at time $t_2, \ldots, \Delta T_n$ at time t_n) which are sufficiently small to ensure the applicability of the linearity condition. Thus, eq. (1.15) will be written as [35]

$$T_f(t) = T' + \Delta T_1(1 - \delta_1) + \Delta T_2(1 - \delta_2) + \cdots + \Delta T_n(1 - \delta_n)$$

$$= T' + \sum_{i=1}^{n} \Delta T_i(1 - \delta_i) \tag{1.16}$$

Heating or cooling at a constant rate $q^\pm = dT/dt$ may be considered as a series of incremental temperature jumps ΔT_i with subsequent isothermal storage during $dt = dT/q^\pm$. Thus, it is possible to go from the isothermal relaxation conditions over to a continuous heating and/or cooling regime, changing the summation in eq. (1.16) by integration. In this way the final expression has the following form [35–37]:

$$T_f(T) = T' + \int_{T_1}^{T} \left\{ 1 - \left[\exp - \left(\int_{T'}^{T} dT''/q\tau \right)^{\beta} \right] \right\} dT' \qquad (1.17)$$

where the increments dT' and dT'' are chosen depending on the required accuracy of calculations.

The parameters of eq. (1.17) determined from rate-heating and rate-cooling experiments with DSC for PS filled with GP [38] and with silica powder SP (aerosil A-175) [39] are listed in Table 1.2. As is clear from the data obtained, both fillers appear to hinder the thermal mobility of the polymer segments (Δh increases), to broaden the spectrum of relaxation times (β decreases) and to increase further the non-linearity of relaxation (X decreases), the more so the higher is the filler content.

The observed changes in the kinetic parameters of structural relaxation in the glass transition interval of filled PS are qualitatively consistent with the structural model discussed above (Fig. 1.4). In fact, the assumed modulated pattern of the polymer PD profile with distance from the filler surface suggests different relaxation kinetics in microregions with a different PD; thus, both apparent broadening of the relaxation times spectrum and more pronounced non-linearity of the relaxation process are quite natural consequences of this model. Moreover, the increase in Δh may be reconciled with the apparent loosening of the overall PD provided the chain segment mobility depends on the 'kinetic', rather than 'geometrical', free volume, as already pointed out above.

Table 1.2 Kinetic parameters of enthalpy relaxation at T_g of filled PS

Filler	w	$\Delta h/k \times 10^{-4}$ (K)	$-\ln \tau_0$ (s)	X	β
GP	0	8.2525	220.59	0.51 ± 0.02	0.50 ± 0.02
	0.01	8.0318	214.76	0.49 ± 0.02	0.49 ± 0.02
	0.20	8.4255	225.89	0.48 ± 0.02	0.45 ± 0.01
	0.50	9.9105	246.39	0.36 ± 0.02	0.43 ± 0.02
	0.70	9.8325	246.31	0.28 ± 0.01	0.38 ± 0.01
SP	0.10	9.0112	236.12	0.65 ± 0.05	0.45 ± 0.05
	0.20	9.3150	244.30	0.58 ± 0.05	0.40 ± 0.05
	0.30	10.2220	267.05	0.56 ± 0.05	0.38 ± 0.05

1.2.2 QUASI-EQUILIBRIUM ASPECTS

In so far as the properties of the glassy state depend on its formation history (see above), the melt–glass transition at T_g cannot be characterized in the familiar terms of thermodynamics of equilibrium phase transitions; in this case, the properties of a glass are associated with the phenomenological parameter $Z(T, P)$ frozen in below T_g in the course of either isobaric cooling $[Z_g(T_g, P)]$ or isothermal compression $[Z_g(T, P_g)]$. Quite a few experimentally measurable quantities were tested as Z_g for pure polymers; now, a similar analysis for filled polymers will be made.

Free Volume Fraction at T_g

By definition

$$f_g = v_f/v_g = 1 - v^O/v_g \tag{1.18}$$

where $v_f = v_g - v^O$ is the free volume, and v^O and v_g are the 'occupied' and real volumes of a liquid at T_g, respectively. Depending on the method of estimation of v^O, the following definitions of the free volume fraction may be identified [14].

(i) $v^O = v_l(0)$, where $v_l(0)$ is obtained by a linear extrapolation of the melt-specific volume to 0 K. Assuming $v^O = $ const. [i.e. $\mathrm{d}\ln v^O/\mathrm{d}T = 0$] we obtain

$$f'_g = \alpha_l T_g \tag{1.19}$$

where $\alpha_l = \mathrm{d}\ln v_l/\mathrm{d}T$ is the expansion coefficient of the melt. If, however, $\mathrm{d}\ln v^O/\mathrm{d}T \cong \mathrm{d}\ln v_g/\mathrm{d}T = \alpha_g$ (expansion coefficient of the glass), then

$$f''_g = \Delta\alpha T_g \tag{1.20}$$

where $\Delta\alpha = \alpha_l - \alpha_g$.

(ii) $v^O = v_\infty$ from HM (see above); then

$$f'''_g = 1 - v_\infty/v_g \tag{1.21}$$

(iii) $v^O = v_g[1 - \alpha_l(T_g - T_0)]$, where T_0 is the characteristic temperature of the Vogel–Tammann equation for liquid viscosity [29]

$$\eta = \eta_0 \exp[B/(T - T_0)] \tag{1.22}$$

η_0 is the pre-exponential constant and $B \cong \alpha_l^{-1}$. In this case

$$f''''_g = \alpha_l(T_g - T_0) \tag{1.23}$$

As could be expected, the values of f_g exhibit a considerable scatter depending on both the definition of v^O and the nature of the polymer and filler content (Table 1.3). Broadly speaking, f_g in all cases tends to increase with filler loading; therefore the possibility that a 'universal' value of the free volume fraction is frozen in at T_g [40, 41] should be ruled out both for pure [14, 29] and for filled [42] polymers.

Table 1.3 Free volume fraction at T_g after different definitions

System	f_g^i	Filler content, w					
		0	0.01	0.05	0.10	0.20	0.50
PS/GP	f_g'	0.237	0.211	0.194	—	0.239	0.259
	f_g''	0.115	0.093	0.091	—	0.152	0.151
	f_g'''	0.085	0.041	0.099	—	0.127	0.255
PMMA/GP	f_g'''	0.102	0.125	0.120	—	0.110	0.138
Oligoester/ SP	f_g''''	0.032	0.031	0.033	0.043	—	—

Excess entropy at T_g

The excess (with respect to the crystalline state) or 'configurational' entropy of a liquid may be defined as

$$\Delta S = \Delta S_m - \int_T^{T_m} \Delta C_p \, \mathrm{d}\ln T = \int_{T_2}^{T} \Delta C_p \, \mathrm{d}\ln T \qquad (1.24)$$

where T_m and ΔS_m are the crystal melting temperature and entropy, respectively; $\Delta C_p = C_1 - C_c \cong C_1 - C_g$; C_1, C_c and C_g are polymer heat capacities in the liquid, crystalline and glassy states, respectively.

According to eq. (1.24), $\Delta S \to 0$ as $T \to T_2$ (where T_2 is the hypothetical temperature of the second-order phase transition). Thus, the configurational entropy at T_g may be defined as

$$\Delta S_g = \Delta C_p \ln(T_g/T_2) \qquad (1.25)$$

Assuming that both terms on the right-hand side of eq. (1.25) are 'universal' constants (i.e. $\Delta C_p \cong 11.7 \, \mathrm{J/deg.mol}$ of main chain 'beads' [43] and $T_g/T_2 \cong 1.3$ [44]), it follows from eq. (1.25) that $\Delta S_g \cong 3.1 \, \mathrm{J/deg.mol}$ 'beads' should also be a universal constant for pure polymers [45]. Subsequent analysis of more extensive experimental data proved, however [46], that it is only ΔC_p that may be regarded as approximately constant, while the values of ΔS_g increase with T_g/T_2. Nevertheless, it is still possible that the 'partial' (i.e. per polymer) ΔS_g of a filled sample will be the same as in pure polymer, provided the eventual change in either of two terms of eq. (1.25) will be compensated by a concomitant reverse change of the second term (Fig. 1.6).

While the values of ΔC_p and T_g may be readily obtained by calorimetry, the situation with T_2 is not that simple. Making a seemingly reasonable assumption that T_2 from eq. (1.25) corresponds to T_0 from eq. (1.22) [47–49], it was possible to show by melt viscosity and heat capacity studies [50] that an increase in the ratio T_g/T_0 for oligoesters filled with 10% SP (due to a decrease in T_0 from

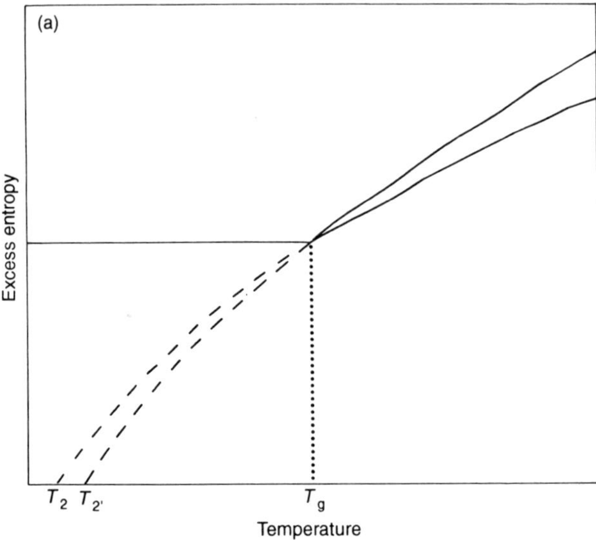

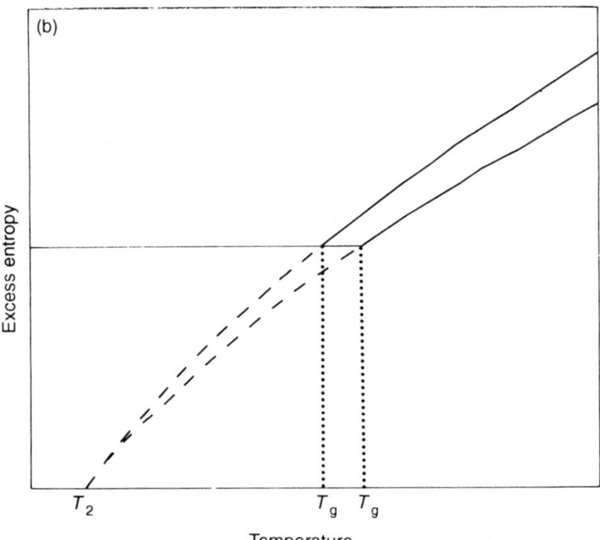

Fig. 1.6. Schematic of the excess entropy vs. temperature plots for unfilled (upper curves) and filled (lower curves) polymer melts illustrating the possibility of keeping $\Delta S_g = \text{const}$ by either shifting T_2 at $T_g = \text{const}$ (a) or shifting T_g at $T_2 = \text{const}$ (b)

175 K to 155 K at $T_g = 217$ K) was apparently compensated by an about 30% decrease in ΔC_p, so that ΔS_g remained approximately unchanged.

On the other hand, in case of PS/GP and PMMA/GP no measurable change in either T_g or ΔC_p in filled samples was detected [51], whereas the values of T_2 estimated from HM [9, 10] tended to decrease for the former system (hence, ΔS_g increased) but remained unchanged (as also did ΔS_g) for the latter. Once again, the notion of structural differences between polymer components in filled samples should be invoked to rationalize the different behavior of these two latter systems. It seems more appropriate, however, to conclude that the amount of experimental information available is insufficient to decide whether the excess entropy frozen at T_g is a 'universal' parameter for all polymer systems.

Viscosity at T_g

On cooling, the shear viscosity η of any non-crystallizable liquid is known to increase smoothly up to 10^{13}–10^{15} P in the glassy state [26–28]. This was the basis for one of the earliest empirical definitions of T_g as that temperature at which the shear viscosity of all glass-forming systems reaches a 'universal' value, $\eta_g = 10^{13}$ P [52]. So far, the maximum Newtonian viscosity of filled polymers in the glass transition interval, apparently has not been measured; hence, the applicability of the above criterion of T_g to filled systems has to be assessed from indirect data.

According to the limited experimental evidence available [53–55], in the range of fairly high temperatures ($T \gg T_g$) the shear viscosity of polymer melts increases with filler content, while the activation energy of viscous flow, ΔE, remains essentially unchanged. From these observations two alternative conclusions may be drawn.

(i) The criterion $\eta_g = 10^{13}$ P may apply to filled polymers in so far as it does not contradict the trend for an increase in T_g in filled polymers claimed for several systems [56–58]. In that case, however, T_g should have steeply increased with filler content, whereas in experiments, on the contrary, a levelling off is observed [56–58].

(ii) η_g increases with filler content (i.e. it is not longer a universal constant) if T_g is not appreciably affected by fillers.

This conclusion may be questioned, too, since ΔE measured in a sufficiently broad temperature interval is not constant, as implied, but, as is clear from eq. (1.22), inversely depends on temperature, i.e.

$$\Delta E = R[\mathrm{d} \ln \eta/\mathrm{d}(1/T)] = RB[T/(T - T_0)^2] \qquad (1.26)$$

Moreover, in highly filled polymers the eventual breakdown of filler aggregates under shearing stresses may also lead to ever-increasing deviations from the Newtonian flow pattern. Under such circumstances, experimental tests of the applicability of the 'isoviscosity' criterion to filled polymer melts must satisfy the following conditions:

(a) viscosity should be measured in conditions ensuring Newtonian flow;

(b) the temperature interval of the measurements must be sufficiently broad for a meaningful application of eq. (1.22).

These conditions were apparently met in experimental studies of oligoether and oligester filled with SP [50]. As can be seen from Table 1.4, for both oligomers the values of $\ln \eta_0$ and T_0 decrease with filler content w, while B increases. Given the apparently constant $T_g = 217$ K for all samples, the values of η_g for filled oligoester calculated from eq. (1.22) decreased with w (it proved impossible to measure T_g and thus to make a similar estimate for filled oligoether). It is pertinent to remark here, however, that the above result was deduced assuming the applicability of eq. (1.22) with the same set of parameters in the whole temperature interval of measurements. This assumption may be valid only for non-crystallizable liquids free from any structural rearrangements in the course of cooling [59, 60], whereas for both pure crystallizable oligomers two temperature intervals of non-Arrhenius flow separated by T_m, each obeying eq. (1.22) with a different set of parameters, were found [61]. It appears that the above conclusions, arrived at from extrapolation to T_g over the rather extended temperature interval $T > T_g$, are not unambiguous.

As already emphasized, the presence of fillers drastically limits the range of shear stresses in which polymer melts behave as Newtonian fluids. In view of the standard definition of viscosity as a measure of resistance to irreversible changes of shape (i.e. structure), it seemed preferable to measure the bulk viscosity, $\eta_g^v = \tau_g/\Delta\beta$ [29] (where τ_g is the relaxation time and $\Delta\beta$ is the compressibility jump at T_g), rather than the shear viscosity of filled polymers at T_g, since the probability of any structural changes under the action of three-dimensional, low-amplitude deformations should definitely be lower than in the case of one-dimensional shearing stresses.

As follows from studies of volume relaxation kinetics of PS/GP after sudden pressure jumps [62], all samples, regardless of filler content, had about the same volume viscosity, $\eta_g^v = (2-5) \times 10^{14}$ P, while the apparent activation energy tended to decrease with w. At first sight these data do not correlate with those

Table 1.4 Parameters of eq. (1.22) for oligomers filled
with SP

Oligomer	w	T_0 (K)	B (K)	$\ln \eta_0$	$\ln \eta_0$
Oligoester	0	175	1260	−4.76	25.2
	0.01	175	1280	−4.80	25.6
	0.05	165	1330	−4.11	21.5
	0.10	155	1420	−2.87	20.2
Oligoether	0	150	1380	−4.46	—
	0.01	150	1400	−4.54	—
	0.05	130	1490	−3.50	—
	0.10	120	1510	−2.27	—

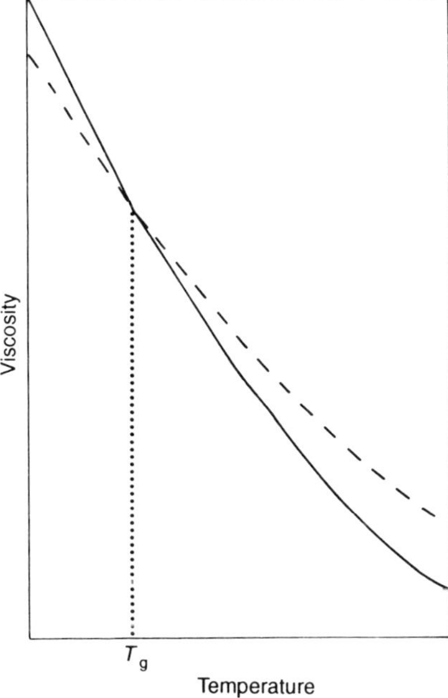

Fig. 1.7. Schematic of the viscosity vs. temperature plots for unfilled (solid line) and filled (broken line) polymer melts illustrating the possibility of $n_g = \mathrm{const}$

from shear viscosity studies of filled oligomers (see above); however, this inconsistency can be easily resolved by taking into consideration the temperature dependence of ΔE as embodied in eq. (1.26). In fact, assuming that the apparent decrease in ΔE observed for PS/GP is also caused by an increase in T_0, the volume viscosity at high temperatures ($T \gg T_g$) will be substantially higher for the filled sample, whereas the apparent ΔE will be similar (Fig. 1.7). As the temperature moves closer to T_g, however, similar values of η_g^v for pure and filled samples may become possible owing to the different temperature dependences of the respective ΔE.

Order Parameter at T_g

As documented above, the notion of freezing at the glass transition of a universal value of the appropriately chosen order parameter Z_g (e.g. free volume fraction or excess entropy) proved invalid both for pure and filled polymers. A further insight into this problem may be gained from an analysis of the pressure dependence of T_g.

According to theory [27], the case $Z_g(T_g, P) = Z_g(T, P_g)$ corresponds to the validity of Ehrenfest-like equations for equilibrium second-order phase transitions

$$dT_g/dP = \Delta\beta/\Delta\alpha \tag{1.27}$$

$$\Delta\beta/\Delta\alpha = T_g v_g \Delta\alpha/\Delta C_p \tag{1.28}$$

whereas in the opposite case, $Z_g(T_g, P) \neq Z_g(T, P_g)$, the right-hand sides of eqs. (1.27) and (1.28) should become larger than the corresponding left-hand sides.

It is the latter case that seems to hold for the majority of glass formers studied so far [63–65]. This means, in effect, that at least two, rather than a single, order parameters are involved in the glass transition event. As soon as the first equality, eq. (1.27), can be derived assuming a universal value of the free volume fraction frozen at T_g [29, 41], it follows that f_g is not a unique order parameter controlling the glass transition. On the other hand, the apparent validity of eq. (1.27a)

$$dT_g/dP = T_g v_g \Delta\alpha/\Delta C_p \tag{1.27a}$$

for pure polymers [66–68] suggests that the glassy state may be characterized by a unique value of either entropy and/or enthalpy regardless of sample prehistory.

The pressure dependence of T_g for PS/GP was studied in two different regimes (i.e. quasi-isochoric cooling of the melt and isobaric heating of the glass) [69, 71]. As can be seen from Table 1.5, the derivatives dT_g/dP in both regimes increase with filler content w. The right-hand sides of eqs. (1.27) and (1.27a) vary in a similar fashion, the former equation being valid for all samples within the limits of experimental uncertainty. In light of the arguments cited above, this result means that the glass structure of PS in filled systems is determined by temperature, pressure and a single order parameter Z_g. According to theory [27], this situation may be visualized as the generation in a liquid during cooling of negative pressure, the magnitude of which is sufficient to keep the liquid structure (literally, its free volume) unchanged. The similarity of this theoretical picture to the experimental conditions of quasi-isochroic cooling is obvious.

As can be seen from Table 1.5, for all filled samples the right-hand side of eq. (1.27a) is smaller compared with the left-hand side, which may be attributed to the stronger effect of the fillers on $\Delta\beta$ than on $\Delta\alpha$, v_g and ΔC_p. In so far as $\Delta\beta$ depends mainly on the packing density change at T_g, while $\Delta\alpha$ and ΔC_p characterize the concomitant changes in molecular mobility, it may be inferred from the above data that as a result of retardation of the relaxation processes, cooling of the filled PS from the melt state to T_g affects the PD of the polymer more than its segmental mobility. This is the likely reason for an unexpected failure of eq. (1.27a).

Summarizing, while no universal value of any order parameter is proved to be frozen at T_g, it appears, however, that the glassy state structure of each filled sample of PS may still be characterized by its own single value of f_g.

Table 1.5 Parameters of eqs. (1.27) and (1.28) for PS/GP[a]

Parameter	Filler content, w			
	0	0.05	0.20	0.50
v_g (cm^3/g)	0.972	0.935	1.042	1.094
$\Delta\alpha \times 10^4$ (K^{-1})	3.17 ± 0.05	2.50 ± 0.05	4.18 ± 0.05	4.14 ± 0.05
$\Delta\beta \times 10^4$ (MPa^{-1})	1.51 ± 0.15	2.02 ± 0.15	3.15 ± 0.15	4.00 ± 0.25
ΔC_p (J/g·K)	0.29 ± 0.03	0.27 ± 0.03	0.33 ± 0.04	0.25 ± 0.04
$dT_g/dP \times 10^2$ (K/MPa)	$(37 \pm 4)/(52 \pm 6)$	$(97 \pm 9)/(74 \pm 8)$	$(56 \pm 6)/(89 \pm 9)$	$(98 \pm 12)/(98 \pm 10)$
$\Delta\beta/\Delta\alpha \times 10^2$ (K/MPa)	47 ± 4	81 ± 8	76 ± 5	97 ± 8
$T_g v_g \Delta\alpha/\Delta C_p \times 10^2$ (K/MPa)	39 ± 6	33 ± 5	49 ± 8	66 ± 10

[a] Values in the numerator and denominator refer to quasi-isochoric and isobaric regimes, respectively.

1.3 THE CONCEPT OF A BOUNDARY INTERPHASE

As already pointed out in section 1.1, the idea of structural changes in a polymer invoked to rationalize the deviation of various thermodynamic and kinetic properties of filled polymers from additivity, was intended to apply to the part of the polymer located sufficiently close to the filler surface (cf. Fig. 1.4) rather than to the whole quantity of the polymer available. In other words, we have to introduce explicitly the concept of a specific 'boundary interphase' (BI) [19, 71].

1.3.1 THICKNESS OF A BI

The thickness of a BI, Δr, may be defined (Fig. 1.4) as the characteristic distance from the solid surface at which the selected structure-sensitive property of a BI becomes indistinguishable from that of a pure polymer (in this context, it should be clear that Δr is, in fact, not so much an intrinsic property of each combination of two components, as rather an 'effective parameter', the numerical value of which will also heavily depend on the sensitivity and accuracy of the experimental technique used to measure the property of interest). Current methods of estimation of Δr will be reviewed below.

(i) As frequently observed (cf. [50, 53, 54]), the reduced Newtonian melt viscosity of filled polymers, $\eta/\eta^0 = f(\varphi)$ (where η^0 is the viscosity of a pure polymer and φ is the filler volume fraction), increases with φ considerably faster than predicted by any of the hydrodynamic theories of flow of suspensions. Formally, this behavior may be attributed to an apparent increase in the volume fraction of the disperse phase due to a BI which forms a non-deformable shell around the filler particles. The relative amount of a BI, v_{BI}, can be estimated from geometrical considerations assuming $\varphi^* = \varphi + v_{BI}$ (where φ^* is the apparent volume fraction of the disperse phase calculated by hydrodynamic theories from experimental values of η/η^0). Values of v_{BI} estimated in this fashion for filled oligomers [50] correlated with φ, the apparent thickness of a BI being roughly similar for all samples studied ($\Delta r \cong 12\,\text{nm}$).

This approach may also be applied to estimate v_{BI} for systems exhibiting non-Newtonian flow; in this case, however, the values of φ^* have to be estimated from the limiting viscosity, $\eta_{\min}$, obtained by extrapolation to infinite shear stress [71]. It is likely that the 'effective' values of φ^* (hence v_{BI} and Δr) estimated from maximum and minimum Newtonian viscosities, for the same filled system, will be different.

(ii) The characteristic feature of specific volume vs. filler content plots (cf. Fig. 1.2) is the occurrence of an initial non-linear portion and a final linear portion which may be extrapolated to $\varphi = 1$ to yield the filler specific volume, v_f (provided aggregation of the filler particles was avoided). Assuming the linear portion of such plots at the highest filler loadings corresponds to the transition of all polymer available into a BI with specific volume v_{BI}, it is possible to estimate v_{BI} from an obvious additive relationship [72]

$$v = v_f \varphi + [v_{BI} v_{BI} + (1 - v_{BI})v_p](1 - \varphi) \tag{1.29}$$

where v_p is the specific volume of a pure (unperturbed) polymer.

The limitation of this method is that it does not account for the assumed modulations of the PD within a BI (Fig. 1.4).

(iii) Assuming additivity of the thermal expansion coefficients of the filler, pure polymer and BI (α_f, α_p and α_{BI}, respectively), i.e.

$$\alpha^* = \alpha_f \varphi + \alpha_p \varphi_p + \alpha_{BI} v_{BI} \tag{1.30}$$

(where $\varphi + \varphi_p + v = 1$), the BI fraction may be obtained from the following relationship [73]:

$$v_{BI} = (\alpha^* - \alpha)/(\alpha + \alpha_p - \alpha^*) \tag{1.31}$$

where α is the thermal expansion coefficient of the filled polymer.

Values of v_{BI} calculated by eq. (1.31) from the relevant experimental data for a metal-filled epoxy resin [73] either increased with φ (for an aliminum filler) or decreased (for an iron filler). Moreover, these values turned out to increase by nearly an order of magnitude as samples were cooled from the melt state to below T_g. It appears that the starting assumptions of the model employed were not met in the experiment (a more detailed analysis of the factors affecting the thermoelastic properties of filled polymers will be made in Chapter 5).

(iv) Assuming that the deficit in the 'partial' (per polymer component) heat capacity jump at T_g of filled polymers [56–58, 73] reflects immobilization of the segmental mobility of the polymer in a BI, we can write [74]

$$v_{BI} = 1 - \Delta C_p / \Delta C_p^0 \tag{1.32}$$

where ΔC_p and ΔC_p^0 are the heat capacity jumps at T_g of filled and pure polymers, respectively.

As might be expected, the values of v_{BI} calculated by eq. (1.32) for several filled polymers increased with φ [57, 75], which is consistent with the physically reasonable assumption of a constant Δr regardless of filler content. However, the validity of the starting assumption about complete inhibition of the segmental mobility of polymer chains in a BI needs further experimental verification.

(v) It follows from simple geometrical considerations that if the quantity of polymer in a BI, $v_{BI}(1 - \varphi)$, is distributed as shells of thickness Δr around filler particles of specific surface area s, then the following relationship should apply:

$$v_{BI}(1 - \varphi)v_{BI} = \Delta r.s.\varphi$$

which may be rewritten as

$$v_{BI} = S\varphi/(1 - \varphi) \tag{1.33}$$

where $S = \Delta rs/v_{BI}$ should be a characteristic constant for each polymer–filler combination. Therefore, among the variety of experimental methods used to estimate v_{BI}, preference should be given to those that yield the same values of Δr regardless of filler content.

1.3.2 THE PROPERTIES OF A POLYMER IN A BI

Glass Transition Temperature

Making the seemingly reasonable assumption that the eventual elevation of T_g in a filled polymer above that for a pure polymer, T_g^0, is caused by the formation of a BI, it follows that the increment observed, $\Delta T_g = T_g - T_g^0$, should depend on v_{BI} as

$$\Delta T_g = Af(v_{BI}) \tag{1.34}$$

where A is an empirical constant accounting for polymer–filler 'affinity'. A linear dependence of ΔT_g on v_{BI} calculated by eq. (1.32) was observed [i.e. $f(v_{BI}) = v_{BI}$ in eq. (1.34)] for PS, PMMA, poly(vinyl chloride), poly(vinyl acetate) (PVAc), polyurethane and poly(dimethyl siloxane) filled with SP [57, 75], as well as for PS and PVAc filled with cellulose powder [76], the slopes of the straight lines [i.e. the parameter A in eq. (1.34)] roughly correlating with the cohesion energies of polymers. On the other hand, for PS and PMMA filled with GP and mica flakes the values of ΔT_g determined from a shift in the mechanical loss maxima exhibited an apparently linear dependence on $\ln S$ [77, 78] [i.e. $f(v) = \ln v_{BI}$ in eq. (1.34)], whereas the values of A correlated with heats of filler wetting by low-molecular weight analogs of PS and PMMA. Similar data were also obtained for polymers filled with surface-treated SP [79].

Assuming the validity of eqs. (1.30) and (1.31), eq. (1.34) should be replaced by the following [73]:

$$\Delta T_g = \Delta \alpha_{BI} v_{BI}(T_g^{BI} - T_g^0)/(\Delta \alpha_{BI} v_{BI} + \Delta \alpha_p \varphi_p) \tag{1.35}$$

where $\Delta \alpha_{BI}$ and $\Delta \alpha_p$ are the thermal expansivity jumps at T_g of a polymer in a BI and a pure polymer, respectively, and T_g^{BI} is the glass transition temperature of a BI.

As can be seen from eq. (1.35), the sign of ΔT_g depends on the absolute values of T_g^{BI} and T_g^0. Thus, the apparent decrease in T_g of epoxies filled with some metallic powders [73, 80] may be attributed to $T_g^{BI} < T_g^0$; however, as is clear from model calculations [81, 82], the latter inequality may also result in an increase, rather than a decrease, in T_g.

Energy State of Macromolecules

It has been tacitly assumed so far that the experimentally observed changes in the 'partial' properties of polymers in filled systems reflect mainly changes in the polymer PD in a BI. Any change in the packing pattern of long-chain molecules, however, may be only a consequence of certain conformational changes within the latter as a result of interactions with a solid surface. In thermodynamic language, the total excess free energy of a polymer in a BI may be generally expressed as

$$G_{ex} = G_{ex}^{intra} + G_{ex}^{inter} \qquad (1.36)$$

where G_{ex}^{intra} is the intramolecular contribution due to assumed loss of conformational degrees of freedom of macromolecules which are in direct contact with the filler surface, and G_{ex}^{inter} is the intermolecular contribution arising from inevitable difficulties of the efficient packing of macromolecules, the conformation of which depends on the distance to the solid surface. For obvious reasons, direct experimental evaluation of each of the above contributions (especially the former) seems hardly possible; useful indirect information, however, may be extracted from studies of the conformation-dependent properties of filled polymers.

In the case of pure polymers, a feeling for the conformational state of macromolecules in an equilibrium melt may be obtained from an analysis of the contributions to the energy of interaction with a solvent, ΔH_1, according to eq. (1.37) [83, 84].

$$\Delta H_1 = \Delta H_r + \Delta H_v + \Delta H_{conf} \qquad (1.37)$$

where $\Delta H_r = v_M(\delta_p - \delta_s)^2 \varphi_p(1 - \varphi_p)$ is the heat of mixing from regular solution theory, in which v_M is the specific volume of the solution, and δ_p and δ_s are the solubility parameters of the polymer and solvent, respectively; $\Delta H_v = v_{ex}P_1$ is the contribution from volume change on mixing; $\Delta H_{conf} = [4\Delta\varepsilon\sigma^2/3(\sigma^2 + 1)][(1 - \alpha^2)/(1 + \alpha^2\sigma^2)]$ is the 'conformational' contribution due to a change in the size of the macromolecular coil (i.e. swelling or contraction) as it is transferred from the bulk melt into a dilute solution, in which $\alpha^2 = \langle h^2 \rangle / \langle h_0^2 \rangle$ is the square of the coil swelling parameter, $\sigma^2 = \langle h_0^2 \rangle / \langle h_f^2 \rangle$ is the square of the stiffness parameter of the macromolecules, $\langle h^2 \rangle, \langle h_0^2 \rangle$ and $\langle h_f^2 \rangle$ are the mean square, end-to-end distances for a real macromolecule in the bulk state and in a dilute solution, and that for an equivalent model chain with free internal rotation, respectively; and $\Delta\varepsilon$ is the energy difference between the stable rotational isomers of the chain.

This approach was applied to characterize the energy state of the macromolecules in a BI of atactic PS ($\langle M_w \rangle = 3.5 \times 10^5, \langle M_n \rangle = 1.2 \times 10^5$) filled with SP [85–87]. Special precautions (i.e. adding a carefully outgassed filler to dilute solutions of PS in benzene or cyclohexane and several day-long gentle stirrings of suspensions) were undertaken to ensure the formation of a BI by the mechanism of initial surface adsorption and subsequent deposition of individual macromolecules in the course of slow solvent evaporation and final sample evacuation to a constant weight.

The experimental heat of a solution of glassy PS in methylene chloride at $T = 303$ K, $\Delta H_s = (-30.1 \pm 1.0)$ J/g, was converted into the required value of

ΔH_1 for a hypothetical equilibrium melt using eq. (1.38)

$$\Delta H_s = \Delta H_1 + \Delta H_g \tag{1.38}$$

where

$$\Delta H_g = \int_T^{T_g} \Delta C_p dT \cong \langle \Delta C_p \rangle (T - T_g) \cong -23.0 \pm 0.4\,\text{J/g}$$

is the exothermic contribution of excess (with respect to the equilibrium melt) enthalpy of glass and $\langle \Delta C_p \rangle = (0.340 \pm 0.006)\,\text{J/g.K}$ is the mean heat capacity difference between the melt and the glass in the temperature interval from $T = 303\,\text{K}$ to $T_g = 370\,\text{K}$.

Next, substituting $\Delta H_r \cong 0$ (since δ_s for methylene chloride and δ_p for PS are approximately equal) and $\Delta H_v = -4.9\,\text{J/g}$ (calculated from $v_{ex} = v'_1 - v_1 = -0.013\,\text{cm/g}$ and $P_i = 375\,\text{J/cm}^3$ [88], where $v'_1 = 0.925\,\text{cm}^3/\text{g}$ and $v_1 = 0.938\,\text{cm}^3/\text{g}$ are the partial specific volume of PS in solution and the specific volume of the equilibrium melt obtained by extrapolation) into eq. (1.37), we obtain $\Delta H_{conf} = (-2.2 \pm 1.5)\,\text{J/g}$, which corresponds to $\alpha = 1.08 \pm 0.05$ (in calculations, $\sigma^2 \cong 6$ and $\Delta \varepsilon = 9\,\text{kJ/mol}$ [14] were assumed). Thus, we may conclude from these estimates that PS coils slightly expand (as far as $\alpha > 1$) as they are transferred from the bulk melt into a dilute solution in methylene chloride.

As can be seen from Fig. 1.8, the room temperature specific volume v vs. filler content w plot consists of two straight segments intersecting at $w = 0.3$, the

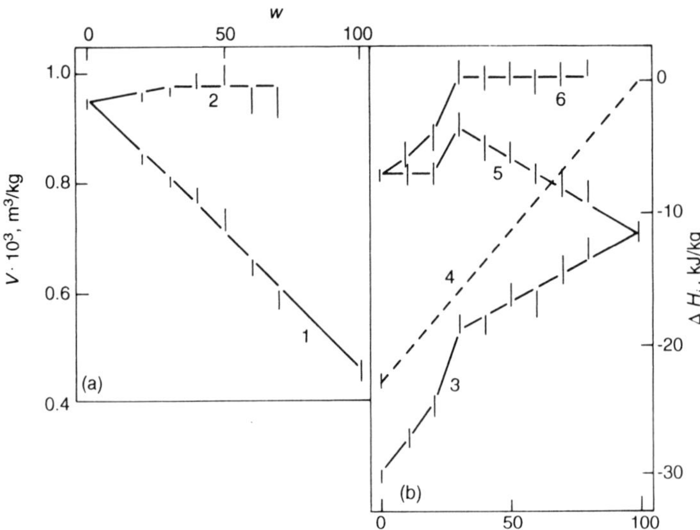

Fig. 1.8. Composition dependence of the specific volume of filled PS-2 (1), partial specific volume of PS (2), ΔH_s (3), ΔH_g (4), ($\Delta H_s - \Delta H_g$) (5) and ΔH_1 (6)

segment at $w > 0.3$ ending at $w = 1.0$ by $v_s = (0.460 \pm 0.03)\,cm^3/g$, which is the specific volume of SP. This means that the polymer specific volume slightly increases from $v_p = (0.953 \pm 0.010)\,cm^3/g$ at $w = 0$ to $v_p = 0.975 \pm 0.035\,cm^3/g$ at $w = 0.3$ and then remains essentially unchanged (broken line 2 in Fig. 1.8(a)).

Initially, the absolute values of (exothermic) heats of interaction with the solvent (curve 3 in Fig. 1.8(b)) fall sharply from $\Delta H_s = (-30.1 \pm 1.0)\,J/g$ at $w = 0$ to $\Delta H_s = (-18.3 \pm 1.0)\,J/g$ at $w = 0.3$ and then decrease linearly to $\Delta H_f = (-11.8 \pm 0.5)\,J/g$ at $w = 1.0$, which is the heat of wetting of SP by methylene chloride.

Assuming the ΔH_g of PS to be independent of filler content (which is consistent with approximate additivity of the heat capacity both in the glassy and melt states) and subtracting the corresponding straight line ΔH_g (line 4) from ΔH_s (line 3), the heats of the interaction of filled PS melts with solvent, $\Delta H_1(w)$, were recovered (line 5). Finally, subtracting the contribution from the heats of the filler wetting according to one more additive relationship

$$\Delta H_1(w) = \Delta H_1'(1 - w) + w\Delta H_f \tag{1.39}$$

the 'partial' heats of solution of the polymer melt at each filler content, $\Delta H_1'$, were obtained (curve 6).

It turned out that the values of $\Delta H_1'$ decreased in magnitude from $\Delta H_1 = (-7.1 \pm 1.0)\,J/g$ for pure PS ($w = 0$) down to an apparently constant limiting value $\Delta H_1' = (0.6 \pm 1.3)\,J/g$ above $w = 0.3$. This means that at a characteristic filler content $w^* = 0.3$, the heat of the polymer–filler interaction, $\Delta H_{pf} = \Delta H_1 - \Delta H_1'$, reaches the limiting value $\Delta H_{pf} = (-7.7 \pm 1.3)\,J/g$, which corresponds to saturation of the interfacial interactions between PS and SP (i.e. to the maximum possible fraction of chain segments v in direct contact with adsorption-active sites on the filler surface). Stated otherwise, at $w > w^*$ all available polymer transforms into a BI.

Substituting $\Delta H_f' = (-21.0 \pm 0.5)\,J/g$ [89] for the heat of the wetting of SP by benzene, and $V_p = 96\,cm^2/mol$ and $V_s = 89\,cm^3/mol$, respectively, for molar volumes of a PS chain repeating unit and benzene molecule, we obtain $v = \Delta H_{pf} V_p / \Delta H_f' V_s \cong 0.37$ for the fraction of phenyl radicals of PS chains which are presumably adsorbed on SP. A similar figure was obtained from spectroscopic studies of PS adsorption onto SP from carbon tetrachloride solutions [90]. This means, in effect, that less than half of the phenyl radicals of PS chains in a BI is directly anchored to the SP surface while the larger part remains in more remote loops.

Judgement on the conformational state of PS macromolecules in a BI may be made from the analysis of the $\Delta H_{conf}'$ contribution to $\Delta H_1'$ by eq. (1.37). Assuming, to a good approximation, 'partial' ΔH_r and ΔH_v to be the same as for a pure polymer, we obtain $\Delta H_{conf}' = (4.9 \pm 1.5)\,J/g$ for $w > w^*$, which corresponds to $\alpha' = 0.85 \pm 0.05$; hence, macromolecular coils contract ($\alpha < 1$) on going from a BI into a dilute solution. It can thus be concluded that macromolecules in a BI of filled PS are in a more extended ('swollen') conformation than either in solution or in a pure polymer, which is the price to be paid for maximum saturation of all possible polymer–filler interactions.

Similar studies for filled samples of two other PS fractions with different molecular masses ($\langle M_w \rangle = 0.5 \times 10^5$, $\langle M_n \rangle = 0.3 \times 10^5$, and $\langle M_w \rangle = 8.7 \times 10^5$, $\langle M_n \rangle = 4.1 \times 10^5$) [91,92] yielded essentially similar results, except for the values of w^* which turned out smaller the higher was the polymer $\langle M_w \rangle$ (Fig. 1.9). As can be seen from Fig. 1.10, the double log plot of the thickness of a BI (calculated as the interparticle distance at w^*, $\langle L^* \rangle$, assuming $\varphi_M = 0.8$ as the maximum packing density of polydisperse filler particles) vs. $\langle M_w \rangle$ is reasonably linear (solid straight line), which can be expressed by the following empirical equation:

$$\langle L^* \rangle (\text{nm}) = 5.25 \times 10^{-3} \langle M_w \rangle^{0.64} \qquad (1.40)$$

Also shown in Fig. 1.10 (broken line) is the dependence of the gyration radius of an unperturbed PS chain, $\langle R_g \rangle$, on $\langle M_w \rangle$ calculated by [93]

$$\langle R_g \rangle (\text{nm}) = 2.65 \times 10^{-2} \langle M_w \rangle^{0.5} \qquad (1.41)$$

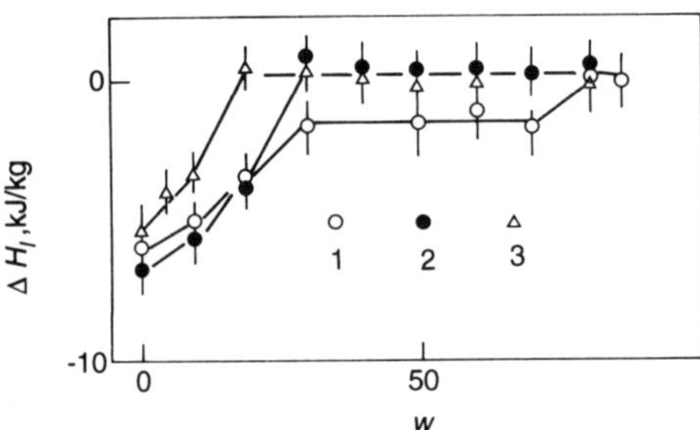

Fig. 1.9. Dependence of ΔH_1 on filler content for PS-1 (1), PS-2 (2) and PS-10 (3)

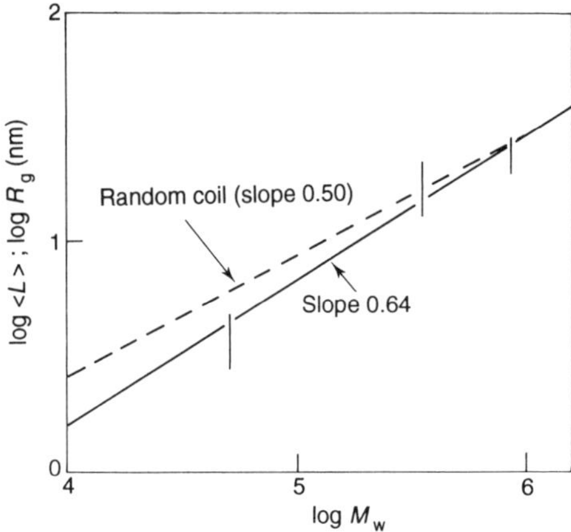

Fig. 1.10. Dependence of $\langle L \rangle$ and $\langle R_g \rangle$ on the molar mass of PS

Despite the difference in exponents at $\langle M_w \rangle$ in eqs. (1.40) and (1.41) (in fact, the exponent 0.64 in the former implies extended conformation, which seems quite reasonable in view of the above discussion), the rather close numerical correspondence between the values of $\langle L^* \rangle$ and $\langle R_g \rangle$ warrants the conclusion that complete saturation of interfacial polymer–filler interactions corresponds to the transformation of all polymer component of a filled sample into a BI of thickness comparable with the average dimensions of a macromolecular coil. This result is in gratifying agreement with theoretical predictions [94, 95].

The last point of concern is the structure of a BI above w^*. Approximately equal, constant values $\Delta H_1' \cong 0\,\text{J/g}$ (per polymer) for all systems studied suggest that the fraction of phenyl radicals in PS chains anchored to a SP surface which reached its limiting value at w^* is the same (i.e. conformations of individual macromolecules in a BI remain essentially unchanged) even at the highest filler loadings. This situation may be qualitatively explained in the following terms.

Limiting values of v_{BI} at w^* characterize the optimum conformation of macromolecular coils in a BI corresponding to the balance between the thermo-dynamic driving force for coverage of all adsorption-active sites on a filler surface by the maximum possible number of phenyl radicals of each macro-molecule (this requires extended conformation of the latter), and a concomitant thermodynamically unfavorable loss of conformational entropy. The balance between these two opposite driving forces should hold regardless of filler content; hence, the macromolecules in a BI will be essentially in the same conformation (i.e. the thickness of a BI, $\Delta r \cong \langle L^* \rangle$, will be the same) throughout the interval $w > w^*$. However, this latter requirement can be met only on condition that with w increasing above w^*, the larger portion of the filler surface available will remain uncovered, so that macrovoids will form in a BI. This conclusion is completely consistent with experimental data [85–87].

1.4   CRYSTALLIZATION AND MELTING OF FILLED CRYSTALLIZABLE POLYMERS

1.4.1   PROPERTIES OF THE CRYSTALLINE PHASE

The solid state properties of crystallizable polymers are controlled by two major factors: the crystallinity, X, and the conformational entropy, S_{conf}, of the 'tie chains' in the interstitial space between chain-folded crystalline lamellae [14, 96]. Experimental evidence of the filler effect (if any) on these factors will be reviewed below.

Crystallinity

It follows from the standard definition of X as a measure of the relative content of the crystalline phase in semi-crystalline polymers that the difference, $(1 - X)$, is the fraction of disordered material rejected outside the crystalline lattice. For pure polymers, the term 'disordered material' should be understood to mean chain folds on basal lamellar planes, tie chains and unattached chain ends, uncrystallizable low molecular weight fractions, etc. In the case of filled polymers, to the above list we should also add polymer chains in a BI which lost crystallizability under topological or kinetic constraints. Hence, the relative crystallinity of a polymer component in a filled sample, $\alpha' = X/X^0$ (where X and X^0 are the crystallinities of a polymer in a filled sample and in the pure state, respectively) may serve as a measure of the BI content, i.e. [97]

$$1 - \alpha' = v_{BI} \tag{1.42}$$

Substitution of eq. (1.33) for v_{BI} yields

$$\alpha' = 1 - S\varphi(1 - \varphi) \tag{1.43}$$

As can be seen from Figs. 1.11(a, b), the predicted linear dependence of α' on the ratio $\varphi/(1 - \varphi)$ is reasonably well obeyed by the relevant data obtained in calorimetric studies of several filled polymers [98–101]. As implied in eqs. (1.33) and (1.43), the slope of the straight line $S = \Delta rs/v_{BI}$ depends on the BI thickness, Δr, which is presumably greater the higher is the polymer–filler 'affinity'. This is qualitatively consistent with empirical correlation between S for Nylon 6/SP and the heats of SP wetting by polar liquids [99].

It is remarkable that in several cases extrapolation of the observed linear plots to $\varphi/(1 - \varphi) = 0$ gives $\alpha' > 1$ (cf. Fig. 1.11). This result means that the same filler may act differently depending on the filler content; namely, it acts as a nucleating agent promoting polymer crystallization ($\alpha' > 1$) at fairly low filler loadings, while with increasing filler content more and more polymer becomes involved in the BI, so that the level of crystallinity drastically decreases down to zero when, presumably, all polymer transforms into a BI.

The absence of appreciable changes in α' up to the highest filler contents ($w = 0.7$–0.8) observed for high density polyethylene filled with SP and chalk [101] implies $\Delta r \cong 0$, which may be the result of either an extremely poor affinity between filler and polymer melt, or an extremely high driving force for crystallization which exceeds by far the thermodynamic driving force for BI formation in the melt.

Thermodynamic Stability of Polymer Crystals

Classical theory [15] predicts that the thermal stability of a crystalline substance, as expressed by its melting point T_m, should decrease with crystal size D according to the Gibbs–Thomson equation, eq. (1.44):

$$T_m(D) = T_m(\infty)[1 - 2\sigma_{sl}v_s/(\Delta H_m D)] \tag{1.44}$$

where $T_m(D)$ and $T_m(\infty)$ are the melting points of a crystal of size D and of a macroscopic crystal, respectively, v_s is the specific volume of a crystalline phase,

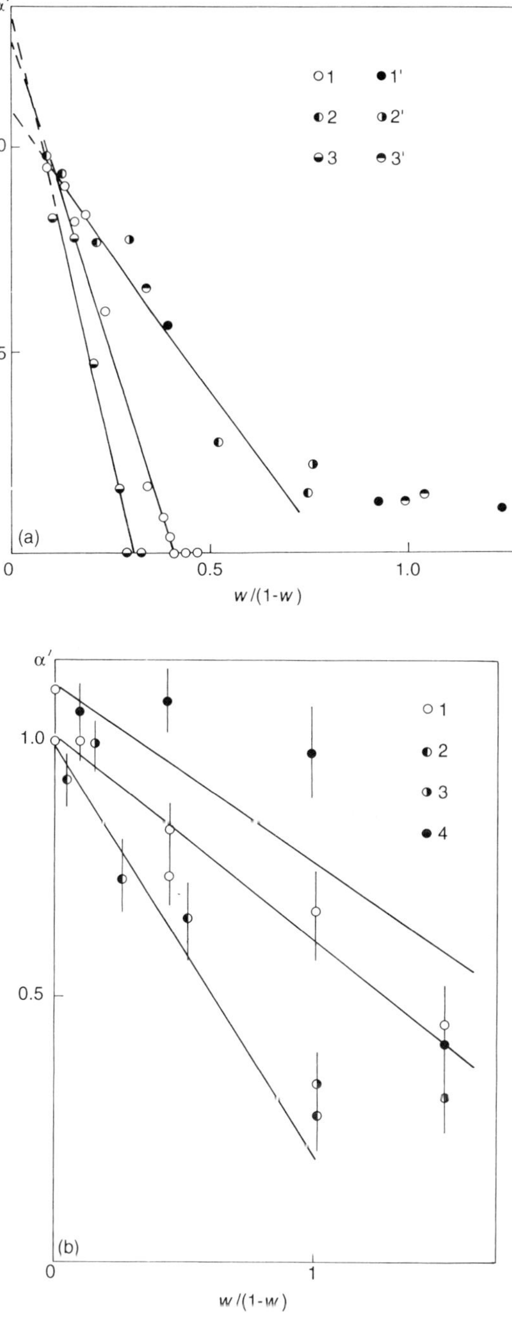

Fig. 1.11. dependence of relative crystallinity on the ratio $w/(1-w)$ for poly(butylene terephthalate) (1), Nylon 6 (2), isotactic polypropylene (3) and low-density polyethylene (4) solution-filled with untreated silica (a) and for poly(ethylene oxide) filled with silica (1–3) and carbon black (1'–3') from dilute solutions in benzene (1, 1'), methanol (2, 2') and water (3, 3')

σ_{sl} is the specific free energy at the solid–liquid interface and ΔH_m is the melting enthalpy of a macroscopic crystal (per unit of mass).

It was experimentally established [102–104] that eq. (1.44) holds for polymer crystals, too, provided D is understood to be the highest of a folded-chain lamella, l, and σ_{sl} is the specific free energy at the interface between the polymer melt and the basal face of a lemmella, σ_e. Moreover, it turned out that the ever-increasing contribution of the excess free energy of the basal faces accompanying lamella thinning manifests itself as a continuous decrease in the melting enthalpy, so that [104, 105]

$$\Delta H_m(l) = \Delta H_m(\infty) - 2h_e v_s/l \tag{1.45}$$

where $h_e = \sigma_e - s_e T$ is the enthalpic, and s_e is the entropic contribution to the excess free energy σ_e.

The applicability of eqs. (1.44) and (1.45) to highly filled crystallizable polymers in which the longitudinal dimensions of the crystalline lamellae will be limited not only by the thermodynamic conditions of crystal nucleation but also by spatial restrictions (i.e. thickness of the gap between filler particles, $\langle L \rangle$), was tested experimentally [99, 101, 106, 107] for a series of linear polymers solution-filled with SP, as described above for PS/SP (cf. section 1.3.2).

In the case of high molecular weight polymers, the values of the 'equilibrium' melting temperature, T_m^0, and the 'thickening rate' parameter, γ, from eq. (1.46) [108]

$$T_m = T_m^0(1 - \gamma) + \gamma T_0 \tag{1.46}$$

(where T_m and T_c are the experimental values of the melting and crystallization temperatures, respectively), exhibited a sudden drop at a characteristic filler content w^* (Table 1.6). As is clear from the standard definition of the parameter

Table 1.6 Parameters of eq. (1.46)

w	PE (low density)		PP		Nylon 6		Poly(butylene terephthalate)	
	T_m^0 (K)	γ	T_m^0 (K)	γ	T_m^0 (K)	γ	T_m^0 (K)	γ
0	394	0.589	475	0.514	511	0.538	505	0.582
0.01	390	0.523	480	0.561	—	—	504	0.535
0.03	390	—	—	—	499	0.345	—	—
0.10	393	0.555	438	0.255	—	—	505	0.574
0.20	—	—	—	—	500	0.378	—	—
0.30	394	0.576	440	0.492	—	—	478	0.508
0.33	—	—	—	—	502	0.399	—	—
0.50	386	0.384	—	—	499	0.429	484	0.475
0.60	382	0.340	—	—	—	—	—	—
0.70	381	0.250	—	—	—	—	—	—

γ [99, 108, 109]

$$\gamma = 0.5(l^*/l)(\sigma_e/\sigma_e^*) \tag{1.47}$$

(where asterisks are used to differentiate the 'kinetic' values for crystal nuclei from the 'equilibrium' values for mature crystals) the observed drop in γ for polar polymers may be attributed to an increase in the energetic barrier to crystallization (i.e. the parameter σ_e^* in the denominator of the ratio in the second parentheses), while for non-polar polymers the crystal thickening effect (i.e. an increase in the ratio in the first parentheses) seems more probable [97, 99].

Similar to the PS/SP case (Fig. 1.9), the thickness of the gap between filler particles at w^*, $\langle L^* \rangle$, correlates with the dimensions of the macromolecular coil for different polymers, $\langle R_g \rangle$ (Fig. 1.12). Literally, this means that in the case of filled, high molecular weight crystallizable polymers, the parameter T_m^0 estimated by (1.46) should be understood as the melting temperature of crystalline entities with thickness corresponding not to a complete chain extension as implied for pure polymers [102–105, 108], but rather to the limiting thickness of the polymer layer in the gap between filler particles, which is comparable with the dimensions of a macromolecular coil.

This 'size effect' was studied in more detail for two SP-filled fractions of oligoester with $\langle M_n \rangle = 1.7 \times 10^3$ (OE-2) and $\langle M_n \rangle = 4.0 \times 10^3$ (OE-4), thus bracketing the critical 'entanglement' molecular mass $\langle M_c \rangle = 2.5 \times 10^3$ [106, 107]. Empirically, $\langle M_c \rangle$ separates 'polymers' ($\langle M \rangle > \langle M_c \rangle$) existing as random coils in the melt and crystallizing by the chain-folding mechanism, from 'oligomers' ($\langle M \rangle < \langle M_c \rangle$) which are in a more extended conformation in the melt (i.e. the end-to-end distance is comparable with the chain 'contour' length) and form extended-chain crystals on cooling [14, 110].

As can be seen from Fig. 1.13, for both oligomers the increase in filler content w is accompanied by a smooth decrease in the apparent crystal melting

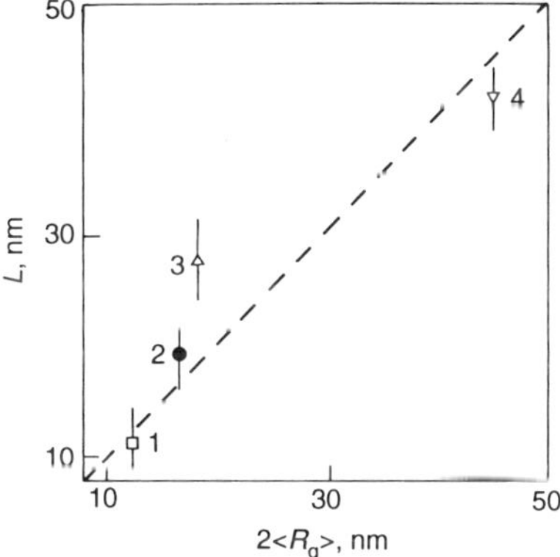

Fig. 1.12. Correlation between $\langle L \rangle$ and $2\langle R_g \rangle$ for filled polymers from Fig. 1.11(a)

temperature T_m and a slight upturn in the glass transition temperature of the amorphous phase T_g, while both the melting enthalpy, ΔH_m, and the crystallinity, $X = 1 - \Delta C_p^a/\Delta C_p^q$ (where the superscripts a and q refer to annealed and quenched samples, respectively) after an initial moderate decrease exhibit sudden drops above $w^* \cong 0.5$. In so far as the corresponding mean width of the gap between filler particles, $\langle L^* \rangle \cong 10\,\text{nm}$, is comparable with the contour length of OE-2 chains ($\langle L' \rangle \cong 12\,\text{nm}$), the observed suppression of crystallizability at $w > w^*$ may be explained as follows. In the range of low filler contents (i.e. when $\langle L \rangle < \langle L' \rangle$) each macromolecule in a BI will be anchored by either one, or by both of its end hydroxyl groups to the same SP particle, while at $w > w^*$ there will be a much higher probability of simultaneous adsorption of both end hydroxyls of the same macromolecule in an extended conformation on the surfaces of two neighboring solid particles. Therefore, the hindrances to crystal nucleation by such macromolecular 'bridges' with both ends fixed would be

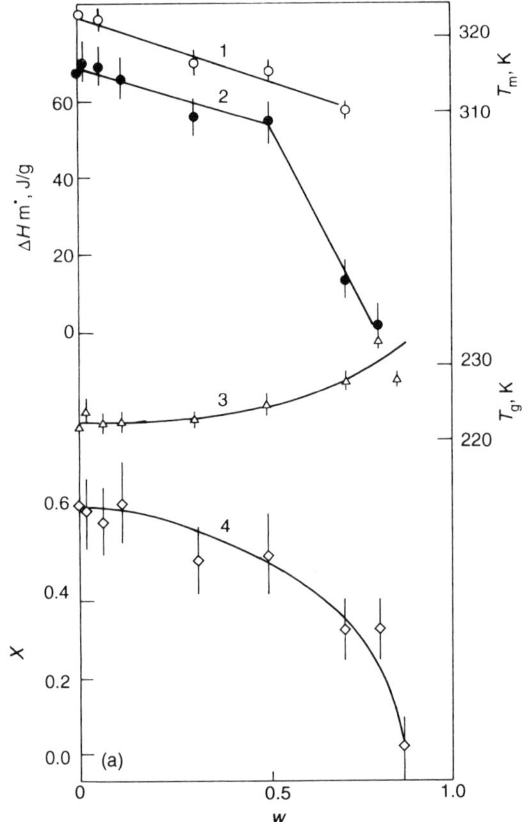

Fig. 1.13. Dependence of the melting point (1), melting heat (2), glass transition temperature (3) and crystallinity (4) on filler content for EA-2 (a) and EA-4 (b)

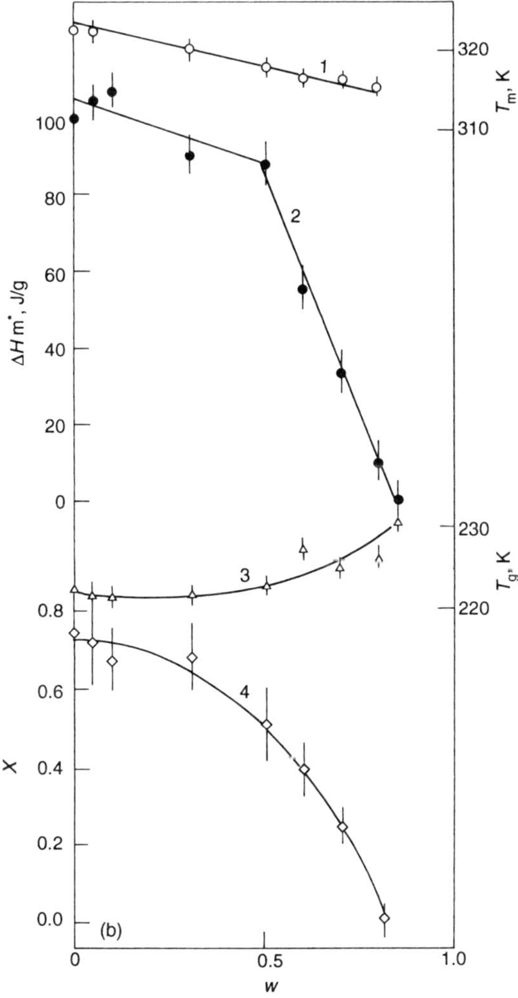

Fig. 1.13. *(Continued)*

much too severe for crystallization to occur. This conclusion is confirmed by a sudden increase in the polymer–filler interaction energy above $w^* = 0.5$ [89, 101].

These arguments also should apply to explain the similar behavior of OE-4 (Fig. 1.13(b)) which is believed to be in a randomly coiled conformation in the melt. As might be expected, in this latter case the characteristic thickness, $\langle L^* \rangle$, is severalfold smaller than the chain contour length ($\langle L \rangle \cong 28$ nm), whereas it is close to the diameter of the corresponding random coil, $2\langle R_g \rangle = 8$ nm. Thus, a sudden drop in ΔH_m and X for filled OE-4 above w^* is likely to be caused by adsorption on the filler surface of remote segments on the periphery of the

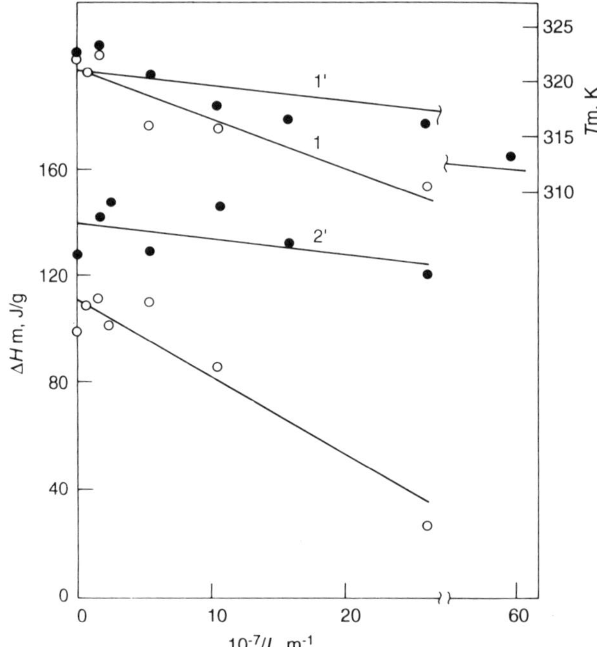

Fig. 1.14. Dependence of the melting point (1, 1′) and true melting heats (2, 2′) on the reciprocal interlayer thickness for EA-2 (1, 2) and EA-4 (1′, 2′)

macromolecular coils, rather than by anchoring of the end groups, as was apparently the case with OE-2.

More insight into this problem provides the plots of T_m and the 'true' (i.e. independent of crystallinity) melting enthalpy $\Delta H_m/X$ vs. $1/\langle L \rangle$ (Fig. 1.14). In apparent agreement with eqs. (1.44) and (1.45), these plots are reasonably linear for both systems, although the slopes (i.e. the specific free energy, σ_e, in the former equation and h_e in the latter) for OE-4 are severalfold smaller. This is clearly a manifestation of the weaker affinity to the polar SP surface of middle-chain segments on the periphery of macromolecular coils of OE-4 compared with the hydroxyl end groups of extended chains of OE-2.

The results obtained imply that the probability of crystal nucleation and/or subsequent growth in filled melts of crystallizable polymers drastically decreases as the thickness of the polymer interlayer between neighboring filler particles becomes comparable with the chain dimensions. This problem will be dealt with in more detail in the next section.

1.4.2 KINETICS OF MELT CRYSTALLIZATION

According to classical concepts [15, 103], liquid–crystal transformation, in the simplest case, may be visualized as filling of the initial liquid phase of volume

V with N spherical crystal entities ('spherulites') each of volume $v_s = 4\pi R^3/3$ (where $G_R = dR/dT$ is the radial growth rate and $R = G_R t$ is the spherulite radius). Hence, the bulk crystallization rate, G_B, will increase with the surface area of spherulite, S_R, the radial growth rate, G_R, and the concentration of spherulites, N/V, i.e.

$$G_B \sim S_R G_R N/V \tag{1.48}$$

while the transient transformation degree will be defined as

$$\alpha = Nv/V = 4\pi N(G_R t)^3/3V$$

or, generalizing to n-dimensional growth,

$$\alpha = K_n t^n \tag{1.49}$$

where K_n is the appropriate rate constant.

Differentiation of eq. (1.49) yields the transformation rate for unrestricted growth, $d\alpha/dt = nK_n t^{n-1}$; however, assuming that the contribution of sterical restrictions imposed on the growth rate by eventual impingement of crystals above a certain value of α may be empirically accounted for by the difference $(1 - \alpha)$, i.e. $d\alpha/dt = nK_n t^{n-1}(1 - \alpha)$, we obtain, after integration, the well-known Kolmogorov–Avrami equation, eq. (1.50)

$$\alpha = 1 - \exp(-K_n t^n) \tag{1.50}$$

The treatment of experimental data for polymer crystallization from the melt using eq. (1.50), however, invariably resulted in non-integer values of the shape parameter n ⌊111⌋, which was clearly at variance with the starting assumptions of the theory; moreover, the applicability of eq. (1.50) was limited by the earliest stage of crystallization. Inadequate conditions of the experiment (e.g. failure to detect the onset of crystallization [111, 112], the effect of impurities [103], etc.) were invoked as an immediate explanation for this empirical finding; however, the assumption of the occurrence of two overlapping, kinetically different mechanisms of crystallization seemed physically more appealing. It was argued [113–115] that the first (fast) stage of crystallization consists of initiation and radial, three-dimensional growth of randomly branched fibrillar crystallites ('spherulite core'), whereas the second (slow) stage involves final solidification of the melt in interfibrillar space. Assuming the applicability of eq. (1.50) to both stages, the development of the total transformation degree α with time t may be expressed as [113–116]

$$\alpha = K_1 n_1 \int_0^{t+\Delta t} \tau^{n_1 - 1} \exp(-K_1^{\tau^{n_1}})\{1 - C \exp[-K_2(t + \Delta t - \tau)^{n_2}\} \, d\tau \tag{1.50a}$$

where τ is the time of the onset of stage 2, C is the mass fraction of the material involved in that stage and Δt is the fitting parameter. In contrast to eq. (1.50), eq. (1.50a) applies quantitatively to the whole interval of transformation degree α with integer values of the shape parameters n_1 and n_2 [113–117].

The temperature dependence of the transformation rate constant K_n (or, equivalently, the crystal growth rate G_R) is assumed to depend on the probabilities of two independent events, namely formation of a stable crystal nucleus, $\exp(-\Delta G/kT)$, and mass transport across the liquid–crystal interface, $\exp(-\Delta E/kT)$ [where ΔG and ΔE are corresponding energy barriers, the latter conveniently defined by eq. (1.26) from section 1.3]. Hence [103, 105, 116]

$$G_R = G_0 \exp(-\Delta G/kT)\exp(-\Delta E/kT) \tag{1.51}$$

where G_0 is the pre-exponential parameter.

By definition, the upper limit for a thermodynamically stable crystal phase is the 'true' melting temperature, T_m^0, which corresponds to equilibrium between Gibbs free energies of the melt, G_1, and the crystal, G_c (i.e. $\Delta G_v = G_1 - G_c = \Delta H_m - T_m^0 \Delta S_m = 0$, where ΔS_m is the entropy of melting). It might have seemed, therefore, that the crystal phase should grow at a non-negligible rate at whatever 'undercooling', $\Delta T = T_m^0 - T_c$, since in this interval it is the crystal that is a single thermodynamically stable phase ($\Delta G_v < 0$). However, as soon as the crystallization rate in a relatively extended temperature interval below T_m^0 was not experimentally observed, it was argued that the true barrier to crystallization depends not only on the crystallization driving force, $\Delta G_v = v_s \Delta S_m \Delta T$, but also on the contribution of excess free energy of a growing crystal, $\Delta G_s = \Sigma^i \sigma_i S_i$ (where σ_i and S_i is the specific free energy and S_i is the surface area of nucleus ith face), i.e. [103, 105, 116]

$$\Delta G = \Delta G_v + \Delta G_s = Z_m(\Delta G_v)^m \tag{1.52}$$

In eq. (1.52) Z_m accounts for the excess energy at the liquid–nucleus interface and m is an integer. For example, in a typical case of polymer melt crystallization by the mechanism of secondary (surface) nucleation with chain folding, we should use $m = 1$ and $Z_1 = 4b_0\sigma\sigma_e$ [where b_0 is the height of the monomolecular overlayer, or 'chain thickness', and σ and σ_e are the melt–nucleus interfacial energies at the lateral and basal (i.e. fold-containing) faces].

Effect of Artificial Nucleation

It was established experimentally for many polymers [118–123] that at a constant cooling rate even tiny (fractions of per cent) quantities of dispersed solid particles cause a shift in the crystallization exotherm to higher temperatures. In terms of eq. (1.50) this apparent acceleration of crystallization may be attributed to an increase in the rate constant K_n due to an increased concentration of the nuclei, N, and/or a decrease in the nucleation barrier, ΔG.

Qualitatively, the first conclusion agrees with the trend for an increase in the crystallization rate with concentration of the 'active' additive [121]. On the other hand, lowering ΔG may reflect a decrease in the polymer–substrate interfacial energy due to either a crystallographic match of components (i.e. epitaxy) or wetting of the substrate by the polymer melt. The overwhelming majority of documented cases apparently agree with the latter situation in so as far as the substrates on which epitaxial crystallization was observed were not necessarily effective nucleants [124–127].

Lowering of the energetic barrier to nucleation by the active additive may be formally accounted for as [15, 116, 128]

$$\Delta G^* = \Delta G f(\theta) = [Z_m/(\Delta G_v)^m] f(\theta) \tag{1.53}$$

where $f(\theta) = [(2 + \cos\theta)(1 - \cos\theta)^2]/4$; θ is the angle of substrate wetting by the polymer melt. It follows therefore that wetting will always result in accelerated crystallization as soon as $f(\theta)$ in eq. (1.53) varies from 0 to 0.5 as θ increases from $0°$ to $90°$.

Equation (1.53) was first derived to describe heterogeneous nucleation for the crystallization of metals [128] assuming a lense-shaped drop as the appropriate nucleus model. Hence, eq. (1.53) cannot be applied in a straightforward way to polymers for which the asymmetric shape of the nucleus requires us to distinguish between the surface energies of the lateral and basal faces [103, 116]. In this case, eq. (1.54) is preferable [129]

$$\Delta G^* = Z_m/(\Delta G_v - \Delta\sigma T_m^0/b_0) \tag{1.54}$$

where $\Delta\sigma = \sigma + \sigma_s - \sigma_{ls}$, in which σ_s and σ_{ls} are the surface free energies on the interface between the solid substrate and the lateral nucleus face and melt phase, respectively. According to eq. (1.54), the nucleating activity of a solid will be higher the larger is the denominator on the right-hand side due to a decrease in $\Delta\sigma$ (provided $\Delta\sigma > 0$).

Analysis of the data of model experiments on the crystallization of isotactic polypropylene (PP), polybutene-1 (PB), polycaprolactone (PCL) and poly(ethylene oxide) (PEO) on low-energy polymeric substrates [130–132] proved that an increase in the crystallization rate (in the absence of an appreciable change in the concentration of nucleation centers) may be quantitatively accounted for by substituting eq. (1.54), with $\Delta\sigma/\sigma = 0.1$–$0.5$, into eq. (1.51). In the case of PP crystallization on high surface energy substrates, a similar analysis yielded $\Delta\sigma < 0$ [133, 134]

A nearly unchanged value of the parameter n from eq. (1.50) obtained in bulk crystallization rate studies of polylauryllactam with 0.1% talc [135], PEO and isotactic poly(propylene oxide) with untreated and treated SP [136, 137], as well as poly(ethylene terephthalate) containing about 1% of talc, kaolin, titanium dioxide [138] or barium sulphate, corundum, calcium and magnium oxides [139], suggests that acceleration of crystallization results from a decrease in the nucleation barrier ΔG as expressed by eqs. (1.53) or (1.54). On the other hand, changes in the apparent values of n concomitant with a sharp narrowing of the transformation interval of constant n observed for the systems Nylon 6/kaolin [123], OE/PE and OE/SP [140] and PPO/indigo [141], clearly indicate changes in the crystallization mechanism. A similar conclusion was arrived at in a low-angle light scattering study of heterogeneous nucleation of PP crystallization [142].

It is pertinent to mention here that sometimes additives caused significant slowing down of spherulitic crystallization as was the case for Nylon 66/surfactants [143] and for isotactic PS/SP modified with trimethylsilane [144]. Since polymer melts proved not to wet those additives, the nucleation barrier ΔG in eq. (1.51) is unaffected; hence, the observed crystallization slowing down should be attributed to the increased height of the transport barrier ΔE, due to either 'screening' of the growing crystal face by the surfactant layer [143] or formation of a BI with decreased molecular mobility [144]. The latter effect may be formally accounted for by eq. (1.55) [145]

$$\Delta E^* = \Delta E + aH_w \tag{1.55}$$

where H_w is the heat of wetting of the additive by the polymer melt, and a is a numerical constant.

Summarizing, minute quantities of dispersed solid particles may cause either acceleration of crystallization due to a lowering of the thermodynamic barrier to nucleation in polymer melt wetting of the additive, or a slowing down of crystallization due to an increase in the kinetic barrier to molecular transport across the melt–nucleus interface. This purely phenomenological conclusion implies a certain change in melt microstructure at the BI near the solid surface, although the fraction of polymer at the BI in this case should be fairly small. Crystallization in conditions when a substantial portion of the polymer transforms into the BI will be treated in the next section.

Crystallization of 'Thick' Layers

Direct evidence for BI in crystallizable polymers is provided by the observation of 'transcrystalline' (TC) morphology near solid substrates. It was originally assumed [146, 147] that TC layers were only formed in cases when the polymer melt was in contact with crystalline, high-energy solids (e.g. pure metals like gold or inorganic crystals like quartz, etc.). Later studies [131, 148–152] revealed, however, that this assumption was not strictly correct. Moreover, it turned out that, sometimes, TC morphology is induced by polymeric substrates (e.g. Teflon) even better than by metals [151, 152]. According to the experimental data available [127, 152, 153], the formation of TC layers is favored by preliminary 'overheating' of the melt to lower the activity and/or concentration of pre-existing centers of heterogeneous nucleation, and subsequent quenching of the melt to the temperature of maximum rate of crystallization.

Fairly thick TC layers are often observed around the particles of artificial nucleants [118, 154]. Because the thickness of a TC layer presumably depends on the relative rates of polymer crystallization near the solid–melt interface (i.e. in a BI) and in the more remote 'bulk' phase, it may be expected that crystallization kinetics will be affected by an eventual overlap of the TC layers in the course of a continuous increase in the nucleant content.

As follows from DSC study of isothermal crystallization kinetics from the melt of Nylon 6 filled with GP [155], narrowing of the gap between filler particles, $\langle L \rangle$, below typical spherulite dimensions (ca. 5 µm) was accompanied by a decrease in the bulk crystallization rate G_B (referred to the same undercooling ΔT). Since analysis of the temperature dependence of G_B by eq. (1.51) did not reveal any difference in nucleation barriers for unfilled and filled samples, it was concluded that the observed retardation of crystallization ought to be attributed to contribution from either N or S_R in eq. (1.48). The latter suggestion seems more appropriate in so far as in the course of continuous thinning of the interparticle polymer layer $\langle L \rangle$, as the filler content increases, the spherulites should change shape from spherical to disc-like; hence, the contribution of surface area to eq. (1.48) will diminish from $S_R = 4\pi R^2$ for the former to $S_R = 2\pi RD$ (where D is the disc thickness) for the latter [156]. This conclusion allows rationalization of the observed linear decrease of G_B with $\langle L \rangle$ (assuming $\langle L \rangle \cong D$) [155].

Comprehensive studies of crystallization of both low molecular weight liquids [157] and polymers [158–161] in thin layers revealed, however, that thinning of the layer below about 30 µm brings about a decrease in not only the bulk crystallization rate G_B but also the rate of spherulite growth G_R, complete loss of crystallizability (i.e. $G_R \cong 0$) being observed at $\langle L \rangle < 1$ µm.

A quantitative interpretation of this phenomenon in terms of the standard eq. (1.51) means, in effect, seeking a possible dependence on $\langle L \rangle$ of the nucleation and transport barriers (ΔG and ΔE, respectively). Although such a dependence may in fact be expected for the former barrier through that observed for both T_m and ΔH_m (cf. section 1.4.1), numerical estimates readily show that the magnitude of this effect for layers as thick as several microns may be safely neglected. The same conclusion also applies to the transport barrier defined by eq. (1.26). Thus, closer scrutiny of the state of the liquid or melt phase prior to crystallization in thin layers on solid substrates is required.

According to theory [19], the phase pressure of low molecular weight liquids confined within thin capillaries differs from the external pressure by the 'capillary pressure' $\Delta P = 2\gamma_l/r$ (where γ_l is the surface tension at the liquid–air interface and r is the radius of the meniscus curvature). Hence, the volume contraction needed for crystal nucleation and subsequent growth in an undercooled liquid will be opposed by a negative (in the case of capillary walls wetting by liquid) capillary pressure. That means, in effect, that the probability of the formation of a stable crystal nucleus would depend on a thermodynamic barrier comprising additional contributions spent on the elastic deformation of such a 'pre-stressed' liquid, ΔG_d, i.e.

$$\Delta G = \Delta G_v + \Delta G_S + \Delta G_d \tag{1.56}$$

Qualitatively, this situation is completely similar to solid-state transformations (like the martensite transformation in metals) in which the contribution of ΔG_d is minimized owing to a change in the nucleus shape from spherical to disc-like. In this case the energetic barrier to nucleation may be expressed as [156]

$$\Delta G = Z_d/(\Delta G_v)^4 \qquad (1.57)$$

where $Z_d = C_4 A^2 (\sigma^*)^3$, $C_4 = 32\pi/3$, A is the energy density of elastic deformation at the interface (per unit of volume) and σ^* is the corresponding surface free energy. Assuming $A \cong \Delta P$ and the capillary radius r_c as the radius of the meniscus curvature r, we obtain

$$Z_d = C_4 (2\gamma_l/r_c)^2 \sigma^{*3} \qquad (1.58)$$

Thus, if the observed change in G_R with r_c is caused exclusively by the mechanism postulated (i.e. assuming G_0 and ΔE unchanged), then $\ln G_R$ vs. $1/r_c^2$ plots should be linear with a slope $Y = C_4 (2\gamma_l)^2 \sigma^{*3}/(\Delta G_v)^4 kT$. As can be seen from Fig. 1.15 (insert), this suggestion is consistent with the experimental data for the crystallization of diphenylamine at $\Delta T = 16\,\mathrm{K}$ [157].

This approach was extended [163] to the treatment of experimental data on melt crystallization of thin polymer films between parallel glass plates [158–161]. As expected, the rates of spherulitic crystallization for gutta-percha (GP) and isotactic PS plotted vs film thickness d, as described above, were also reasonably linear (Fig. 1.16). Moreover, it proved possible to explain in a similar fashion the case of crystallization when only one face of the polymer film was in contact with the high-energy substrate (Fig. 1.15). In fact, in the case of substrate wetting by a thin film of polymer melt, the volume of the latter at any undercooling will remain essentially unchanged owing to a fairly large (at least one order of

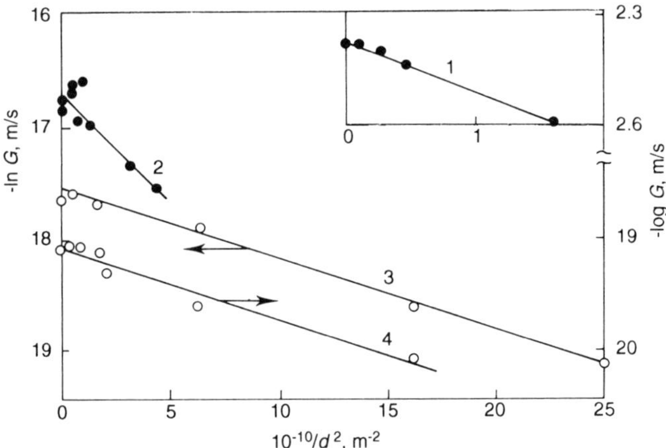

Fig. 1.15. Dependence of the spherulitic growth rate on the inverse squared layer thickness for diphenylamine (1), gutta-percha (2, 3) and isotactic polystyrene (4). Substrates: glass–glass (1, 3, 4); air–glass (2)

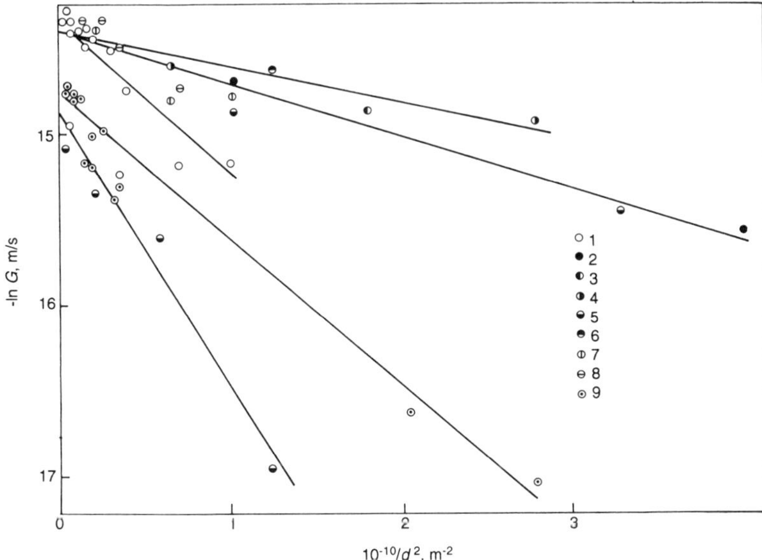

Fig. 1.16. Dependence of the spherilitic growth rate on the inverse squared layer thickness for PP. Substrates: NaCl, (111) plane (1); solid alloy 1 (2); solid alloy 2 (3); NaCl, (110) plane (4); copper (5); CaF_3 (6); liquid alloy 1 (7); liquid alloy 2 (8); glass (9)

magnitude) mismatch between the thermal expansivities of the solid substrate and the polymer melt. The resulting extensional stress (in other words, the negative 'disjoining pressure' [191]) will hinder volume contraction during crystallization of a thin polymer film by exactly the same mechanism as a negative capillary pressure hinders crystal nucleation and growth in liquids confined within capillaries or between parallel plates. As can be seen from Fig. 1.15, a linear decrease in $\ln G_R$ plotted vs. $1/d^2$ is observed for GP crystallization as a thin film either deposited over one glass plate or sandwiched between two glass plates; it is remarkable, however, that the slope of the straight line for the former case is fourfold higher than that for the latter. Taking into consideration that crystallization in both cases was carried out at equal ΔT, whereas both γ_l and σ^* should be considered as characteristic material constants for each polymer–substrate combination, the difference of the 'effective' radii of the curvature is likely to be the only reasonable cause of the different slopes observed. In fact, an exactly fourfold difference between the slopes is predicted by eq. (1.57) if we assume $r \cong d$ for the case of a polymer film between two glass plates, and $r \cong d/2$ for the alternative case of a polymer film on one glass plate.

Slopes of the straight lines Y derived from the treatment of relevant experimental data [158–161] for a wide variety of polymer–substrate pairs (Table 1.7) proved to depend on the substrate surface energy σ_s. Calculated values of σ^* for all studied pairs are also unusually high (in fact, approaching σ_S). Taking into

Table 1.7 Energetic properties of polymers and substrates

Substrate	σ_s (J/m^2)	Polymer	ΔT (K)	$\gamma_1 \times 10^3$ (J/m^2)	$Y \times 10^{10}$ (m^2)	σ^* (J/m^2)
Glass–glass	0.70	GP	35	28.1	0.060	0.325
Glass–glass		PS	90	32.1	0.060	0.325
Glass–air		GP	40	28.1	0.215	0.325
Glass–air		PP	40	23.5	0.850	0.595
Glass–air		PPO	45	29.1	0.170	0.325
Glass–air		PMO	50	36.4	0.074	0.395
Copper–air	1.07	PP	40	23.5	1.640	0.735
NaCl–air:						
(111)	0.87	PP	40	23.5	0.880	0.605
(110)	0.38	PP	40	23.5	0.017	0.350
Metallic						
alloys–air	0.45	PP	50	23.5	0.285	0.410

consideration that in the case of the same substrate (e.g. glass) the values of σ^* tend to decrease as the polymer polarity increases (cf. the data for non-polar PP and for polar PPO and polymethylene oxide), it appears that this parameter may be understood as the surface energy at the polymer–solid interface, σ_{cs}. This suggestion is consistent (on the order-of-magnitude basis) with the values of σ_{cs} estimated by eq. (1.59) [164]

$$\sigma_{cs} = \sigma_c + \sigma_s - 2\Phi(\sigma_c\sigma_s)^{1/2} \tag{1.59}$$

assuming $\Phi = 1.0$ to be typical for wetting.

It follows from the above analysis that the apparent retardation of spherulitic growth in 'thick' polymer layers may be reasonably interpreted assuming an inverse dependence of the nucleation barrier on the square of the layer thickness, as expressed by eq. (1.58). The validity of this approach, however, rests on the assumption of the unusual disc-like shape of the crystallization nucleus, which needs to be verified experimentally. On the other hand, the sweeping claim for complete loss of crystallizability in polymer films thinner than about 1 μm [158–161] seems unnecessarily restrictive because this thickness exceeds by far any expected critical nucleus size (roughly, a few nanometers). It would be more appropriate to conclude that on a micron-large structural scale the spherulites are no longer visible, while on a finer scale crystallization still may remain possible (cf. section 1.4.1). Additional experimental evidence in favor of this conclusion will be discussed in the next section.

Crystallization of 'Thin' Layers

According to the data obtained in dilatometric studies of isothermal crystallization kinetics under elevated pressures of oligoester OE-2 filled with GP and SP

[165], crystallization slows down with increased content of both fillers; however, the effect of the fillers gradually diminishes as the pressure increases. In agreement with other data for this system [106, 107] (cf. section 1.4.1), no traces of crystallinity could be found at SP content above $w^* = 0.5$ (i.e. when the thickness of the layer $\langle L \rangle$ becomes smaller than the chain contour length, $\langle L' \rangle$).

In the framework of the standard eq. (1.51), retardation of crystallization may be caused either by an increase in the barrier to molecular transport ΔE due to strong interactions (wetting) between the filler surface and the oligomer melt, or an increase in the thermodynamic barrier to nucleation ΔG due to structural changes at the oligomer–filler interface. The former effect looks less probable since one should have assumed a highly improbable decrease in the contribution from heat of wetting H_w into ΔE^* [cf. eq. (1.55)] to explain the experimentally observed vanishing of the filler effect at high pressures. Therefore, for all systems ΔE was assumed invariant.

As can be seen from Fig. 1.17, the calculated values of the surface energy of the lateral faces of the crystal nucleus σ for all filled samples are roughly the same regardless of filler content (i.e. $\langle L \rangle$) and exceed those for unfilled samples at the lowest pressure; however, at maximum pressure this difference vanishes.

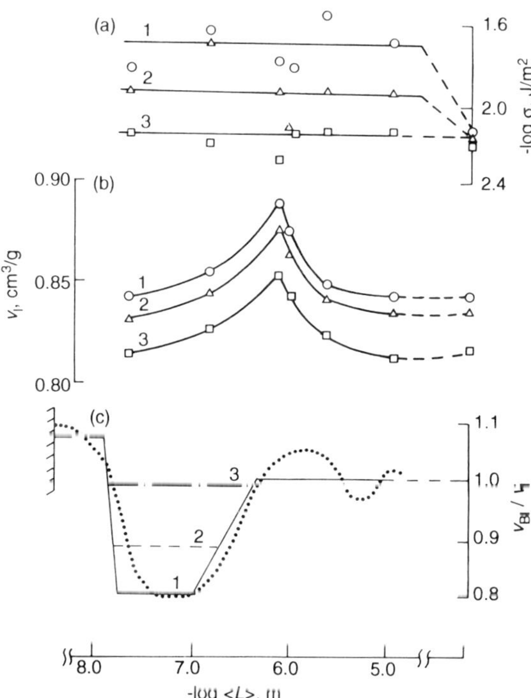

Fig. 1.17. Dependence of σ (a), v_1 (b) and the ratio v_{BI}/v_1 (c) on the interlayer thickness for silica-filled EA-2. Pressures: 15.8 (1); 42.7 (2); 112.4 MPa (3).

This finding may be rationalized in terms of an empirical correlation [14, 166]

$$\sigma = 0.15(v_1/v_c)^4 \tag{1.60}$$

as resulting from the pressure dependence of the ratio v_1/v_c (essentially, the pressure dependence of the melt specific volume v_1 in so far as the compressibility of the crystalline phase is obviously much lower). Therefore, higher-than-'normal' values of σ for filled samples (Fig. 1.17) may be attributed to the formation of a BI with specific volume v_{BI} higher compared with that for a pure oligomer v_1^0. Making use of eq. (1.60), the numerical estimates may be readily made by eq. (1.60), i.e.

$$v_{BI}/v_1^0 = (\sigma_{BI}/\sigma)^{1/4} \tag{1.61}$$

It follows from the above data that all filled samples OE-2, regardless of $\langle L \rangle$, should contain micro regions with similar packing density (i.e. v_{BI}); hence, we might have expected that the relative content of the latter (as well as the overall specific volume of the oligomer, v_1) will increase, the smaller is $\langle L \rangle$. This argument, however, is not supported by experiment (Fig. 1.17) in which the values of v_1 after an initial increase pass through the apparent maximum at $\langle L \rangle \cong 400\,\text{nm}$ and then decrease again to $v_1 < v_1^0$ at $\langle L \rangle < 13\,\text{nm}$. Thus, the oligomer in filled samples of OE-2 must comprise structurally different micro regions with lower-than-normal and higher-than-normal PD, the relative content of the latter inversely dependent on $\langle L \rangle$.

These results may be generalized in the framework of the following model (Fig. 1.17(c)). The formation of macromolecular 'bridges' with both ends anchored to different filler particles at $\langle L \rangle < \langle L' \rangle$ hinders crystal nucleation due to unfavorable loss of conformational entropy; however, the PD in this thin, non-crystallizable surface layer may be higher than normal (presumably, due to complete coverage of the adsorption-active sites on the SP surface by end hydroxyl groups). As $\langle L \rangle$ increases above $\langle L' \rangle$, a certain fraction of the oligomer molecules available will be adsorbed on the SP by one end hydroxyl group, whereas the unbonded molecules will remain in unrestricted conformation. It is likely that the maximum of v_1 observed at $\langle L \rangle = 400\,\text{nm}$ reflects the local 'conformational incompatibility' of bonded and unbonded macromolecules referred to above.

In contrast to oligomers ($\langle M \rangle < \langle M_c \rangle$) for which complete loss of crystalliz-ability is observed at $\langle L \rangle < \langle L' \rangle$, polymers of higher molecular mass ($\langle M \rangle > \langle M_c \rangle$) crystallizing by the chain-folding mechanism, retain crystallizability even at $\langle L \rangle < 2\,\langle R_g \rangle$, although the ultimate degree of crystallinity is drastically decreased. For example, crystallization of Nylon 6 ($2\langle R_g \rangle \cong 20\,\text{nm}$), albeit with drastically changed kinetics, was still observable in the both isothermal [155] and non-isothermal [167] DSC experiments at very high SP loadings ($\langle L \rangle < 8\,\text{nm}$). No evidence for a crystallization exotherm was found during rate-cooling experiments at the highest filler content ($\langle L \rangle \cong 3\,\text{nm}$); however, isothermal storage for several hours at extremely high undercoolings resulted in the gradual development of crystallinity, apparently by the diffusion mechanism [$n = 0.5$ in eq. (1.50)].

These data suggest that the crystallizability loss of high molecular mass polymers in extremely thin layers may be the result of a relatively limited fraction of segments of macromolecular coils which are in direct contact with filler particles. It seems obvious that for these polymers the effect of conformational entropy loss on the crystallization driving force would be much less severe than in the case of anchoring of both ends of the same oligomer molecule with neighboring filler particles. These arguments allow us to explain the different crystallization behaviors of highly filled oligomers and polymers at comparable thicknesses of the interparticle space, $\langle L \rangle$.

It is pertinent to emphasize here once again that the experimental data analyzed in this section refer mainly to highly filled polymers prepared by slow evacuation of the solvent from suspensions of filler particles in dilute polymer solutions during the course of gentle stirring to ensure optimum conditions for the adsorption interaction of each individual macromolecule with the filler surface [100, 101, 106, 107, 155, 165]. Meanwhile, the filler effect on polymer crystallization is fundamentally different in the case of melt blending. Namely, the crystallization rate is unchanged in samples of high-density PE melt-blended with SP and chalk [101]; melt blending of polychlorotrifluoroethylene with graphite, carbon black, aluminum and SP, on the other hand, accelerated polymer crystallization [168]. In the case of PP filled with *tert*-butylbenzoic acid the crystallization rate turned out to be higher in melt-blended samples compared with those prepared from solution [169]. The origin of these differences remains unclear; presumably, acceleration of crystallization in melt-filled samples might deal with the effect of stabilization of heterophase fluctuations in the boundary regions of the polymer melt near the filler surface, thus decreasing the nucleation barrier for crystallization.

1.5 CONCLUSIONS

As amply documented in this chapter, the incorporation of filler into linear polymers invariably results in the formation, near the filler surface, of a boundary interphase (BI) structurally different from the pure polymer. However, the properties of the BI (and therefore the properties of the filled polymers) crucially depend on the method of preparation employed. In the case when the filler is introduced into the dilute polymer solution and stored for a sufficient time to complete saturation of the adsorption interactions of individual macromolecules with the filler surface before evacuation of the solvent, all kinetic and thermodynamic properties of the filled polymers exhibit drastic changes with a characteristic filler content, w^*, at which the effective thickness of the polymer interlayer between neighboring filler particles becomes comparable with the molecular dimensions of the polymer (contour length $\langle L \rangle$ in the case of oligomers, or random coil dimensions $2\langle R_g \rangle$ in the case of high polymers). In this context, the intuitive classification of filled polymers into 'low-loaded' and 'high-loaded', on the basis of the absolute values of the filler content w, should be revised in so far as the final judgement will depend on the relationship between w and w^*. Anyway, w^* may be regarded as a theoretical upper limit to filler content in polymers corresponding to the onset of thermodynamic and mechanical instability of a filled system [170].

In the case when the filler is introduced by melt blending with the polymer, the interaction with the filler surface at the level of individual macromolecules is replaced by the interaction at the level of a continuous phase of entangled macromolecules. Thus, it should be no surprise that the effective thickness of a BI (and hence its influence on the macroscopic properties of a filled system) will be greatly diminished.

REFERENCES

1. Griffiths R. B. (1982) An introduction to the thermodynamics of surfaces, in *Phase Transitions in Surface Films*, ed. by J. G. Dash and J. Ruvalds, Plenum Press, New York and London, pp. 1–27.
2. Ninham B. W. (1980) Long-range vs. short-range forces. The present state of play, *J. Phys. Chem.*, **84**, 1423–1430.
3. Agrafonov Yu. V., Nauchitel V. V. and Sarkissov G. N. (1979) A unary correlation function of a liquid near a solid surface: an atomistic approach, *Koll. Zhurn.*, **41**, 963–964.
4. Flory P. J. (1953) *Principles of Polymer Chemistry*, Cornell Univ. Press, Ithaca, New York.
5. Napper D. H. (1983) *Polymeric Stabilization of Colloidal Dispersions*, Academic Press, London–New york–Paris.
6. Allen G., Sims D. and Wilson G. J. (1962) Intermolecular forces and chain flexibilities in polymers. III. Internal pressures of polymers below their glass transition temperatures, *Polymer*, **3**, 375–382.
7. Kwei T. K. and Arnheim W. M. (1965) Diffusion of gases through filled polymers, *J. Polymer Sci.*, Part C, **10**, 103–110.
8. Ismaylov T. M., Sagalayev G. V., Gabrielyan I. A. *et al.* (1974) PVT properties of thermoplasts in the fluid state, *Plastich. Massy*, **6**, 45–47.
9. Privalko V. P., Besklubenko Yu. D., Lipatov Yu. S. *et al.* (1977) Thermodynamics of filled polystyrene, *Vysokomol. Soed.*, Ser. A, **19**, 1744–1755.
10. Besklubenko Yu. D., Privalko V. P. and Lipatov Yu. S. (1978) Thermodynamics of filled poly(methyl methacrylate), *Vysokomol.* Soed., Ser. A, **20**, 1309–1314.
11. Privalko V. P., Lipatov Yu. S., Besklubenko Yu. D. *et al.* (1977) Compressibility of filled poly(methyl methacrylate) in the rubbery state, in *Physical Chemical Properties and Structure of Polymers*, ed. by Yu. S. Lipatov, Naukova Dumka, Kiev, pp. 33–44 (in Russian).
12. Besklubenko Yu. D. (1980) Thesis, Institute of Macromolecular Chemistry, Academy of Sciences of Ukraine, Kiev, Ukraine.
13. Besklubenko Yu. D., Lipatov Yu. S. and Privalko V. P. (1983), Influence of filler nature on thermodynamic properties of molten polystyrene and poly(methyl methacrylate), *Komposits. Polim. Mater.*, **16**, 72–75.
14. Privalko V. P. (1986) *Molecular Structure and Properties of Polymers*, Khimia, Leningrad (in Russian).

15. Frenkel Ya. I. (1975) *Kinetic Theory of Liquids*, Nauka, Leningrad (in Russian).
16. Hirai N. and Eyring H. (1958) Bulk viscosity of liquids, *J. Appl. Phys.*, **29**, 810–815.
17. Hirai N. and Eyring H. (1959) Bulk viscosity of polymeric systems, *J. Polymer Sci.*, **37**, 51–70.
18. Smith R. P. (1970) Equation of state and the thermodynamic properties of liquid polymers: application of the Hirai–Eyring model, *J. Polymer Sci.*, Part A-2, **8**, 1337–1360.
19. Deryaguin B. V., Churayev N. V. and Muller V. M. (1987) *Surface Forces*, Nauka, Moscow (in Russian).
20. Lipatov Yu. S., Moisya E. G. and Semenovich G. M. (1975) The study of packing density of macromolecules in surface layers at the interface, *Polymer*, **16**, 582–584.
21. Popov V. Ya., Remizov N. A. and Lavrentyev V. V. (1976) On the diffusive properties of polymeric films in adhesive contact with solid substrates, *Koll. Zhurn.*, **38**, 1212–1213.
22. McKinney J. E. and Simha R. (1976) Configurational thermodynamic properties of polymer liquids and glasses. Poly(vinyl acetate). II, *Macromolecules*, **9**, 430–441.
23. Roe R.-J. and Tonelli A. E. (1979) Specific heat difference between liquid and glass: the contribution of the conformational and torsional specific heat of polymer main chains and side groups, *Macromolecules*, **12**, 878–883.
24. Robertson R. E. (1979) The aging of glassy polymers as determined by scanning calorimeter measurements, *J. Appl. Phys.*, **49**, 5048–5054.
25. Tool A. Q. and Eichlin C. G. (1931) Variations caused in heating curves of glass by heat-treatment, *J. Am. Ceram. Soc.*, **14**, 276–308.
26. Kobeko P. P. (1952) *Amorphous Substances*, Izdat. Acad. Nauk USSR, Leningrad (in Russian).
27. Davies R. O. and Jones G. O. (1953) Thermodynamic and kinetic properties of glasses, *Adv. Phys. (Phil. mag. Suppl.)*, **2**, 370–410.
28. Bartenev G. M. (1960) *Mechanical Properties and Thermal Treatment of Glasses*, Stroiizdat, Moscow (in Russian).
29. Kovacs A. J. (1963) Transition vitreuse dans les polymeres amorphes. Etude phenomenologique, *Fortshcr, Hochpolym.-Forsch.*, **3**, 394–507.
30. Rehage G. and Borchard W. (1973) The thermodynamics of the glassy state, in *The Physics of Glassy Polymers*, ed. by R. N. Haward, Applied Science Publishers Ltd, London, chapter 1.
31. Leaderman H. (1943) *Elastic and Creep Properties of Filamentous Materials and Other High Polymers*, The Textile Foundation, Washington.
32. Williams G., Watts D. C., Dev S. B. and North A. M. (1971) Further considerations of non-symmetrical dielectric relaxation behavior arising from a simple empirical decay function, *Trans. Faraday Soc.*, **67**, 1323–1335.
33. Reklison S. M., Bulaeva A. V. and Mazurin O. V. (1971) Change in linear dimensions and viscosity of window glass during stabilization, *Izv. Acad. Nauk USSR (Nerog. Mater.)*, **7**, 714–715.
34. Narayanaswamy O. S. (1971) Model of structural relaxation in glass, *J. Am. Ceram. Soc.*, **54**, 491–498.
35. Moynihan C. T., Easteal A. J., DeBolt M. A. and Tucker J. (1976) Dependence of the fictive temperature of glass on cooling rate, *J. Am. Ceram. Soc.*, **59**, 12–16.
36. DeBolt M. A., Easteal A. J., Macedo P. B. and Moynihan C. T. (1976) Analysis of structural relaxation in glass using rate heating data, *J. Am. Ceram. Soc.*, **59**, 16–21.
37. Moynihan C. T., Macedo P. B., Montrose C. J. *et al.* (1976) Structural relaxation in vitreous materials, *Ann. New York Acad. Sci.*, **279**, 15–35.
38. Lipatov Yu. S., Demchenko S. S. and Privalko V. P. (1983) Enthalpy relaxation in the glass transition process of filled polystyrene, *Dokl. Acad. Nauk USSR*, **273**, 128–131.

39. Privalko V. P., Lipatov Yu. S., Demchenko S. S. and Titov G. V. (1986) Kinetics of enthalpy relaxation during glass transition of highly filled polystyrene, *Vysokomol. Soed.*, Ser. B, **28**, 425–427.
40. Simha R. and Boyer R. F. (1962) On a general relation involving the glass temperature and coefficients of expansion of polymers, *J. Chem. Phys.*, **37**, 1003.
41. Ferry J. D. (1970) *Viscoelastic Properties of Polymers*, 2nd edn Wiley, New York–London–Sydney–Toronto.
42. Privalko V. P. and Lipatov Yu. S. (1986) Glass transition in filled polymers, in *Physical Chemistry of Multi-Component Polymer Systems*, Vol. 1, ed. by Yu. S. Lipatov, Naukova Dumka, Kiev, pp. 83–105.
43. Wunderlich B. (1960) Study of the change in specific heat of monomeric and polymeric glasses during the glass transition, *J. Phys. Chem.*, **64**, 1052–1056.
44. Adam G. and Gibbs J. H. (1965) On the temperature dependence of cooperative relaxation properties in glass-forming liquids, *J. Chem. Phys.*, **43**, 139–146.
45. Bestul A. B. and Chang S.-S. (1964) Excess entropy at glass transformation, *J. Chem. Phys.*, **40**, 3731–3733.
46. Privalko V. P. (1980) Excess entropies and related quantities in glass-forming liquids, *J. Phys. Chem.*, **80**, 3307–3312.
47. Greet R. J. (1966) Comparison of expressions for cooperative properties in glass-forming liquids, *J. Chem. phys.*, **45**, 2479–2481.
48. Miller A. A. (1968) Polymer melt viscosity and the glass transition: an evaluation of the Adam–Gibbs and the free-volume models, *J. Chem. Phys.*, **49**, 1393–1397.
49. Dworkin A. (1974) Sur la variation de viscosite des liquids surfondus en fonction de la temperature, *J. Chim. Phys.*, **71**, 929–930.
50. Lipatov Yu. S., Privalko V. P. and Shumsky V. F. (1973) Studies of the melt viscosity of filled oligoesters, *Vysokomol. Soed.*, Ser. A, **15**, 2106–2109.
51. Privalko V. P. and Titov G. V. (1979) Heat capacity of filled amorphous polymers, *Vysokomol. Soed.*, Ser A, **21**, 348–354.
52. Tammann G. (1930) *Der Glaszustand*, Leopold Voss Verlag, Leipzig.
53. Kraus G. (1965) Interactions of elastomers and reinforcing fillers, *Rubber Chem. Technol.*, **38**, 1070–1114.
54. Bartenev G. M. and Zakharenko N. V. (1962) On the viscosity and flow mechanism of filled polymers, *Koll. Zhurn.*, **24**, 121–127.
55. Hsich H. S.-Y. (1982) Composite rheology. I. Elastomer–filler interaction and its effect on viscosity, *J. Mater. Sci.*, **17**, 438–446.
56. Solomko V. P. and Pas'ko S. P. (1970) Studies of specific heat of filled polystyrene and poly(methyl methacrylate), *Vysokomol. Soed.*, Ser. A, **12**, 681–686.
57. Lipatov Yu. S. and Privalko V. P. (1972) Glass transition in filled polymer systems, *Vysokomol. Soed.*, Ser. A, **14**, 1643–1648.
58. Howard G. I. and Shanks R. A. (1982) The influence of filler particles on the mobility of polymer molecules, *J. Macromol. Sci.*, Ser. B, **17**, 287–295.
59. Barlow A. J., Lamb J. and Matheson A. J. (1966) Viscous behavior of supercooled liquids, *Proc. Roy. Soc. (London)*, **292**, 322–342.
60. Carpenter M. R., Davies D. B. and Matheson A. J. (1967) Measurement of glass transition temperature of simple liquids, *J. Chem. Phys.*, **46**, 2451–2454.
61. Privalko V. P., Shumsky V. F. and Lipatov Yu. S. (1972) Viscous properties of molten oligoethers and oligoesters, *Vysokomol. Soed.*, Ser. A, **14**, 764–771.
62. Privalko V. P., Demchenko S. S., Besklubenko Yu. D. and Lipatov Yu. S. (1977) Volume relaxation of filled polystyrene, *Vysokomol. Soed.*, Ser. A, **19**, 1763–1769.
63. O'Reilly J. M. (1964) Effect of presure on amorphous polymers: in *Modern Aspects of the Vitreous State*, Vol. 3, ed. by J. P. MacKenzie, Butterworths, London, pp. 59–89.

64. Goldstein M. (1963) Some thermodynamic aspects of the glass transition: free volume, entropy and enthalpy theories, *J. Chem. Phys.*, **39**, 3369–3374.
65. Angell C. A. and Sichina W. (1976) Thermodynamics of the glass transition: empirical aspects, *Ann. New York Acad. Sci.*, **279**, 53–67.
66. Ichihara S., Komatsu A., Tsujita Y. *et al.* (1971) Thermodynamic studies on the glass transition and the glassy state of polymers. I. Pressure dependence of the glass transition temperature and its relation to other thermodynamic properties of polystyrene, *Polymer J.*, **2**, 530–534.
67. Oels H.-J. and Rehage G. (1977) Pressure–volume–temperature measurements on atactic polystyrene. A thermodynamic view, *Macromolecules*, **10**, 1036–1043.
68. McKinney J. E. and Goldstein M. (1977) Thermodynamics of the densification process for polymer glasses, *J. Res. Natl. Bur. Stand.*, **81A**, 283–297.
69. Lipatov Yu. S., Privalko V. P., Besklubenko Yu. D. and Demchenko S. S. (1977) Pressure dependence of the glass transition temperature of filled polystyrene, *Vysokomol. Soed.*, Ser. A, **19**, 1756–1762.
70. Privalko V. P., Besklubenko Yu. D., Demchenko S. S. and Titov G. V. (1981) Glass transition in filled polystyrene under elevated pressure, *Vysokomol. Soed.*, Ser. A, **23**, 116–120.
71. Lipatov Yu. S. (1977) *Physical Chemistry of Filled Polymers*, Khimia, Moscow (in Russian).
72. Rafikov M. N., Guzeev V. V. and Malysheva G. P. (1971) Estimation of the thickness of polymer layer adsorbed on filler, *Vysokomol. Soed.*, Ser. A, **13**, 2625–2626.
73. Papanicolaou G. C. and Theocaris P. S. (1979) Thermal properties and volume fraction of the boundary interphase in metal-filled epoxies, *Colloid & Polymer Sci.*, **257**, 239–246.
74. Privalko V. P., Lipatov Yu. S., Kercha Yu. Yu. *et al.* (1971) Calorimetric studies of filled linear polyurethanes, *Vysokomol. Soed.*, Ser. A, **13**, 103–110.
75. Lipatov Yu. S. and Privalko V. P. (1973) On the glass transition temperature of filled polymers, *Vysokomol. Soed.*, Ser. B, **15**, 749–753.
76. de Ruvo A. and Alfthan E. (1978) Shifts in glass transition temperature of synthetic polymers filled with microcrystalline cellulose, *Polymer*, **19**, 872–874.
77. Iisaka K. and Shibayama F. (1978) Effect of filler particles size on dynamic mechanical properties of poly(methyl methacrylate), *J. Appl. Polymer Sci.*, **22**, 1321–1330.
78. Iisaka K. and Shibayama K. (1978) Mechanical α-dispersion and interaction in filled polystyrene and poly(methyl methacrylate), *J. Appl. Polymer Sci.*, **22**, 3135–3143.
79. Yim A., Chahal R. S. and St.-Pierre L. E. (1973) The effect of polymer–filler interaction energy on the T_g of filled polymers, *J. Colloid & Interface Sci.*, **43**, 583–590.
80. Papanicolaou G. C., Paipetis S. A. and Theocaris P. S. (1977) Thermal properties of metal-filled epoxies, *J. Appl. Polymer Sci.*, **21**, 689–701.
81. Lipatov Yu. S. and Babich V. F. (1982) Features of thermomechanical behavior of simple models of composite materials with interfacial layers, *Mekh. Komposit. Mater.*, **2**, 225–232.
82. Babich V. F. and Lipatov Yu. S. (1982) On shift and resolubility of relaxation maxima with change in properties of the boundary layer in composite materials, *J. Appl. Polymer Sci.*, **27**, 53–62.
83. Bianchi U., Cuniberti C., Pedemonte E. and Rossi C. (1969) Conformational energy contribution to the heat of solution in polymer–solvent systems. Part I. Theory, *J. Polymer Sci.*, Part A-2, **7**, 845–853.
84. Bianchi U., Cuniberti C., Pedemonte E. and Rossi C. (1969) Conformational energy contribution to the heat of solution in polymer–solvent systems. Part II. Experimental results, *J. Polymer Sci.*, Part A-2, **7**, 855–866.

85. Privalko V. P., Demchenko S. S. and Titov G. V. (1985) Calorimetric study of interactions between polystyrene and Aeiosil, *Komposits. Polim. Mater.*, **27**, 72–75.
86. Lipatov Yu. S., Privalko V. P., Demchenko S. S. and Titov G. V. (1985) Energy state of macromolecules in boundary layers of highly filled polystyrene, *Dokl. Acad. Nauk USSR*, **284**, 651–654.
87. Lipatov Yu. S., Titov G. V., Demchenko S. S. and Privalko V. P. (1987) Energetics of interphase interactions in highly filled polystyrene, *Vysokomol. Soed.*, Ser. A, **29**, 604–610.
88. Manzini G. and Crescenzi V. (1974) Simple accurate method of calculation of the thermal pressure coefficient of non-polar liquids and a possible estimate of their excess volumes of mixing, *Gazz. Chim. Ital.*, **104**, 51–61.
89. Lipatov Yu. S., Privalko V. P., Nedrya N. L. *et al.* (1983) Calorimetric study of the thermodynamic state of macromolecules in boundary layers of a highly filled oligoester, *Polymer Commun.*, **24**, 359–361.
90. Kawaguchi M., Maeda K., Kato T. and Takahashi A. (1984) Preferential adsorption of monodisperse polystyrene on silica surface, *Macromolecules*, **17**, 1666–1671.
91. Lipatov Yu. S., Privalko V. P., Titov G. V. and Demchenko S. S. (1987) Molecular mass dependence of the thickness of adsorption layers in highly filled polystyrene, *Dokl. Acad. Nauk USSR*, **295**, 146–149.
92. Lipatov Yu. S., Privalko V. P., Titov G. V. and Demchenko S. S. (1987) Scaling relationship for thickness of adsorption layers in highly filled polystyrene, *Vysokomol. Soed.*, Ser. B, **29**, 163–164.
93. Nesterov A. E. (1984) Properties of polymer solutions and blends, in *Handbook on Physical Chemistry of Polymers*, Vol. 1, ed. by Yu. S. Lipatov, Naukova Dumka, Kiev (in Russian).
94. Fleer G. I. and Scheutjens J. M. H. M. (1982) Adsorption of interacting oligomers and polymers at an interface, *Adv. Colloid & Interface Sci.*, **16**, 341–359.
95. Skvortsov A. M. and Gorbunov A. A. (1986) Conformation of macromolecules in filled polymers, *Vysokomol. Soed.*, Ser. A, **28**, 1941–1946.
96. Marikhin V. A. and Myasnikova L. P. (1977) *Supermolecular Structure of Polymers*, Khimia, Leningrad (in Russian).
97. Privalko V. P. (1986) Melting and crystallization of filled polymers, in *Physical Chemistry of Multi-Component Polymer Systems*, Vol. 1, ed. by Yu. S. Lipatov, Naukova Dumka, Kiev, pp. 106–129 (in Russian).
98. Howard G. J. (1971) Melting behavior of poly(ethylene oxide) in the presence of solid particles, *Kolloid-Z. & Z. Polymere*, **244**, 213–217.
99. Privalko V. P., Kawai T. and Lipatov Yu. S. (1979) Crystallization of filled Nylon 6. I. Heat capacities and melting behavior, *Polymer J.*, **11**, 699–709.
100. Lipatov Yu. S., Nedrya N. L. and Privalko V. P. (1982) On the criteria of optimum filler content in highly loaded crystallizable polymers, *Dokl. Acad. Nauk USSR*, **267**, 127–132.
101. Rymarenko N. L. (1984) Thesis, Institute of Macromolecular Chemistry, Academy of Sciences of Ukraine.
102. Geil P. H. (1963) *Polymer Single Crystals*, Interscience, New York and London.
103. Mandelkern L. (1964) *Crystallization of Polymers*, McGraw-Hill Co., New York– San Francisco–Toronto–London.
104. Wunderlich B. (1973) *Macromolecular Physics. Vol. 1: Crystal Structure, Morphology, Defects*, Academic Press, New York and London.
105. Hoffman J. D., Davis G. T. and Lauritzen J. I. (1976) The rate of crystallization of linear polymers with chain folding, in *Treatise on Solid State Chemistry*, Vol. 3, ed. by N. B. Hannay, Plenum Press, New York, pp. 497–614.
106. Privalko V. P., Nedrya N. L., Khmelenko G. I. and Lipatov Yu. S. (1983) Thermal

stability of oligoester crystallized in thin layers, *Vysokomol. Soed.*, Ser. A, **25**, 2550–2555.

107. Privalko V. P., Lipatov Yu. S., Nedrya N. L. *et al.* (1984) Chain length effect on the crystal thermal stability of oligoesters crystallized in thin layers, *Colloid & Polymer Sci.*, **262**, 9–14.

108. Hoffman J. D. and Weeks J. J. (1962) Rate of spherulitic crystallization with chain folds in polychlorotrifluoroethylene, *J. Chem. Phys.*, **37**, 1723–1741.

109. Matsumoto T., Ikegami N., Ehara K. *et al.* (1970) The structural change on annealing poly(ethylene terephthalate), *Kogyo Kagaku Zasshi*, **73**, 2441–2446.

110. Bartenev G. M. and Frenkel S. Ya. (1990) *Physics of Polymers*, Khimia, Leningrad (in Russian).

111. Godovsky Yu. K. (1977) *Thermophysical Methods of Polymer Characterization*, Khimia, Moscow (in Russian).

112. Ueberreiter K. and Steiner K. (1966) Kristallisationskinetik von Polymeren. 2. Mitteilung: Untersuchungen der Anwendbarkeit der AVRAMI-Gleichung auf den Kristallisationsverlauf von Polyaethylen-succinat, *Makromol. Chem.*, **91**, 175–194.

113. Hoshino S., Meinecke E., Powers J. *et al.* (1965) Crystallization kinetics of polypropylene fractions, *J. Polymer Sci.*, Ser. A, **3**, 3041–3065.

114. Hillier I. H. (1965) Modified Avrami equation for the bulk crystallization kinetics of spherulitic polymers, *J. Polymer Sci.*, Ser. A, **3**, 3067–3078.

115. Price F. P. (1965) A phenomenological theory of spherulite crystallization: primary and secondary crystallization processes, *J. Polymer Sci.*, Ser. A, **3**, 3079–3097.

116. Wunderlich B. (1976) *Macromolecular Physics. Vol. 2. Crystal Nucleation, Growth, Annealing*, Academic Press, New York–San Francisco–London.

117. Rymarenko N. L., Voitenko A. I. and Privalko V. P. (1990) Kinetic analysis of the bulk crystallization of polymers from the melt, *Vysokomol. Soed.*, Ser. B, **32**, 485–489.

118. Kargin B. A., Sogolova T. I. and Shaposhnikova T. K. (1965) Mechanism of nucleation of high-melting solid particles in crystalline polymers, *Vysokomol. Soed.*, **7**, 385–388.

119. Beck H. N. (1966) Effect of molecular weight upon the heterogeneous nucleation of crystallization in polypropylene, *J. Polymer Sci.*, Part A-2, **4**, 631–638.

120. Binsbergen F. L. (1970) Heterogeneous nucleation in the crystallization of polyolefins. Chemical and physical nature of nucleating agents, *Polymer*, **11**, 253–267.

121. Yim A. and St.-Pierre L. E. (1970) The effect of interfacial energy on heterogeneous nucleation in the crystallization of polydimethylsiloxane, *J. Polymer Sci.*, Part B, **8**, 241–245.

122. Johnsen U. and Spielgies G. (1972) Keimbildung und Kristallisation von nucleirtem und nicht nukleirtem Polypropylen bei konstanter Abkuelgeschwindigkeit, *Kollod Z. & Z. Polymore*, **250**, 1174–1181.

123. Gurato G., Gaidano D. and Zanetti R. (1978) Influence of nucleating agents on the crystallization of 6-polyamide, *Makromol. Chem.*, **179**, 231–245.

124. Kargin V. A., Sogolova T. I. and Rapoport-Molodtsova N. Ya. (1965) Mechanism of nucleation in crystalline polymers, *Dokl. Acad. Nauk USSR*, **163**, 1194–1197.

125. Moisya E. G., Semenovich G. M., Menzheres G. Ya. *et al.* (1975) Spectroscopic studies of epitaxial crystallization of polymers, *Teoret. Eksper. Khim.*, **11**, 709–712.

126. Mauritz K. A., Baer E. and Hopfinger A. J. (1978) The epitaxial crystallization of macromolecules, *J. Polymer Sci.*, Ser. D, **13**, 1–61.

127. Moisya E. G. (1978) Epitaxy of high polymers, in *Structural Features of Polymers* ed. Yu. S. Lipatov, Naukova Dumka, Kiev, pp. 104–125.

128. Uhlman D. R. and Chalmers B. (1965) The energetics of nucleation, *Ind. Eng. Chem.*, **57**, 19–31.

129. Price F. P. (1969) Nucleation in polymer crystallization, in *Nucleation*, ed. by A. C. Zettlemoyer, Marcel Dekker, New York.
130. Chatterjee A. M., Price F. P. and Newman S. (1975) Heterogeneous nucleation of crystallization of high polymers from the melt. 1. Substrate-induced morphologies, *J. Polymer Sci.: Polymer Phys. Ed.*, **13**, 2369–2383.
131. Chatterjee A. M., Price F. P. and Newman S. (1975) Heterogeneous nucleation of crystallization of high polymers from the melt. 2. Aspects of transcrystallinity and nucleation density, *J. Polymer Sci.: Polymer Phys. Ed.*, **13**, 2385–2390.
132. Chatterjee A. M., Price F. P. and Newman S. (1975) Heterogeneous nucleation of crystallization of high polymers from the melt. 3. Nucleation kinetics and interfacial energies, *J. Polymer Sci.: Polymer Phys. Ed.*, **13**, 2391–240.
133. Goldfarb L. (1978) Heterogeneous nucleation of isotactic polystyrene, *Makromol. Chem.*, **179**, 2297–2303.
134. Goldfarb L. (1979) Heterogeneous nucleation of isotactic polypropylene—surface energy considerations, *Makromol. Chem.*, **180**, 511–512.
135. Manescalchi F., Rossi R. and Mattiussi A. (1973) Cinetique de crystallization du polylauryllactame, *Eur. Polymer J.*, **9**, 601–608.
136. Cole J. H. and St.-Pierre L. E. (1975) Heterogeneous nucleation of isotactic poly(propylene oxide) crystallization, *Am. Chem. Soc. Symp. Ser.*, **6**, 58–69.
137. Cole J. H. and St.-Pierre L. E. (1978) The role of interfacial energy in the heterogeneous nucleation of polyether crystallization, *J. Polymer Sci.: Polymer Symp.*, **63**, 205–235.
138. Groeninckx G., Bergmans H., Overbergh N. *et al.* (1974) Crystallization of poly(ethylene terephthalate) induced by inorganic compounds. I. Crystallization behavior from the glassy state in a low-temperature region, *J. Polym. Sci.: Polymer Phys. Ed.*, **12**, 303–316.
139. Przygocki W. (1982) Nucleation in poly(ethylene terephthalate), *Acta Polymer*, **33**, 729–735.
140. Privalko V. P., Khmelenko G. I. and Lipatov Yu. S. (1975) Crystallization kinetics of a filled oligoester, *Sintez Fiz. Khim. Polim.*, **15**, 93–97.
141. Slonimsky G. L. and Godovsky Yu. K. (1966) Calorimetric study of melting and crystallization of polypropylene with artificial nucleants, *Vysokomol. Soed.*, **8**, 718–722.
142. Johnsen U. and Spielgies G. (1973) Kleinwinkel-Lichtstreuung an nukleirtem Polypropylen, *Angew. Markomol. Chem.*, **31**, 123–137.
143. Vassilevskaya L. P., Bakeev N. F., Lagun L. G. *et al.* (1964) Influence of minute amounts of surfactants on properties of crystalline polymers, *Dokl. Acad. Nauk USSR*, **159**, 123–137.
144. Turturro G. and St.-Pierre L. E. (1978) The heterogeneous nucleation of crystallization in isotactic polystyrene, *J. Coll. Interface Sci.*, **67**, 349–354.
145. Kennedy M., Turturro G., Brown G. R. and St. Pierre L. E. (1980) Silica retards radial growth rate of spherulies in isotactic polystyrene, *Nature*, **287**, 316–317.
146. Schonhorn H. (1964) Transcrystalline growth at a polymer–metal interface, *J. Polymer Sci.*, Part B, **2**, 465–467.
147. Schonhorn H. (1968) Heterogeneous nucleation of polymer melts on high-energy surfaces: II. Effect of substrate on morhpology and wettability, *Macromolecules*, **1**, 145–148.
148. Fitchmun D. R. and Newman S. (1969) Surface morphology in semicrystalline polymers, *J. Polymer Sci.*, Part B, **7**, 301–305.
149. Giniewski C. and Moore R. S. (1969) Deformation and structure of cylindrical 'spherulites' in transcrystalline polyethylene. Detection and characterization of the pseudomonoclinic crystalline component, *Macromolecules*, **2**, 385–394.

150. Fitchmun D. R. and Newman S. (1970) Surface crystallization of polypropylene, *J. Polymer Sci.*, Part A-2, **8**, 1545–1564.

151. Henke S. J., Smith C. E. and Abbott R. F. (1975) Effect of mold temperature on the morphology and properties of polypropylene. I. A propylene–ethylene copolymer, *Polymer Eng. Sci.*, **15**, 79–83.

152. Goldfarb L. (1980) Transcrystallization of isotactic polypropylene, *Makromol. Chem.*, **181**, 1757–1762.

153. Privalko V. P., Deberdeev R. Ya., Rymarenko N. L. *et al.* (1986) Transcrystallization in polyethylene coatings, *Dokl. Acad. Nauk UkrSSR*, Ser. B, **11**, 44–48.

154. Gutzow I., Dochev V., Pancheva E. and Dimov K. (1978) Induced crystallization of poly(ethylene terephthalate) on small metallic nucleating particles, *J. Polymer Sci.: Polymer Phys. Ed.*, **16**, 1155–1168.

155. Privalko V. P., Kawai T. and Lipatov Yu. S. (1979) Crystallization of filled Nylon 6. II. Isothermal crystallization, *Colloid & Polym. Sci.*, **257**, 841–854.

156. Stein R. S. and Powers J. (1962) Microscopic measurements of spherulite growth rate, *J. Polymer Sci.*, **56**, S9–S10.

157. Mezhidov V. Kh. and Mezhidov S. Kh. (1983) Method to study properties of boundary layers in undercooled melt, *Koll. Zhurn.*, **45**, 349–351.

158. Malinsky Yu. M., Epelbaum I. V., Titova N. M. *et al.* (1968) Features of structure formation in thin polymer films, *Vysokomol. Soed.*, Ser. A, **10**, 786–798.

159. Malinsky Yu. M. (1970) On the influence of solid surface on processes of relaxation and structure formation in boundary layers of polymers, *Usp. Khim.*, **39**, 1511–1535.

160. Malinsky Yu. M., Prokopenko V. V., Titova N. M. *et al.* (1973) Some peculiarities of relaxation and structure-forming processes in polymer boundary layers, *J. Polymer Sci.: Polymer Symp. Ed.*, **42**, 679–691.

161. Malinsky Yu. M., Areshidze M. G. and Bakeev N. F. (1973) On the influence of surface energy and phase state of substrates on linear rate of spherulitic crystallization in boundary layers of polymers, *Dokl. Acad. Nauk USSR*, **208**, 1142–1149.

162. Raghavan V. and Cohen M. (1976) Solid-state phase transformations, in *Treatise on Solid State Chemistry*, Vol. 5, ed. by N. B. Hannay, Plenum Press, New York, pp. 67–127.

163. Privalko V. P. and Lipatov Yu. S. (1980) On the mechanism of polymer crystallization in thin films, *Dokl. Acad. Nauk USSR*, **251**, 151–155.

164. Good R. J. (1977) Surface free energy of solids and liquids: thermodynamics, molecular forces and structure, *J. Coll. & Interface Sci.*, **59**, 398–419.

165. Lipatov Yu. S., Privalko V. P. and Sharov A. N. (1982) On the packing density of macromolecules in thin polymer layers, *Dokl. Acad. Nauk USSR*, **263**, 1403–1406.

166. Privalko V. P. (1978) Dependence of nucleation parameters for polymer crystallization from the melt on chemical nature of a macromolecule, *Polymer J.*, **10**, 607–617.

167. Privalko V. P., Kawai T. and Lipatov Yu. S. (1979) Crystallization of filled Nylon 6. III. Non-isothermal crystallization, *Colloid & Polymer Sci.*, **257**, 1042–1048.

168. Privalko V. P., Tarara A. M. and Lipatov Yu. S. (1983) Crystallization of filled polychlorotrifluoroethylene under elevated pressures, *Dokl. Acad. Nauk UkrSSR*, Ser. B, **5**, 52–55.

169. Johnsen U., Spielgies G. and Zachmann G. (1970) Abhaengigkeit der heterogenen Keimbildung in Polypropylen von der Kristallisationstemperatur und von der Art der Baimengung der Fremdsubstanz, *Kolloid Z. & Z. Polymere*, **240**, 762–766.

170. Lipatov Yu. S. and Privalko V. P. (1984) On the criteria of the concept 'highly filled polymer', *Vysokomol. Soed.*, Ser. B, **26**, 257–260.

2 Polymer Blends

2.1 THERMODYNAMIC CHARACTERIZATION

According to the general thermodynamic equation, eq. (1.8), the excess Gibbs free energy of a binary system is the difference between that measured experimentally, G, and the corresponding additive value, i.e.

$$G_{ex} = G - [G_1\varphi + (1 - \varphi)G_2] \tag{2.1}$$

where G_i is the Gibbs free energy of the ith component, and φ is the volume fraction of component 1. In the case when two components form a homogeneous, single-phase system, $G_{ex} \leqslant 0$; the opposite case, $G_{ex} > 0$, corresponds to heterogeneous, phase-separated systems.

Values of G_{ex} at any temperature may be calculated in a straightforward manner by eq. (2.1) using absolute (i.e. measured from 0 K) heat capacities of pure components and their blends [1]. If, however, this information proves unavailable, the usual procedure is to estimate G_{ex} by eq. (2.2)

$$G_{ex} = H_{ex} - TS_{ex} \tag{2.2}$$

substituting enthalpic (H_{ex}) and entropic (S_{ex}) contributions calculated with the aid of appropriate theoretical models.

Within the framework of the Flory–Huggins lattice model (FHLM) [2–4], H_{ex} is assumed to be the enthalpy of formation of a regular solution

$$\Delta H_r = RTV_M\chi\varphi(1 - \varphi) \tag{2.3}$$

where

$$\chi = BV_1/RT \tag{2.4}$$

is the parameter of thermodynamic interactions between components 1 and 2, $B = (\delta_1 - \delta_2)^2$ is the corresponding interaction energy density, δ_1 and δ_2 are solubility parameters of components 1 and 2, and V_M and V_1 are molar volumes of a binary system and of component 1, respectively.

Accordingly, only the combinatorial contribution to S_{ex}, i.e.

$$\Delta S_{comb} = -RV_M[\varphi \ln \varphi/p_1 + (1 - \varphi)\ln(1 - \varphi)/p_2] \tag{2.5}$$

(where p_i is the degree of polymerization of the ith component), is accounted for.

Because the theoretical ΔH_r is always positive, thermodynamic stability of a single-phase state may be ensured only by contribution from ΔS_{comb}. This contribution is higher the lower are the p_i values of the components; however, for the more general case when both components are 'high polymers' ($p_1, p_2 \Rightarrow \infty$, $\Delta S_{comb} \Rightarrow 0$), a single-phase state ($G_{ex} \Rightarrow 0$) is predicted to occur only on the condition $\Delta H_r \Rightarrow 0$ (i.e. $\delta_1 \Rightarrow \delta_2$).

This prediction proved to apply reasonably well for blends of two non-polar components with similar values of δ_i, cis-1,4-polybutadiene and poly(vinyl ethylene), for which the additivity of specific volumes, heat capacity jumps at T_g, etc. were also observed [5]. However, vanishingly small values of all excess thermodynamic quantities presumed for the system cited should be regarded as an extremely rare occurrence, rather than a rule. In fact, it is believed [cf. eq. (1.37)] that even in the case of a mismatch between δ_i values (i.e. $\Delta H_r > 0$), high molecular mass polymers may still have the chance to the 'thermodynamically compatible' (i.e. $G_{ex} < 0$) due to the contribution of a negative excess volume, $v_{ex} < 0$, to the 'volume' component of the excess enthalpy, $\Delta H_v \cong P_i v_{ex}$ (making the seemingly reasonable assumption of additivity of internal energies).

In simple terms, the occurrence of non-zero values of v_{ex} may result from a redistribution of the free volume within the blend composed of polymers with different intrinsic free volume fractions. In fact, assuming a 'universal' value of the free volume fraction to be frozen at the glass transition temperature T_g for all polymers, $f_g \cong$ const [where f_g may be defined for example, by eqs. (1.19) and (1.20)], it becomes evident that two polymers with different $T_{g,i}$ values heated to the same temperature T, will have different free volume fractions, $f_i \sim (T - T_{g,i})$. At the molecular level, mixing of these two polymers may be visualized as a sort of 'squeezing out' of the excess free volume from the inter-segmental space of a more expanded, low-T_g polymer, by segments of a high-T_g component, thus leading so $v_{ex} < 0$ [3].

Quantitative justification of these admittedly qualitative arguments was provided, for example, by treatment of melt PVT data for blends of PS and poly(2,6-dimethyl-1,4-phenylene ether) (PPO) [6] within the framework of a Simha–Somcynsky scaled theory (SSST) [7] extended to binary mixtures [8]. It appears that the main source of compatibility of PS/PPO blend in the melt state is indeed the contribution of relatively small (0.1–0.5% of the actual volume) $v_{ex} < 0$.

Qualitatively, the occurrence of a volume contraction on mixing ($v_{ex} < 0$) suggests 'stronger-than-normal' interactions between the components. This problem was studied in detail by PVT measurements in the melt state of solution-blended samples of PMMA (viscosity-average molecular mass $\langle M_v \rangle = 500 \times 10^3$) and two fractions of PEO, PEO-2 ($< M_v > = 2 \times 10^3$) and PEO-125 ($\langle M_v \rangle = 125 \times 10^3$), as well as of blends of the same PMMA and poly(vinylidene fluoride) PVDF) ($\langle M_v \rangle = 550 \times 10^3$) [9–13].

As can be seen from Fig. 2.1(a), both melt specific volumes at normal pressure v_1^0, as well as their temperature and pressure derivatives (i.e. thermal expansivity α_1 and compressibility β_1) of PMMA/PEO-2, exhibit strong negative deviations from additivity in the whole composition range, whereas these deviations are considerably reduced for PMMA/PEO-125. More complex behavior was obser-

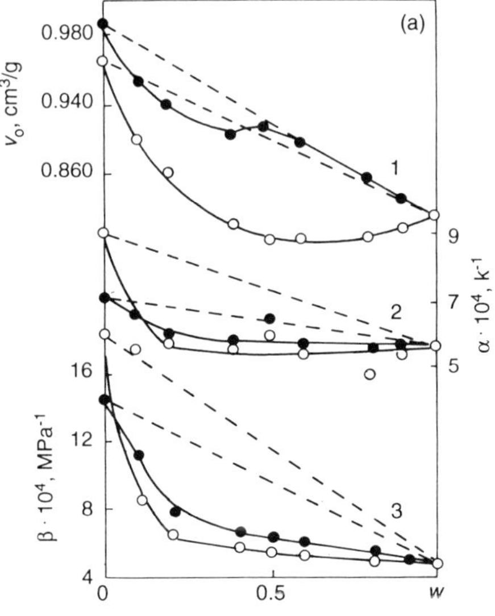

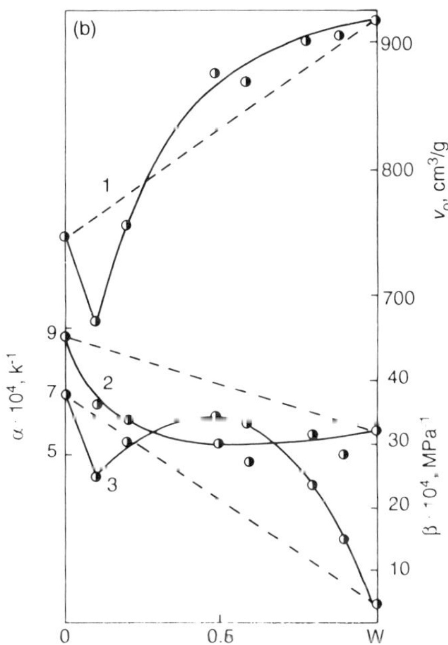

Fig. 2.1. Composition dependences of the melt-specific volume (1), thermal expansivity (2) and isothermal compressibility (3) for PEO/PMMA at 393 K (a) and PVDF/PMMA at 493 K (b). Broken lines: linear additivity behavior; ○: blends with PEO-2; ●: blends with PEO-125 ◑:

ved for PMMA/PVDF (Fig. 2.1(b)), namely, positive deviations from additivity of v_1^0 occurred at PMMA content $\varphi > 0.5$, in essential agreement with other data [14], while at $\varphi < 0.5$ the situation was reversed, a similar reversal being observed for β_1.

Treatment of the experimental PVT data with the aid of the lattice fluid model (LFM) reduced equation of state [15, 16]

$$\tilde{\rho}^2 + \tilde{P} + \tilde{T}[\ln(1 - \tilde{\rho}) + (1 - 1/p)\tilde{\rho}] = 0 \qquad (2.6)$$

(where $\tilde{\rho} = \rho/\rho^*$, $\tilde{P} = P/P^*$ and $\tilde{T} = T/T^*$ are reduced density, pressure and temperature, respectively, $\rho^* = M/pv^*$, $P^* = \varepsilon^*/v^*$ and $T^* = \varepsilon^*/k$ are the corresponding characteristic parameters, M is the molecular mass of chain molecule of close packed volume pv^* and ε^* and v^* are the energy and volume parameters of the intermolecular potential) allowed calculation of the parameter of binary interactions

$$\chi = [\varphi T_1^* + (1 - \varphi)T_2^* - T^*]/(1 - \varphi)T \qquad (2.7)$$

The negative sign of the normal-pressure values of χ for all systems (Fig. 2.2) may be regarded qualitatively as an indication of strong 'specific' interactions between the components. It is likely that dipole–dipole interactions are involved in PMMA/PEO-125 and PMMA/PVDF pairs, while a larger magnitude of negative χ for PMMA/PEO-2 suggests an additional contribution from stronger interactions (presumably, hydrogen bonding between end hydroxyls of the PEO-2 oligomer and carbonyls of PMMA chains). In fact, a 'dip' on the χ curve for this pair occurs at the composition corresponding to saturation of such hydrogen bonds.

Thus, it is specific interactions between polar groups of the components which may be thought of as a major origin of negative v_{ex} and (presumably) H_{ex} in most documented cases of compatible polymer blends (assuming the contributions to the cited excess quantities from a relative mismatch of characteristic volumes and/or temperatures of pure components [16] to be of secondary importance). Contributions of such interactions to the excess enthalpy of the blend may be accounted for by the following empirical equation [17]:

$$H_{ex} = \Delta H_r + n\Delta H_{sp} \qquad (2.8)$$

where $\Delta H_{sp} < 0$ is the enthalpy gain due to the formation of a single specific contact, and n is the number of such contacts per molecule.

According to eq. (2.8), any initially incompatible ($H_{ex} > 0$) binary system may be transformed into a compatible one ($H_{ex} < 0$) by chemical incorporation into the chain repeating unit of one or both components of a sufficient number of groups of specific interactions (say, donor–acceptor, dipole–dipole, hydrogen bonding, etc.), so that their contribution, $n\Delta H_{sp} < 0$, will dominate over $\Delta H_r > 0$. In fact, this empirical approach is now widely used to improve the compatibility of a large variety of polymer blends [18–21]. More rigorous, quantitative justification of this approach is provided by recent theoretical models [22, 23].

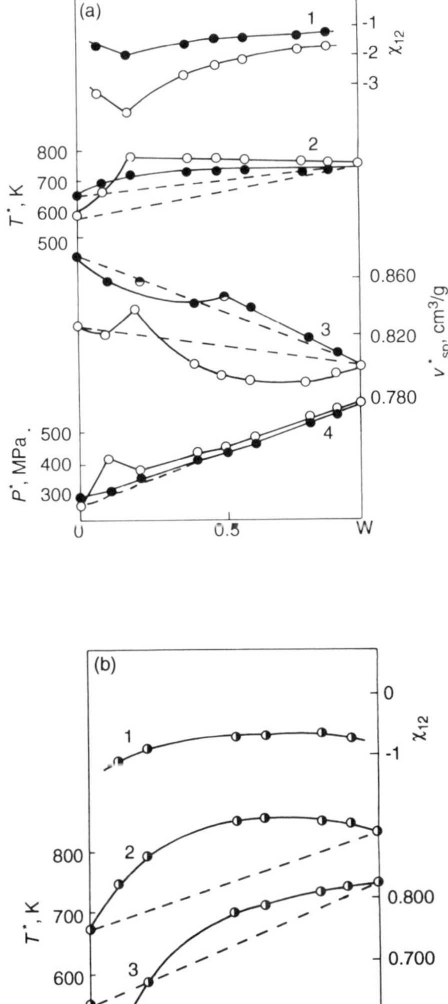

Fig. 2.2. Composition dependences of the thermodnamic interaction parameter (1), characteristic temperature (2), characteristic specific volume (3) and characteristic pressure (4) for PEO/PMMA (a) and PVDF/PMMA (b). Broken lines: linear additivity behavior

The occurrence of specific interactions between definite sites on the chain backbone of compatible components has two consequences: namely, the exothermic contribution $\Delta H_{ap} < 0$ lowers G_{ex} and thus promotes melt homogenization, whereas a network of specific bonds will hinder molecular mobility, which is basically a violation of a postulate of perfectly random mixing implicit in all theoretical models. The effect of non-randomness manifests itself, for example, as the excess intensity of small-angle X-ray scattering (SAXS) at zero angle, $I_0 \sim kT\beta_1$, for PMMA/PEO-2, PMMA/PEO-125 and PMMA/PVDF in the melt state [12, 13, 24]. Non-random distribution of segments also became evident for the former system from an electron spin resonance study of motion of nitroxide radicals attached to PEO [25], and from polarized light scattering measurements at room temperature (i.e. in the solid state) for the latter pair [26]. Essentially similar data were obtained for other apparently compatible polymer blends [27].

It is pertinent to emphasize here that non-zero values of v_{ex} in the equilibrium melt state should not be considered as necessary and sufficient conditions for compatibility, because such situations may occur under certain conditions even in blends of incompatible polymers. This was observed, for example, in the PVT studies of melt-blended samples of PS/bisphenol. A polycarbonate (PC) [28, 29], on plots of the composition dependence of melt-specific volume at normal pressure, v_1^0, estimated by application of the Tait equation

$$1 - v_1/v_1^0 = 0.0894 \ln(1 + P/B) \tag{2.9}$$

(where B is the 'material constant') to high-pressure data. In this case the values of v_{ex} apparently say more about the 'pressure-induced' thermodynamic state of the blend than about the true compatibility of components at normal pressure.

2.2 PHASE SEPARATION AND STRUCTURE FORMATION

2.2.1 EQUILIBRIUM ASPECTS

Blends of Linear Polymers

As pointed out in the preceding section, in the framework of FHLM [2, 4] the main criterion of a single-phase state of a binary polymer mixture in the equilibrium melt is the perfect matching of solubility parameters of two components, $\delta_1 = \delta_2$ (i.e. vanishingly small excess enthalpy H_{ex}) in so far as the non-zero value of the excess entropy S_{ex} at any finite temperature T ensures $G_{ex} < 0$ in eq. (2.2). Generally speaking, a change in temperature T or pressure P will also be accompanied by a change in enthalpic contribution to G_{ex} owing to a mismatch of δ_i values arising from differences in their temperature and pressure derivatives $(\partial \ln \delta_i/\delta T)_P \cong -\alpha_i$ and $(\partial \ln \delta_i/\delta P)_T = \beta_i$. However, while heating is expected to stabilize the single-phase state even further owing to increasing contribution of the $-TS_{ex}$ term, a decrease in $-TS_{ex}$ in the course of cooling (or, equivalently, hydrostatic compression), on the contrary, may eventually lead to $G_{ex} > 0$ (i.e. to separation of the melt into individual phases of each component).

These qualitative considerations apply to the lower portion of a schematic phase diagram (Fig. 2.3) where the solid line ('binodal') satisfies the condition of coexistence at equilibrium ($\mu'_1 = \mu''_1, \mu'_2 = \mu''_2$) of a single-phase state with chemical potential of components μ'_1 and μ'_2, and of a two-phase state with chemical potentials μ''_1 and μ''_2. For the FHLM considered, values of μ_i for each state may be calculated by differentiating G_{ex} by the corresponding number of moles N_i, i.e. [4]

$$\mu_1 = \partial G_{ex}/\partial N_1 = \ln \varphi + (1 - p_1/p_2)(1 - \varphi) + \chi p_1(1 - \varphi)^2 \qquad (2.10a)$$

$$\mu_2 = \partial G_{ex}/\partial N_2 = \ln(1 - \varphi) + (1 - p_2/p_1)\varphi + \chi p_2 \varphi^2 \qquad (2.10b)$$

The broken line in Fig. 2.3 ('spinodal') corresponds to the geometrical location of inflection points (i.e. those satisfying the conditon $\partial^2 G_{ex}/\partial \varphi^2 = 0$) which may be defined as

$$\chi_{sp} = 0.5[1/p_1 \varphi_{sp} + 1/p_2(1 - \varphi_{sp})] \qquad (2.11)$$

where the subscript sp means 'spinodal'.

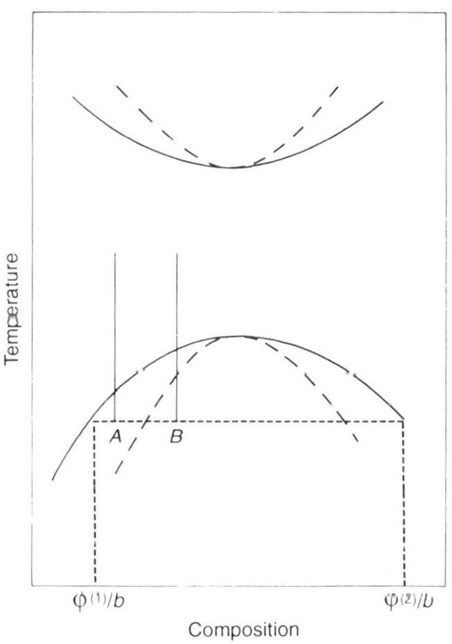

Fig. 2.3. Typical phase diagram for a binary liquid system (schematic)

It is possible now to differentiate between the following phase states of a binary polymer system:

(i) $\partial^2 G_{ex}/\partial\varphi^2 > 0$: absolute stability of a single-phase state (phase space above binodal);

(ii) $\partial^2 G_{ex}/\partial\varphi^2 < 0$: absolute instability of a single-phase state with respect to any small composition fluctuations (phase space 'inside' spinodal); and

(iii) metastability of a single-phase state (phase space in between binodal and spinodal).

Binodal and spinodal have a common 'critical' point at the highest temperature of coexistence of different phase states (the 'upper critical solution temperature', or UCST) which is defined by a corresponding 'critical' interaction parameter as

$$\chi_{cr} = 0.5\,(1/p_1^{1/2} + 1/p_2^{1/2})^2 \tag{2.12}$$

In so far as $\chi_{cr} \Rightarrow 0$ when p_1 and $p_2 \Rightarrow \infty$, eq. (2.12) is an alternative formulation of the vanishingly small excess enthalpy as a criterion of compatibility of high polymers within the framework of FHLM.

As already pointed out, this model predicts an improvement in compatibility of two polymer components in the melt state during continuous heating above the UCST; however, in the course of such heating the mismatch of not only solubility parameters, but also of the free volume fractions of components, becomes so dramatic that segments of more expanded, low-T_g polymers start to 'condense' onto segments of high-T_g components. Since new intersegmental bonds are formed, this process is, of course, exothermic (i.e. a part of the excess enthalpy is lost); however, the concomitant entropy loss is so severe that the Gibbs free energy becomes positive and the system separates into two phases. Relevant to the latter case, a simple phase diagram with a common point for binodal and spinodal at the lower critical solution temperature (LCST) is shown in the upper portion of Fig. 2.3.

It may be expected that the occurrence of both the UCST and the LCST should be regarded as a universal feature of phase diagrams for any combination of linear polymers (provided other factors like glass transition on cooling or thermal degradation on heating do not interfere). However, the shape of the phase diagram as well as the width of the 'miscibilty window' between the UCST and the LCST will be critically affected by the molecular characteristics of components. Phase diagrams constructed according to the spinodal equation of LFM (for the simplest case of identical characteristic volumes of components) [16]

$$\chi_{sp} = 0.5\{[1/p_1\varphi + 1/p_2(1 - \varphi)]/\rho - T\psi^2 P^*\beta_1\} \tag{2.13}$$

(where $\psi = \rho[\varphi/T_1(1 - \varphi) - (1 - \varphi)/T_2\varphi - 1/T(1 - \varphi) - 1/T\varphi] - (1/p_1 - 1/p_2)$ and subscripts sp on the volume fractions has been omitted) may change the shape from the simplest one shown in Fig. 2.3 to 'closed loops', 'hourglass'-like, etc. depending on the relative mismatch between the characteristic parameters of components. In qualitative terms, the major factor ensuring stability of a single-phase state, i.e. a negative value of the excess enthalpy H_{ex}, will be favored both by a negative value of the interaction parameter χ as well as by higher-than-unity ratios of the characteristic parameters v_1^*/v_2^* and T_1^*/T_2^* [16].

Equation (2.13) may also be used to predict the pressure dependence of the phase diagrams. Broadly speaking, for the LCST the initial increase with pressure is expected to change for gradual decrease, whereas the sign of the pressure derivative of the UCST is a more delicate function of the various 'material parameters' in eq. (2.13) [16]. The signs of the temperature and pressure derivatives of the interaction parameter estimated by LFM ($d\chi/dT \geqslant 0$ and $d\chi/dP < 0$ for PMMA/PEO-125 and $d\chi/dT < 0$ and $d\chi/dP > 0$ for PMMA/PVDF [10–13]) suggest the probability of the LCST for the former and the UCST for the latter pair.

Interpenetrating Polymer Networks

Originally, interpenetrating polymer networks (IPN) were conceived as chemically homogeneous (on a sufficiently large structural scale) binary polymer systems which could be conveniently prepared by swelling the first loosely cross-linked component in a suitably chosen monomer and subsequently cross-linking the latter [30, 31] (in this fashion, the 'sequential', seq-IPN, are produced). Conceptually, this procedure of 'chemical' blending offered the possibility to avoiding the hazards of eventual phase separation which were notorious for 'physical' blending of linear polymers (cf. the preceding section), due to topological constraints (permanently entangled, chemically different macrocycles) on IPN unmixing [32, 33].

Invariably, the positive values of G_{ex} estimated by eq. (2.1) from vapor sorption data [34,35] combined with results of morphological studies [36 37] revealed, however, that the overwhelming majority of both 'simultaneous', sim-IPN (i.e. those prepared by simultaneous polymerization of the homogeneous mixture of two different functional monomers) and seq-IPN studied so far were, in fact, phase-separated systems. Qualitatively this result may be attributed to the continuous decrease of the combinatorial mixing entropy S_{ex} [cf. eq. (2.5)] in the course of polymerization, so that the initially homogeneous single-phase mixture becomes unstable ($G_{ex} \geqslant 0$) and separates into two phases.

A more rigorous approach to the phase stability problem of IPN (in the case of sufficiently low cross-link densities) involved evaluation of the following contributions to the Helmholtz free energy of IPN [38].

(i) The elastic contribution (of each component)

$$\Delta F_{el}/kT = [v(\varphi_0)/2](\lambda_x^2 + \lambda_y^2 + \lambda_z^2 - 3) - Bv(\varphi_0)\ln(\lambda_x\lambda_y\lambda_z) \qquad (2.14)$$

where $v(\varphi_0)$ is the number of elastically active chains (i.e., network density), φ_0 is the volume fraction of monomers of the first component in the initial mixture during the cross-linking reaction, λ_x, λ_y and λ_z are the components of the relative chains extension and B is the proportionality factor which is assumed equal to zero, $2/f$ or unity in different theoretical formulations (here f is the cross-link functionality).

(ii) The interaction energy between components [cf. eq. (2.3)].

(iii) An entropy change due to mutual entanglements

$$\Delta S_{\text{ent}} = (1 - \varphi_0)^{1/3}/\varphi_0^{2/3} p_1' p_2'^{1/2} A_1^2 A_2 \tag{2.15}$$

where the A_i values are the chain segment lengths of two components, and primes are used to distinguish network chains from free ones.

Assuming additivity of the elastic contributions from two components and $\Delta S_{\text{ent}} \Rightarrow 0$ (in the limit of low network densities or high p_i values and $\varphi_0 \gg 1 - \varphi^0$), the spinodal curve for sim-IPN may be calculated from eq. (2.16) [38]:

$$\chi_{\text{sp}}(\varphi_0) = -0.5\{[(C_1/3 - B_1)/p_1'(\varphi_0)]/\varphi_0 - [(C_2/3 - B_2)/p_2'](1 - \varphi_0)]\} \tag{2.16}$$

where $C_1 = \varphi_{\text{sp}}/\varphi_0$ and $C_2 = (1 - \varphi_{\text{sp}})/\varphi_0$.

Analysis of eq. (2.16) shows that at $B = 0$ sim-IPN would be thermodynamically unstable at any φ_0, whereas in other cases (i.e. $B \neq 0$) phase stability is predicted at small φ_0. On the other hand, seq-IPN are not expected to be stable at any φ_0 or B [38]. More detailed analysis of issue (iii) [39] has shown that one more factor promoting phase separation in IPN is the change in the primitive paths of subchains in each network component (in other words, a change in the corresponding local densities) due to topological entanglements at a critical wave vector length.

Recently, a series of new sim-IPN of PPO-c/PMMA-c, PMMA-c/poly(carbonate-urethane) (PCU-c), etc. was synthesized [40–42]. These apparently single-phase IPN are unique in the sense that the 'physical' blends of the corresponding linear polymers are immiscible in the melt phase. However, the major result of thermodynamic characterization of IPN (PMMA-c/PCU-c = 46/54) [43] is $G_{\text{ex}} > 0$ in the broad temperature interval encompassing the solid (glassy) and rubbery states (Fig. 2.4). Moreover, inspection of the plots in Fig. 2.4 allows us to identify roughly the following characteristic temperature intervals.

(i) $T < 30$ K (all samples are in the glassy state).

In view of approximately equal heat capacities for all samples, all excess quantities are negligibly small.

(ii) 30 K $< T < 200$ K (all samples are still in the glassy state).

Both H_{ex} and S_{ex} are negative and pass through a shallow minima around 120 K for the former and 160 K for the latter.

(iii) 200 K $< T < 350$ K (PCU-c and IPN pass into the rubbery state; PMMA-c is still glassy).

H_{ex} changes sign to positive; S_{ex} is still negative; both pass through the shallow minima located around 280 K (i.e. about 30 K above the corresponding T_g values).

(iv) 350 K $< T < 450$ K (all samples pass into the rubbery state).

Both H_{ex} and S_{ex} are positive and exhibit an accelerated rise with temperature.

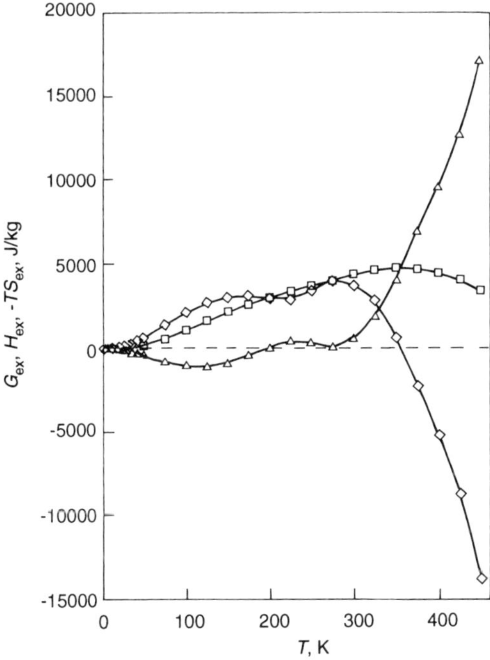

Fig. 2.4. Temperature dependence of the excess enthalpy ($\triangle$), excess entropy ($\diamond$) and excess Gibbs free energy ($\square$) for IPN

It appears from the above data that at the lowest temperatures [i.e. in the interval (i)] the vibration wavelength by far exceeds the dimensions of any structural entity of IPN (say, chain segments, cross-linking sites, etc.); thus, differences in the molecular structure of the components become insignificant and IPN behaves as a single-phase solid body ($G_{ex} \cong 0$). As the temperature rises to the interval (ii), the energetically favorable ($H_{ex} < 0$) interactions between the components are, however, counterbalanced by a concomitant decrease in the vibrational mobility ($S_{ex} < 0$); hence, $G_{ex} > 0$. On further heating up to interval (iii), the increased amplitude of molecular vibrations effectively destroys the interchain coupling, which manifests itself as change in the sign of H_{ex} to positive and an increase in negative S_{ex} close to zero, with G_{ex} still positive. Finally, in the highest temperature interval (iv), the interchain repulsion between different species ($H_{ex} > 0$) leads to still higher molecular mobility ($S_{ex} > 0$). However, the contribution of this latter effect turns out to be smaller compared with the latter; thus, G_{ex} remains positive.

A further insight into the origin of $G_{ex} > 0$ observed is provided by substitution of $H_{ex} = 600 \, \text{J/kg}$ and the excess volume $v_{ex} = -0.11 \, \text{cm}^3/\text{g}$ (at room temperature) into $H_{ex} \cong \pi_{ex} v_{ex}$, which yields $\pi_{ex} \cong -55 \, \text{MPa}$ for the excess internal pressure of IPN. This result is additional experimental evidence suggesting that the apparently single-phase, 'chemically densified' state of IPN is, however, far from equilibrium since the negative π_{ex} may be regarded as the driving force for the transition into another (i.e. phase-separated) state which is closer to the internal equilibrium. As a plausible structural model of IPN, one may invoke the notion of an array of densely packed chains in a somewhat extended (as compared with random coil) conformation. This model is qualitatively consistent with both observations, i.e. the negative signs of v_{ex}, π_{ex} and S_{ex} (under the latter quantity, basically, intramolecular or conformational entropy should be understood), and the positive values of H_{ex} and G_{ex}. Moreover, the improved mechanical properties of IPN [42] now seem a quite natural feature of proposed model because the extended chains provide for a sort of 'self-reinforcement'.

In this respect, the situation with sim-IPN looks conceptually identical to the one encountered with polymer glasses 'physically' densified ($v_{ex} < 0$) from the equilibrium rubbery state above T_g under high hydrostatic pressures into a non-equilibrium, higher-energy ($H_{ex} > 0$) glassy state [44–46]. The fundamental difference between the 'physical' densification of linear polymers and the 'chemical' densification of IPN is that, in the absence of external constraints, the former retain the possibility of relaxing to a more equilibrium state, while the latter are bound to remain in the non-equilibrium, single-phase state indefinitely owing to the permanent topological constraints on the transition into an equilibrium, phase-separated state.

2.2.2 KINETIC ASPECTS

The sudden 'quench' of an initially homogeneous, single-phase binary system above the LCST or below the UCST will have the following two fundamentally different consequences depending on the amplitude of this perturbation (cf. Fig. 2.3).

(i) A moderate amplitude (the system is transferred to point A in the phase space in between binodal and spinodal).

This point corresponds to the thermodynamic stability of a two-phase state with lower chemical potential $\mu'' < \mu'$; however, at least for short enough times, a single-phase state may still be kinetically stable (i.e. 'metastable') against smaller-than-critical composition fluctuations. This situation is conceptually similar to metastability of the melt state of a crystallizable polymer at small undercoolings when the bulk driving force for crystallization, ΔG_v, is insufficient to overcome the unfavorable contribution of the excess interfacial energy ΔG_s (cf. section 1.4 of Chapter 1). Therefore, it is possible to apply the formalism of classical nucleation theory of liquid–vapor of liquid–crystal transformation to the present case of liquid–liquid phase separation assuming $\Delta G_v = V(\mu'' - \mu') \cdot [\varphi_b^{(2)} - \varphi_b^{(1)}]$ and $\Delta G_s = \gamma S$ [47] (where $\varphi_b^{(i)}$ are the coordinates of points on binodal defined by Fig. 2.3, γ is the interfacial energy, and V and S are the volume and surface area of a nucleus). Thus, phase separation in the metastable region of the phase diagram is assumed to start by homogeneous nucleation of, say, component 1 on a limited number of sufficiently large composition fluctuations in a single-phase system and to develop further by the usual mechanism of molecular diffusion across the interface (because $\partial^2 G_{ex}/\partial\varphi^2 > 0$ in the metastable region, the diffusion flux will be 'positive', i.e. oriented opposite to the concentration gradient). In the course of the latter process, the critical nucleus size will increase due to a continuously growing deficit of component 1 in the homogeneous phase; hence, nuclei of subcritical size will dissolve in the latter to be 'swallowed' by growing supercritical nuclei (the 'Ostwald ripening' mechanism of growth) [48].

(ii) Large amplitudes (the sysem is transferred to point B within spinodal).

Corresponding to this case the inequality $\partial^2 G_{ex}/\partial\varphi^2 < 0$ implies absolute instability of a single-phase state against even small composition fluctuations. Therefore, no nucleation barrier will affect the probability of a phase separation event which will thus occur homogeneously throughout the whole system by the diffusion mechanism; however, in view of the above inequality the diffusion flux will be oriented in the 'negative' direction, i.e. along the concentration gradient (the mechanism of 'spinodal decomposition').

In phenomenological terms [49], small-scale concentration fluctuations in a homogeneous (on a large scale) solution give rise to additional contributions to the bulk free energy, namely

$$G = [g(\varphi) + K(\nabla\varphi)^2]\,dV \tag{2.17}$$

where $g(\varphi)$ is the free energy density of a homogeneous system, $K(\nabla\varphi)^2$ is the contribution from the composition gradient and $K > 0$ is constant. Thus, the free energy difference between the initially homogeneous state and the inhomogeneous state may be expressed as

$$G' - G'' = [0.5(\partial^2 g/\partial\varphi^2)(\varphi' - \varphi'')^2 + K(\nabla\varphi)^2]\,dV \tag{2.18}$$

Equation (2.18) predicts that the fate of the solution will depend on the sign of the derivative $\partial^2 g/\partial\varphi^2 = \partial^2 G_{ex}/\partial\varphi^2$: namely, the positive sign will ensure the stability of the homogeneous state, otherwise the latter will be unstable.

It is expected [49] that the phase separation by the spinodal decomposition mechanism will develop due to the growth of the amplitude of the composition fluctuations with wavenumbers β smaller than a certain critical value

$$\beta_{cr} = [(-\partial^2 G_{ex}/\partial\varphi^2)/2K]^{1/2} \tag{2.19}$$

while those with $\beta > \beta_{cr}$ will decay with time and vanish. The corresponding Fourier component of the mean local concentration fluctuations at time t is expressed as

$$\langle \varphi(\beta, t) \rangle = \langle \varphi(\beta, 0) \rangle \exp[R(\beta)t] \tag{2.20}$$

where

$$R(\beta) = -D\beta^2 - 2MK\beta^4 \tag{2.21}$$

is the 'amplification factor' passing through a sharp maximum at $\beta_{\max} = \beta_{\mathrm{cr}}/2^{1/2}$, $D = M(\partial^2 G_{\mathrm{ex}}/\partial\varphi^2)$ is the diffusion coefficient and M is the 'mobility factor'.

The most general predictions of this phenomenological theory which should apply to any binary system (at least at the earliest stage of spinodal decomposition) are the following [49].

(a) Morphologically, the phase-separated system will consist of two continuous, interpenetrating strutures with a characteristic linear dimension (fluctuation wavelength) $\lambda_{\max} = 2\pi/\beta_{\max}$.

(b) The composition of each phase during phase separation [cf. eq. (2.20)] should change exponentially with time, keeping $\beta_{\max} = \mathrm{const}$.

(c) Experimental values of the diffusion coefficient D must be negative.

All these predictions were apparently confirmed for solution-blended samples of styrene-acrylonitrile (SAN) copolymer/PMMA studied by transmission electron microscopy [50], for PS/poly(vinyl methyl ether) (PVME) studied by light microscopy and pulsed NMR [51], light scattering [52, 53] and excimer fluorescence [54, 55], as well as for initial stages of synthesis of IPN based on polyurethanes [56, 57] and epoxy resins [58] studied by the SAXS technique. On the other hand, failure of issue (b) established in light scattering studies of PS blended wth poly(o-chlorostyrene) [59] and with poly(methylphenylsiloxane) [50, 61] is likely the result of an 'overshoot' of the initial stage of spinodal decomposition where the linear theory [49] was intended to apply.

It was also established experimentally that the dimensions of microphases were inversely dependent on temperature [50, 51]. Qualitatively, this finding may be rationalized in the framework of FHLM as related to the temperature dependence of χ [cf. eq. 2.4)], i.e. substituting $\partial^2 G_{\mathrm{ex}}/\partial\varphi^2 \sim (T/T_{\mathrm{sp}}^{-1})$ into eq. (2.22) to obtain

$$\lambda_{\max} = 2\pi l_{\mathrm{D}}[3(1 - T/T_{\mathrm{sp}})]^{-1/2} \tag{2.22}$$

where l_{D} is the characteristic Debye distance of intermolecular interactions. Further progress in the application of spinodal decomposition theory to polymers may be expected by accounting for non-linear effects at the later stages [47, 62–64], as well as by replacement of $\partial^2 G_{\mathrm{ex}}/\partial\varphi^2$ from simple FHLM with more appropriate expressions (e.g. from LFM [16, 23]).

2.2.3 BOUNDARY INTERPHASE

Intuitive expectations that the pure phases of each component of a polymer blend will be recovered after completion of phase separation by whatever mechanism (i.e. the degree of phase separation, α_{sep}, will be unity) were not confirmed experimentally. In fact, the degree of phase separation, α_{sep}, which may be estimated either by a pragmatic application of eq. (1.32), i.e.

$$\alpha_{\mathrm{sep}} = 1 - v_{\mathrm{BI}} = \Delta C_{\mathrm{p}}/\Delta C_{\mathrm{p}}^0 \tag{2.23}$$

(where v_{BI} is the fraction of the boundary interphase, BI, ΔC_{p} and ΔC_{p}^0 are heat capacity jumps at T_g of the polymer in the blend and in the pure state, respectively), or by the theoretically more appropriate eq. (2.24) [65, 66]

$$\alpha_{\text{sep}} = \langle \Delta\rho^2 \rangle / \langle \Delta\rho^2 \rangle_0 \tag{2.24}$$

(where $\langle \Delta\rho^2 \rangle = KQ$ is the experimental value of the mean square of electron density fluctuations, $\langle \Delta\rho^2 \rangle_0 = \varphi(1 - \varphi)(\rho_1 - \rho_2)^2$ is the corresponding theoretical quantity, Q is the small-angle X-ray scattering invariant, K is the calibration constant, and ρ_1 and ρ_2 are the electron densities of the components), invariably, turned out smaller than unity [67, 68]. It can be thus concluded that phase separation is never complete, so that a fraction v_{BI} of material available remains in the boundary interphase (BI). Theoretical and experimental methods to estimate v_{BI} (or, equivalently, the thickness of the BI, Δr) will be reviewed below.

Theoretical Aspects

As can be seen from Fig. 2.5, the local density of a phase-separated system is assumed to vary continuously in the BI of thickness Δr between the microphases of each individual component with different densities. Straightforward application of Gibbs' postulate of local equilibrium (i.e. equality of the chemical potentials in three coexisting microphases, $\mu_1 = \mu_2 = \mu_{\text{BI}}$ [69]), yields [70]

$$\Delta r = (\gamma_2 - \gamma_1) V_2^{\text{BI}} / RT(p_2/p_1)\varphi^{\text{BI}} \tag{2.25}$$

where γ_i is the surface tension of the ith component, and V_2^{BI} and φ^{BI} are the molar volume of component 2 and volume fraction of component 1 in the BI, respectively.

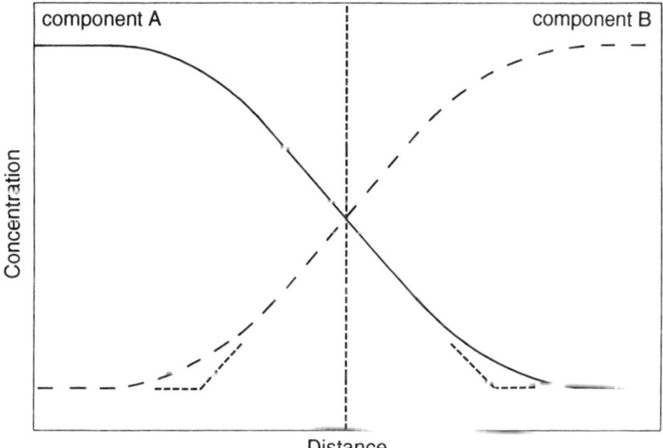

Fig. 2.5. Dependence of the relative concentration of components of a binary system on distance from the interface (schematic)

An apparently better approximation is provided by substitution of eq. (2.4) for χ from FHLM to account for the thermodynamic interactions between the components, i.e. [71]

$$\Delta r = (\gamma - \gamma_2)V/RT[\ln(1 - \varphi)^{\text{BI}} + p\chi(\varphi^{\text{BI}})^2] \qquad (2.26)$$

where γ is the interfacial tension, and V and p are the geometrical means of molar volumes and polymerization degrees of components, respectively.

Alternatively, the properties of the BI may be derived from considerations of possible contributions to gradient terms in the expressions for G [cf. eq. (2.14)]. A simple lattice model treatment of the contribution of conformational entropy of a macromolecule within the BI predicted a sigmoidal pattern of reduced density variation in the BI, namely [72, 73]

$$\rho_i(x)/\rho_0 = [1 + \exp(x/\Delta r)]^{-1} \qquad (2.27)$$

where

$$\Delta r = 2[(R_1^2 + R_2^2)/2\chi)^{1/2} \qquad (2.28)$$

is the BI thickness, $R_i^2 = p_i l_{\text{b}}^2/6$ is the squared gyration radius of the macromolecule of the ith component and l_{b} is the length of the main chain bond.

On the other hand, an account of the enthalpic contribution evaluated by LFM yields [74]

$$\Delta r = [(P_1^* P_2^*)^{1/2} k_0/2\Delta a_{1/2}](v_1^* v_2^*)^{1/6} \qquad (2.29)$$

where $k_0 = (T_1^*/T_2^*)^{1/2}(v_2^*/v_1^*)^{1/6}\rho_1^2 + (T_2^*/T_1^*)^{1/2}(v_1^*/v_2^*)^{1/6}\rho_2^2$, and $\Delta a_{1/2}$ is the maximum (i.e. taken at the half-width of the BI) value of $\Delta a = \varepsilon^* \rho/v^* \{-\rho + T[(v-1)\ln(1-\rho)+(1/p)\ln\rho + (\varphi/p_1)\ln\varphi + (1-\varphi)/p_2 \ln\varphi_2]\} + P - \rho/v^*[\varphi\mu_1^* + (1-\varphi)\mu_2^*]$.

In spite of the different formalisms, all theories cited predict Δr decreasing from 'infinity' (i.e. in the limit of $p_i < 100$) to a finite value of the order of several nanometers (for immiscible high polymers). Qualitatively similar behavior (i.e. a continuous decrease of Δr from infinity at T_{cr} to a finite value at $T \neq T_{\text{cr}}$) is predicted as a result of the inverse proportionality between Δr and the depth of penetration into spinodal space, $(T - T_{\text{cr}})$ [75, 76].

Experimental Aspects

Rough estimate of Δr may be made by combining eqs. (1.32) [e.g. (2.23)] and (1.33); however, in this case the dimensions of the microphases (or, equivalently, the interfacial surface area) should be determined from independent measurements.

For several polymer blends the thickness of the BI determined by electron microscopy and IR spectroscopy was as high as several microns [77–79]; however, these data should be regarded with caution in view of the theoretical arguments suggesting Δr smaller by several orders of magnitude (cf. preceding section). Thus, it seems preferable to use small-angle scattering techniques in the range of scattering angles ($2\theta > 3°$) where Porod's law for microheterogeneous, two-phase systems should apply.

In the case where the interface is sharp (i.e. $\Delta r = 0$) the true intensity of scattering is expressed as [66, 68]

$$J_t(S) = K'S^{-n} \tag{2.30}$$

where $n = 4$ (point collimation) or $n = 3$ (infinite slit collimation) and K' is the structure-dependent parameter. Deviations from eq. (2.30) [i.e. $K' \neq \text{const} = f(S)$] imply a density gradient across the BI; in this case, the thickness of the latter, Δr, may be obtained from the following general expression:

$$\Delta r = C\sigma, \tag{2.31}$$

where C is a numerical constant which depends on the 'smearing' function used to aproximate the shape of the density profile in the BI (Fig. 2.5), and σ is the corresponding dispersion.

A quantitative application of eq. (2.30) requires the following steps in the preliminary treatment of the raw data [68].

(i) Correction for the contribution of background scattering due to thermal density fluctuations, $J_T(S)$, to the intensity on a descending portion ('tail') of the experimental small-angle scattering curve, i.e.

$$J(S) = J_t(S) + J_T(S) \tag{2.32}$$

where $S = 2\sin\theta/\lambda$ is the scattering vector, 2θ is the scattering angle and λ is the radiation wavelength. Both exponential and power series versions of $J_T(S)$ have been proposed; however, the assumption $J_T(S) = \text{const}$ also works well [68].

(ii) Seeking a structural function $K' = f(S)$ yielding the best fit of the corrected values of $J_t(S)$ to eq. (2.30). So far, this criterion was apparently satisfied best by the following function, which accounts empirically for a sigmoidal pattern of density variation within the BI [cf. eq. (2.27)], namely [80, 81]

$$K'(S) \sim \exp[-38(\sigma S)^{1.81}] \tag{2.33}$$

Application of the procedure described above to an analysis of the SAXS data for a PMMA/epoxy resin blend [82] yielded, via eq. (2.31), $\Delta r \cong 2\,\text{nm}$ (assuming $C = 1$), which is consistent with theoretical predictions. On the other hand, $\Delta r = (2 \pm 0.5)\,\text{nm}$ for a PMMA/PS blend determined by the neutron critical reflection technique [83] is considerably smaller than theoretical estimates by both eq. (2.28) [72] and eq. (2.29) [74] (16.0 and 5.7 nm, respectively). Moreover, it turned out that the sigmoidal shape of the interfacial density profile assumed in current theories of polymer–polymer interfaces was less adequate (as concerns the theoretical fit to the experimental data on the kinetics of mutual diffusion of chlorinated PE and deuterated PMMA blend obtained by the latter technique) compared with a more sophisticated one (a linear gradient followed by Gaussian tail) [84]. Nevertheless, it seems safe to conclude that the most salient features of the static and dynamic behavior of polymer–polymer interfaces may be (at least, semi-quantitatively) accounted for by the current theories outlined in the preceding section.

2.3 GLASS TRANSITION IN BLENDS OF NON-CRYSTALLINE POLYMERS

2.3.1 KINETIC ASPECTS

As described in detail in section 1.2.1, in isobaric conditions the study of the relaxation kinetics of a selected structure-sensitive property j of a single-phase glass-forming liquid involves monitoring its evolution from the initial state (j_0) to the final state (j_∞) in functions of either time t (after a sudden temperature jump, ΔT) or temperature T (during heating or cooling at a constant rate $q^\pm$).

In the former case the experimental data at each ΔT are treated first by eq. (1.13) to obtain the relaxation time τ and the non-exponentially factor β; next, the activation enthalpy Δh is estimated by eq. (1.12) and, finally, the non-linearity factor X is derived by eq. (1.14). Enthalpy relaxation data for pure components and a solution-blended, single-phase sample of PS and PVME (50/50) treated in this fashion [85] revealed that while eq. (1.12) did not apply (apparently due to the close proximity of the aging temperatures T_a to $T_g = 256$ K for PVME), eq. (1.13) worked reasonably well yielding $\beta = 0.390 \pm 0.090$ for the blend (i.e. in between 0.350 for PS and 0.572 for PVME). It was concluded that the overall kinetic pattern was dominated by the contribution from PVME, whereas the relaxation rate was decreased by PS [85]. In a similar study of a compatible PMMA/SAN blend [86], the relaxation rate turned out to be slower, the smaller the difference $T_g - T_a$, and the higher the SAN content; at the same time, the experimental data for all compositions at the lowest aging temperatures, $T_a = T_g - 50$, could be fitted to a single master curve with $\beta = 0.732$.

Alternatively, the kinetic parameters of relaxation may be derived by fitting the experimental rate-heating DSC curves to eq. (1.17). As can be seen from Table 2.1 [87], the apparent activation enthalpies, $\Delta h/k$, for all PMMA/SAN blends were slightly in excess of the corresponding additive values, which is qualitatively consistent with relatively small $H_{ex} < 0$ [88, 89]. On the other hand, values of β and X fell somewhere in between those for either of the pure components, the latter parameter exhibiting fairly strong negative deviations from additivity at high SAN contents. The origin of this behavior is not fully understood; it is perhaps pertinent to mention here the results of a pulsed NMR study [90] in which experimental evidence for composition inhomogeneity on a scale of 2–15 nm was obtained; moreover, marginally non-exponential rotating frame relaxation was also noted for the blend with the highest SAN content.

It has been tacitly assumed so far that the above approach was applicable

Table 2.1 Kinetic parameters of enthalpy relaxation at T_g for PMMA/SAN blends

SAN content	$\Delta h/k \times 10^{-4}$ (K)	$-\ln \tau_0$ (s)	X	β
0	13.22	359.8	0.338	0.265
0.2	14.03	377.6	0.227	0.351
0.4	14.62	392.6	0.253	0.346
0.6	13.86	370.8	0.147	0.423
0.8	13.47	358.5	0.137	0.460
1.0	12.50	328.9	0.221	0.492

to compatible, single-phase blends (i.e. with a single T_g). However, it may be extended [91] to study the relaxation behavior of binary systems like poly(vinyl chloride) (PVC)/poly(isopropyl methacrylate) (PiPMA) [71], PS/poly(2-vinyl pyridine) (P2VP) [92], PMMA/PS [93–96], etc. each composed of nearly immiscible polymers but still exhibiting an apparently single T_g owing to a fortuitous overlap of the respective glass transition intervals of the pure components.

In so far as for sufficiently small times of aging t_a at $T_a < T_g$ the fictive temperature T_f should not be too much different from T_g, we obtain from eq. (1.14) [92]

$$\tau \cong \tau_0 \exp[(X \Delta h/k)(1/T_g - 1/T_a)] \tag{2.34}$$

It follows from eq. (2.34) that even in the case of a blend composed of incompatible polymers with relaxation parameters (i.e. $\Delta h, X, \tau_0$, etc.) all identical, with the exception of slightly different $T_{g,i}$ values, aging during the same 'physical' time t_a will bring the component with larger 'effective' time $t_{eff} = t_a/\tau$ closer to equilibrium (i.e. that with a smaller difference between T_a and $T_{g,i}$). In DSC annealing and rate-heating experiments with immiscible PVC/PiPMA [91], PS/P2VP [92] and PMMA/PS [93] blends, this effect manifested itself as a gradual splitting of the initially smooth, broad transition interval into two relaxation peaks, both the location and shape of the latter exhibiting different rates of evolution with t_a. In this context, the experimentally observed slowing down of enthalpic relaxation in PMMA/SAN blends enriched with SAN [86] in which composition inhomogeneity was detected [90] (see above) may now be attributed to the effect of the difference $T_g - T_a$ on t_{eff}.

2.3.2 QUASI-EQUILIBRIUM ASPECTS

Composition Dependence of T_g

The occurrence of a single T_g located in between the $T_{g,i}$ values of pure components is traditionally considered to be one of the most convincing empirical criteria of miscible polymer blends [97, 98]. Theoretical approaches to explain and/or predict the T_g of single-phase blends are usually based on the analysis of composition dependence of an appropriately chosen order parameter, Z_g, frozen below T_g (cf. section 1.2.2). These approaches will be reviewed below.

(i) By definition, the free volume fraction of any amorphous system above T_g is simply

$$f = f_g + \alpha_f (T - T_g) \tag{2.35}$$

where f_g is the free volume fraction at T_g and α_f is the corresponding expansion coefficient. Assuming additivity of the contributions of the components f_i to f of the blend and a universal value $f_g = \text{const}$ for all systems regardless of composition, we obtain from eq. (2.32) for the temperature interval between $T_{g,1}$ and $T_{g,2}$ [99]

$$T_g = \frac{T_{g,1}\varphi + T_{g,2}(1 - \varphi)K_\alpha}{\varphi + (1 - \varphi)K_\alpha}, \tag{2.36}$$

where $K_\alpha = \alpha_{f,2}/\alpha_{f,1}$ (in usual practice, $\alpha_{f,i}$ is identified as $\Delta\alpha_i$).

(ii) The assumption of additivity of the 'configurational' entropies of the components $\Delta S_{g,i}$, frozen at the respective $T_{g,i}$ values, to that of a single-phase blend ΔS_g at its T_g [i.e. neglecting S_{ex} from eq. (2.5)], may be expressed in two alternative ways.

(a) In phenomenological terms, defining ΔS_g by eq. (1.24) and setting the limits of integration for each system in the temperature intervals from respective $T_{g,i}$ to T_g of the blend, we obtain after rearrangement [100]

$$\ln T_g = \frac{\ln T_{g,1}\varphi + \ln T_{g,2}(1 - \varphi)K_s}{\varphi + (1 - \varphi)K_s} \tag{2.37}$$

(where $K_s = \Delta C_{p,2}/\Delta C_{p,1}$).

(b) In terms of a molecular model [101], any polymer system at its T_g is characterized by a universal value of the configurational entropy ΔS_g, which is a function of the chain stiffness factor, $\Delta\varepsilon/kT_g$ (where $\Delta\varepsilon$ is the 'flex energy', or energy difference between rotational isomers), and the hole fraction (in phenomenological terms, the free volume fraction), v_0, i.e. [102]

$$\Delta S_g = A + D(\Delta\varepsilon/kT_g) + CV_0 \tag{2.38}$$

where A, D and C are constants. Assuming additivity of the flex energies and hole fractions

$$\Delta\varepsilon = B\Delta\varepsilon_1 + (1 - B)\Delta\varepsilon_2 \tag{2.39a}$$

$$V_0 = BV_{0,1} + (1 - B)V_{0,2} \tag{2.39b}$$

(where $B = p_1 N_1 \gamma_1/(p_1 N_1 \gamma_1 + p_2 N_2 \gamma_2)$ is the fraction of flexible bonds of component 1, γ_1 and γ_2 are the numbers of flexible bonds per chain repeating unit of components 1 and 2, and N_1 and N_2 are the numbers of corresponding macro-molecules), we obtain [102]

$$T_g \cong T_{g,1}B + T_{g,2}(1-B) + K_m B(1-B)(T_{g,1} - T_{g,2})(V_{0,1} - V_{0,2}), \qquad (2.40)$$

where K_m is an adjustable constant. The assumption of a universal value of the hole fraction frozen in at T_g, i.e., $V_0 = V_{0,1} = V_{0,2}$, reduces eq. (2.40) to a simple rule of additivity

$$T_g = T_{g,1}B + T_{g,2}(1-B) \qquad (2.41)$$

(iii) Explicitly accounting for the contribution of the excess mixing enthalpy, H_{ex}^g, and the enthalpy fractions of the pure components, $\Delta H_{g,i}$, frozen in at respective $T_{g,i}$, to the enthalpy fraction of a single-phase blend, ΔH_g, frozen in at the corresponding T_g, i.e.

$$\Delta H_g = \Delta H_{g,1}\varphi + \Delta H_{g,2}(1-\varphi) + H_{ex}^g \qquad (2.42)$$

(where $\Delta H_{g,i} = \Delta C_{p,i}(T_g - T_{g,i})$ is assumed) and making use of eq. (2.3), we obtain for the temperature interval between $T_{g,1}$ and $T_{g,2}$ (neglecting for the moment the difference between weight fractions and volume fractions)

$$T_g = \frac{T_{g,1}\varphi + T_{g,2}(1-\varphi)K_s - \chi k v_M(T_{g,2} - T_{g,1})\varphi(1-\varphi)}{\varphi + (1-\varphi)K_s} \qquad (2.43)$$

Presumably, more exact versions of eq. (2.43) may be obtained by incorporation of the empirical corrections for the non-additivity of the heat capacity jumps at T_g values and for the composition dependence of the interaction parameter χ [103].

(iv) Recallng that the contributions of the binary, ternary, etc. interactions to the concentration dependence of many properties (i.e. viscosity, osmotic pressure, etc.) of the binary systems (in appropriately reduced form) may be conveniently accounted for by expansion into a power series of concentration, it was postulated [104] that the simple (linear) rule of additivity of the glass transition temperatures expressed as

$$(T_g - T_{g,1})/(T_{g,2} - T_{g,1}) = \varphi \qquad (2.44)$$

may be also regarded as a simplified, truncated version of a more general 'virial' equation, eq. (2.45), i.e. [104, 105]

$$(T_g - T_{g,1})/(T_{g,2} - T_{g,1}) = (1 + K_1)\varphi - (K_1 + K_2)\varphi^2 + K_2\varphi^3 \qquad (2.45)$$

where K_1 and K_2 are the empirical fitting parameters. K_1 is expected to account for differences between the interaction energies of like and unlike contacts (i.e. $K_1 > 0$ suggests 'a predominant contribution of the energetic effects of unlike interactions, whereas $K_1 < 0$ indicates a prevailing effect of the conformational rearrangements accompanied by changes of the free volume', and K_2 considers only effects 'related to redistributions in the contact neighbourhood' [105].

The relative merits of the different approaches described above as concerns the quality of experimental data fit to theoretical predictions will probably best be assessed from the original publications [99–105]. Broadly speaking, it is only the two latter approaches, (iii) and (iv), which explicitly take into account (albeit in a simplified form) the effects of specific interactions and thus seem to be more flexible compared with approaches (i) and (ii) as concerns their intrinsic ability to reproduce diverse patterns of composition dependence of T_g encountered for single-phase polymer blends.

It may also be easily verified that the incorporation of various simplifying assumptions (e.g. universal values of the free volume fraction f_g'' in eq. (1.20) or of the product, $\Delta C_p T_g$, etc.) will reduce nearly all of the above equations to one of many earlier, popular empirical relationships [100, 102, 104–108].

Pressure Dependence of T_g

So far isobaric conditions have been implied (cf. preceding section) in the derivation of all expressions for the composition dependence of T_g. However, when the glassy state is reached by compression of an equilibrium melt in isothermal conditions, it is the glass transition pressure, P_g, which is the relevant variable. It can be shown, however, that the composition dependences of both T_g and P_g are described by nearly identical expressions, provided jumps in the isobaric expansivities, $\Delta\alpha$, are changed for jumps in the isothermal compressibilities, $\Delta\beta$, and assuming pressure-independent heat capacity jumps, ΔC_p, a binary interaction parameter, χ, etc. Theoretically, this symmetry implies the validity of the Ehrenfest equations, eqs. (1.27) and (1.27a), in the case of single-phase polymer blends [109].

Apparently, PS/PPO is the only compatible blend for which this prediction can be tested experimentally [6]. As can be seen from Table 2.2, eqs. (1.27) and (1.27a) apply to both pure components, notwithstanding a more than two-fold mismatch in the free volume fractions at T_g [in fact, $f_g'' = 0.277$ calculated by eq. (1.20) for PPO is unusually high compared with the close to universal value $f_g'' = 0.104$ for PS]. All glass transition quantities of the blends (except the

Table 2.2 Glass transition quantities for PS/PPO blends [6]

	PPO/PS ratio						
Property	0/1	1/4	2/3	1/1	3/2	4/1	1/0
$T_g(O)$ (K)	352.3	377.6	403.6	415.5	424.4	449.0	476.4
v_g(cm^4 cc/g)	0.972	0.974	0.968	0.968	0.964	0.970	0.974
$\Delta\alpha \times 10^4$ (K^{-1})	2.96	3.20	3.74	4.20	4.52	5.34	5.82
$\Delta\beta \times 10^4$ (MPa^{-1})	1.66	1.99	2.25	2.57	2.95	3.82	5.24
$dT_g/dP \times 10^2$ (K/MPa)	42.8	62.2	66.2	67.5	66.2	67.9	8.24
$\Delta\beta/\Delta\alpha \times 10^2$ (K/MPa)	56.1	60.2	60.2	61.2	65.3	71.5	90.0

specific volume, v_g) tended to increase with PPO content, eq. (1.27) being reasonably well obeyed at all compositions. The same conclusion might be valid also for eq. (1.27a), provided the ΔC_p of the blends is an additive sum of components $\Delta C_{p,i}$ values, as is implicit in eq. (2.37). Limited experimental evidence available [110] suggests this latter assumption to be approximately correct fot the PS/PPO case, although it is unlikely to be a general rule for all compatible polymer blends [111, 112]).

Glass Transitions in Phase-separated Blends

As already emphasized, the degree of phase separation, α_{sep}, for many allegedly incompatible binary blends is below unity. This experimental finding might be explained by two alternative models.

(i) The blend is separated into two homogeneous phases each consisting predominantly of one component diluted with a small amount of the second component.

(ii) In phase-separated blends, besides the pure phases of both components, a non-negligible fraction, $v_{BI} = 1 - \alpha_{sep}$, of the total amount of blend material is involved in the BI of intermediate composition φ'.

The former model implies the occurrence of two glass transitions located somewhere above the lower and below the higher $T_{g,i}$ values of pure the components, whereas three glass transitions, two of which coincide with the $T_{g,i}$ values of pure components and the third located somewhere in between, should conform to model (ii). It appears that the PS/PMMA blend [93, 96], as well as blends with one component crystallized, like PMMA/PEO [11–13], PMMA/PVDF [12, 13, 113] and PC/poly(butylene terephthalate) [114] behave according to model (i), whereas model (ii) applies to PVC/butadiene-nitrile rubber blends [115]. In either case, the local composition of each microphase may be estimated by fitting the relevant T_g to an appropriate equation for single-phase systems (see the preceding sections), while the relative content of microphases, v_i, will be obtained by substituting experimental and additive $\Delta C_{p,i}$ values into the numerator and the denominator of eq. (2.23), respectively.

It is pertinent to remark here that the pattern of composition dependence of T_g for 'semi-compatible' blends (i.e. those exhibiting a limited miscibility of components) will crucially depend on the size d of isolated particles of one component dispersed in a continuous matrix of the second component. In fact, smooth variation of T_g with composition typical for single-phase systems was observed for solution-blended, unannealed samples of PMMA/PVC [1, 116], PVC/acrylonitrile-butadiene copolymer [97], etc. although structural micro-heterogeneity at the level of about 10 nm was detected by electron microscopy. On the basis of these observations, it was claimed that two glass transitions coinciding with the $T_{g,i}$ values of pure components or a single T_g located in between should be observed in phase-separated blends at $d > 100$ nm and at $d < 10$ nm, respectively [97].

Because this latter characteristic size was of the same order of magnitude as the expected thickness of the BI (cf. section 2.2.3), it would have been tempting to associate the single T_g observed with the BI. This assumption implies a continuous change in the BI composition with nominal composition of the blend; however, since this is untenable within the framework of equilibrium thermodynamics, it remains to suggest that the experimental effects observed refer to a metastable, non-equilibrium, single-phase state frozen below T_g. In fact, this claim proved inconsistent with the equilibrium data for cross-linked epoxy/diene rubber blends which exhibited two well resolved T_g values even at an extremely small ($d \cong 6{-}7\,\mathrm{nm}$) size of the microheterogeneities [117].

The equilibrium thermodynamic approach will also predict the following two alternative mechanisms by which the phase morphology of phase-separated blends may interfere with the glass transition phenomenon in isolated inclusions of a soft (elastomeric) microphase dispersed in the continuous matrix of a stiff (glassy) component.

(i) When the dimensions of inclusions are not too small (say, $d > 100\,\mathrm{nm}$), its T_g may be decreased below that for a pure component as a result of the onset of three-dimensional extensional stresses σ_{ext} (i.e. negative pressure) owing to a mismatch of the expansion coefficients of two components, that is [118]

$$\sigma_{\mathrm{ext}} \sim E_2(\alpha_1 - \alpha_2)(T_{g,1} - T_{g,2})\varphi \qquad (2.46)$$

where E_2 is the Young's modulus of the continuous phase.

This unusual effect of phase morphology was observed for a 'semi-interpenetrating' polymer blend obtained by solution cross-linking of polyurethane (PU) within a matrix of chlorinated PVC (CPVC) [119, 120]. It turned out that the T_g of oligoether blocks of PU exhibited a sudden drop by $15{-}20\,\mathrm{K}$ in the composition range ($20{-}30\%$ of CPVC) where PU formed isolated inclusions within a continuous matrix of CPVC. Assuming a typical (for elastomers) value $dT_g/dP = 0.3\,\mathrm{K/MPa}$ [121], the observed depression of T_g is equivalent to the action of negative hydrostatic pressure $\Delta P = \sigma_{\mathrm{ext}} = -(50{-}65)\,\mathrm{MPa}$.

(ii) As the dimensions of soft inclusions d (i.e. effective radius of curvature $r \cong d/2$) decrease, their T_g should increase [122] with the increment of inner ('capillary') pressure which may be estimated by the Laplace equaton, eq. (2.47):

$$\Delta P_i = 2_{12}/r \cong 4\gamma_{12}/d \qquad (2.47)$$

where γ_{12} is the interfacial tension.

While the latter effect seems of minor importance [e.g. the expected increment of T_g will not exceed a few degrees at typical values $\gamma_{12} = (3{-}10) \times 10^{-7}\,\mathrm{J/cm^2}$ [123], $d = 5\,\mathrm{nm}$ and $dT_g/dP = 0.3\,\mathrm{K/MPa}$], it does draw attention to the problem of a limiting size $d_{\min}$ below which the isolated microphases of one polymer will not be at internal equilibrium with the surrounding continuous matrix of another polymer. As soon as the maximum of the conformational entropy of a single macromolecule corresponds to its random coil dimensions $2\langle R_g^2 \rangle^{1/2}$ with less than 10% of the total configurational space occupied by its own segments [2], it appears that after completion of phase separation in a binary polymer melt with overwhelming excess of one component (i.e. at $\varphi \Rightarrow 0$ o $\varphi \Rightarrow 1$) the limiting size of the inclusions d_{min} will not be appreciably smaller than $2\langle R_g^2 \rangle^{1/2}$; otherwise (i.e. collapse of a macromolecule into a densely-packed globular state) the unfavorable contribution from entropy loss would be much too severe [124]. In other words, it may be assumed that the notoriously high viscosity of polymer melts will hinder the interpolymer diffusion processes, so that the final result of phase separation will be a few isolated inclusions of size $d \geqslant 2\langle R^2 \rangle^{1/2}$ with about 90% of the total configurational space of each macromolecular coil still occupied by foreign chains. A similar conclusion may be reached from other consideratons [4].

2.4 CRYSTALLIZATION AND MELTING OF CRYSTALLIZABLE COMPONENTS

2.4.1 PROPERTIES OF A CRYSTALLIZED COMPONENT

Crystallinity

As can be seen from Fig. 2.6, the crystallinities $X = \Delta H_m / \Delta H_m^0$ (where ΔH_m and ΔH_m^0 are the melting enthalpies of real and perfect crystals, respectively) of both PEO and PVDF decrease to zero above a definite weight content w' of the non-crystallizable component of the blend (PMMA) [11–13]. These results were in fair agreement with the data from other sources [125–130]; essentially similar behavior was observed for many other compatible polymer blends with one component intrinsically crystallizable [131–135]. In view of the inverse proportionality between X and the maximum crystallization rate G_{max} [136], which is known to decrease steeply the smaller is the difference between the glass transition temperature of the amorphous phase T_g and the melting temperature of the crystalline phase T_m [137], it seems quite natural that the crystallizability of PEO is lost at a PMMA content ($w' \geqslant 0.8$) corresponding to $T_g \cong T_m$. In light of this reasoning, we might have expected that the crystallinity of PVDF would be less affected by PMMA due to a much larger difference between T_m and T_g (Fig. 2.6(b)), whereas PVDF, on the contrary, loses its ability to crystallize at considerably lower PMMA content ($w' \geqslant 0.5$).

This aparent discrepancy may be resolved by the following qualitative arguments. As shown in section 2.1, melt compatibility of both PMMA/PEO and PMMA/PVDF pairs is ensured by the specific interactions responsible for the negative values of the interaction parameters χ. This means that the thermodynamic driving force favoring separation of the crystalline phase from a homogeneous melt, $\Delta G_v \sim \Delta S_m \Delta T$, will be opposed by the contribution from excess melt enthalpy, $H_{ex} < 0$, favoring the single-phase state. Hence, at equal undercoolings ΔT the resulting driving force of crystallization, $\Delta S_m \Delta T - H_{ex}$, will be

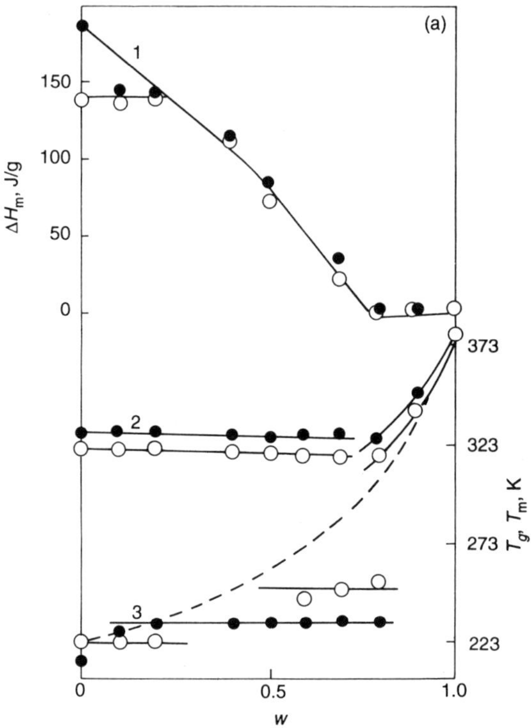

Fig. 2.6. Composition dependence of the melting heats (1), melting (2) and glass transition (3) temperatures for (a) PEO-2/PMMA ($\square$), PEO-125/PMMA ($\bullet$) and (b) PVDF/PMMA

higher the higher is the melting entropy ΔS_m. It is thus the higher ΔS_m for PEO compared with PVDF which should be regarded as responsible for the higher w' of the former (Fig. 2.6).

Therefore, crystallization of component 1 from a homogeneous melt of nominal composition w will have two obvious consequences.

(i) The local composition of the non-crystalline ('amorphous') quasi-phase below T_m will change to $w'' < w$.

(ii) In so far as the amorphous quasi-phase in semi-crystalline polymers is located essentially in the interlamellar space, the enrichment of the latter with chain segments of the second component will move the lamellae apart.

Prediction (i) is confirmed by positive deviations of the experimental T_g values from theoretical curves for compatible blends in the 'crystallizability domains' ($w < w'$) of PMMA/POE [12, 13], PMMA/PVDF [12, 13, 113] and Nylon 6/amorphous aromatic polyamide (AP) [138] pairs, as well as by deviations from additivity of the mean electron density fluctuations for PVC/poly(ε-caprolactone) (PCL) pairs [131]. Analysis of the dipole relaxation process in the non-crystalline

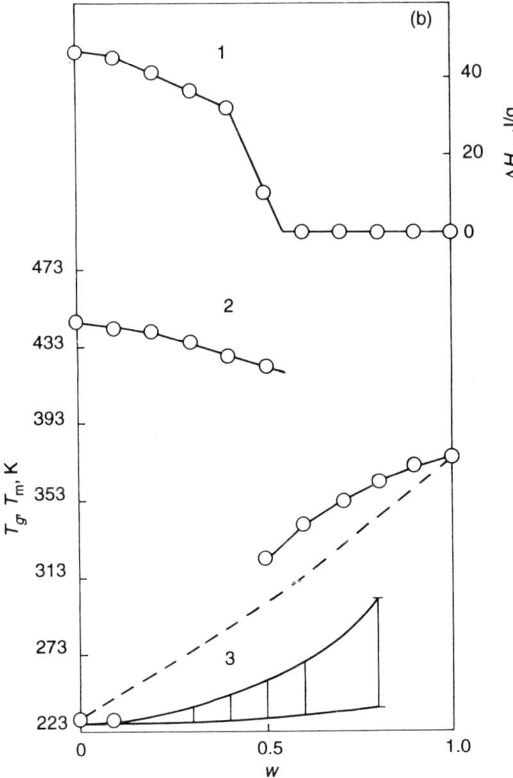

Fig. 2.6. *continued*

phase of PMMA/PVDF blends also suggests the coexistence of microphases of mixed components and of pure PVDF [139].

Prediction (ii) agrees with the gradual coarsening of the spherulite structure and an increase in the extinction spacings between spherulite bands (optical microscopy and light scattering data), as well as with the increase in interlamellar spacings (inferred from electron microscopy and X-ray scattering data) concomitant with the dilution of a crystallizable component with a non-crystallizable polymer, as observed for PMMA/POE [13, 106], PMMA/PVDF [130, 139, 140], PVC/PCL [131, 141] and other [142] compatible pairs.

In this respect the composition-invariant crystallinity, interlamellar spacing and lamella height of each of the two crystallizable aliphatic polyamides, Nylon 6 and Nylon 66, in miscible blends with non-crystallizable AP [143], look quite unusual. Recalling the arguments used to rationalize the different patterns of composition dependence of X for PMMA/POE and PMMA/PVDF pairs (see above), the seemingly unique behavior of polyamide blends may be qualitatively explained by complete rejection of AP from the interlamellar space of both nylons as a result of the negligibly small magnitude of the excess melt enthalpy H_{ex} (since, presumably, the density of the specific interactions in the melts of aliphatic nylons will be hardly changed on addition of AP) compared with the thermodynamic driving force for crystallization ΔG_v. An essentially similar conclusion was arrived at in a lattice model treatment of the effect of H_{ex} on the composition of interlamellar space in crystallizable/amorphous polymer blends (CAPB) [144].

Melting Behavior

A common feature of nearly all compatible CAPB is the more or less significant depression of the melting point T_m of a crystallizable component with increasing content of an amorphous polymeric 'diluent' [12, 13, 97, 125–134]. The classical Flory formula for T_m depression by low-molecular weight diluents [2]

$$1/T_m - 1/T_m^0 = -(R/\Delta H_m^P)(V_1/V_d)[\ln \varphi/p_1 - (1 - 1/p_1)(1 - \varphi) + \chi(1 - \varphi)^2]$$

$$(2.48)$$

(where T_m and T_m^0 are the melting points of the polymer crystal in the presence and absence of a diluent, respectively, V_1 and V_d are the molar volumes of the polymer chain repeating unit and of the diluent, respectively, and φ is the volume fraction of the polymer) may be easily extended to account for a similar effect of a polymeric diluent of polymerization degree p_2 and molar volume V_2, i.e. [128]

$$1/T_m - 1/T_m^0 = -(R/\Delta H_m^0)(V_1/V_2)[\ln \varphi/p_1 + (1/p_1 - 1/p_2)(1 - \varphi) + \chi(1 - \varphi)^2]$$

$$(2.48a)$$

which, after substitution of eq. (2.4) for χ and assuming $p_1, p_2 \Rightarrow \infty$, reduces to

$$(1/T_m - 1/T_m^0)/\varphi = -(BV_1/\Delta H_m^0)(\varphi/T_m)$$
$$(2.48b)$$

Straight lines passing through the origin, as required by eq. (2.48b), were obtained on the plots of the left-hand side of eq. (2.48b) vs. φ/T_m for PMMA/PVDF [13, 128, 130] and poly(ethyl methacrylate)/PVDF [145] pairs, negative values of B confirming the specific interactions between the components in the melt. Similar plots for other CAPB like POE/PMMA [13, 146], PVC/poly-β-propiolactones [147], PVC/PEO [148], PS/PPO [149], etc. were also reasonably linear with $B < 0$; however, non-zero intercepts inconsistent with eq. (2.48b) were invariably observed.

Since thermodynamic equilibrium between infinite crystalline and amorphous phases at T_m was implicitly assumed in the derivation of eq. (2.48b), the latter discrepancy might be attributed to neglecting either of the following effects.

(i) The composition dependence of the interaction parameter B [150].

(ii) The non-equilibrium, finite size of the polymer crystals in melting experiments [149, 151].

Empirically, the latter effect may be eliminated by substitution into eq. (2.48b) of the 'equilibrium' melting temperatures estimated by eq. (1.46) [13, 128, 130, 139]; otherwise, it may be accounted for by eq. (2.49) [149]

$$\Delta H_m^0(T_m^0 - T_m)/\varphi R T_m^0 - T_m/p_1 - \varphi T_m/2p_2 = C/R - b\varphi \qquad (2.49)$$

where $b = \text{const} \cong \chi T$, and C is a fitting 'morphological' parameter. In fair agreement with the expectations, both parameters b and C turned out inversely proportional to the chain length of the polymeric diluent (PS) for the PS/PPO blend [149].

On the other hand, explicit account of both effects (i) and (ii) yielded (in the limit $p_1, p_2 \Rightarrow \infty$) [150]

$$T_m^0/T_m^0 \cong 1 - 2\sigma_e/l\Delta H_m^0 + (RT_m^0/V_2\Delta H_m^0)\chi(1 - \varphi)^2 \qquad (2.50)$$

where $1/l = a\Delta H_m^0(1 - X)(1 - \varphi)^2/2\sigma_e$ is the inverse lamella thickness and σ_e is the specific interfacial energy of its basal (fold-containing) face, $\chi = -aV_1\Delta H_m^0 X/RT_m^0$ and $a = (T_m^0 - T_m)/T_m^0(1 - \varphi)^2$.

Equation (2.50) proved quantitatively applicable [150] to the relevant experimental data for PMMA/PEO blends [126] at normal pressure; similar behavior was also predicted for the PMMA/PVDF pair. On the other hand, experiments under high hydrostatic pressures yielded the following results [13, 152, 153].

(i) Melting temperatures of PEO and PVDF, both in the pure state, as well as in blends with PMMA after an initial increase, tended to level off with increasing pressure.

(ii) With increasing content of PMMA the derivatives dT_m/dP (at normal pressure) tended to decrease for the PMMA/PEO pair, while the reverse effect was observed for PMMA/PVDF.

Apparently, issue (i) may be readily explained in terms of the familiar Clausius–Clapeyron equation, eq. (2.51)

$$dT_m/dP = \Delta v_m/\Delta S_m \qquad (2.51)$$

by a faster (compared with the melting entropy ΔS_m) decrease with pressure in the volume change on melting Δv_m, as observed for other polymers [137].

Issue (ii) will be discussed within the framework of eq. (2.52) derived from eq. (2.48b), i.e.

$$d \ln T_m/dP = (d \ln T_m^0/dP) + (R\varphi/\Delta S_m)(d\chi/dP) \qquad (2.52)$$

According to eq. (2.52), the slope (i.e. the sign of the corresponding derivative by composition) of the plot dT_m/dP vs. φ will depend on the sign of the derivative $d\chi/dP$. In other words, eq. (2.52) predicts negative and positive values of $d\chi/dP$ for PMMA/PEO and PMMA/PVDF, respectively, which is completely consistent with earlier estimates (cf. section 2.2).

It should be emphasized that the depression of the melting temperature in CAPB does not necessarily mean thermodynamic compatibility of components, because a similar effect was observed, for example, for polymer single crystals distorted by a rigid matrix of apparently incompatible component 2 with $T_{g,2} < T_{m,1}$, as was the case with CAPB like PE/PS, PE/PPO, isotactic PS/PMMA, etc. [154, 155]. On the contrary, the melting temperature $T_{m,1}$ of the oriented PE fiber embedded in a rigid matrix with $T_{g,2} > T_{m,1}$ turned out to be some 7 K above that of an identical sample in the unconstrained state [156], which may be tentatively explained in terms of eq. (2.51) by the generation of an excess hydrostatic pressure in PE due to contraction of the matrix during its polymerization.

A final remark is reserved for the case when both components of the blend are crystallizable polymers (CCPB). For a large number of melt-blended CCPB (e.g. mixtures of polyolefins with themselves and with other thermoplastic polymers) two melting endotherms with peaks corresponding to T_m values of pure components were observed [157]. This should be considered as a fairly general phenomenon in view of the extremely low probability of crystal isomorphism for polymers [158], although the appearance of a third peak located somewhere in between and attributed to melting of kinetically entrapped, 'hybrid' crystals was also reported for the first heating run of solution-blended samples of low-density and high-density PEs [157, 159].

2.4.2 KINETICS OF MELT CRYSTALLIZATION

The almost universal trend for the systematic retardation of melt crystallization in compatible CAPB blends with an increase in the relative content of amorphous component 2 (e.g. [133, 145, 157]) (keeping the crystallization temperature T constant) may be readily explained, qualitatively, by a decrease in the crystallization driving force $\Delta G_v \sim \Delta S_m (T_m^0 - T)$ due to the melting point depression effect. Quantitative analysis involves treatment of the growth rate data by an appropriate version of eq. (1.51), i.e. [160, 161]

$$G_R = G_0' \exp(-\Delta G'/kT) \exp(-\Delta E'/kT) \tag{2.53}$$

where $G_0' = G_0 \varphi$ is the pre-exponential factor, $\Delta G' = (Z_m - Z_m')/(\Delta G_v)^m$ is the nucleation barrier, and $\Delta E'/kT = B/(T - T_0')$ is the molecular transport barrier. Equation (2.53) quantitatively described all experimental data on the kinetics of spherulitic crystallization at normal pressure of both PEO [150, 162] and PVDF [150, 163] in blends with PMMA, assuming, as usual, a secondary (surface) nucleation mechanism (i.e. $m = 1$, $Z_1 = 4b_0 \sigma \sigma_e$, $Z_1' = 2\sigma_e \ln \varphi/b_0$) and approximating $T_0' \cong T_g - 50$, $B \cong 1/\alpha_1$.

Similar analysis of the bulk crystallization rates $1/\tau_{0.5} \sim G_R$ (where $\tau_{0.5}$ is the crystallization half-life) of PMMA/PEO and PMMA/PVDF under elevated pressures by eq. (2.53) with proper account taken of the pressure dependences of all material quantities involved [13, 164, 165] showed that the calculated values of σ_e at each pressure, in agreement with other data at normal pressure [130, 162, 163], were approximately the same regardless of blend composition. However, while a pressure increase from 25.2 MPa to 100.6 MPa brought about only a mild (about 10%) decrease in σ, the values of σ_e (in 10^{-3} J/m^2) increased more than two-fold (from about 23 to 80, and from 115 to 240 for PEO and PVDF, respectively). As discussed in detail elsewhere [137], the latter effect may be regarded as a manifestation of increased steric constraints to smoothing of the fold-containing faces of crystallization nuclei under elevated pressures.

In spite of its quantitative applicability, eq. (2.53), however, does not account for the continuous change in composition of the amorphous phase during crystallization (cf. section 2.4.1 above). The experimental evidence for this phenomenon is provided by SAXS data on the melting kinetics of a PEO/PMMA blend of nominal composition 50/50 at $T \geqslant T_m$ [142, 166] in which a SAXS peak appeared, passed through a maximum and finally faded away during isothermal transition from the initial heterogeneous, semi-crystalline state (with PEO/PMMA = 70/30 for the local composition of amorphous interlamellar space) to a homogeneous, single-phase melt (PEO/PMMA = 50/50).

It was suggested [167] that the contribution of such a local reorganization process at the crystallization growth front may be accounted for by the following modification of eq. (2.53):

$$G_R = [G'_0 k_1 k_2/(k_1 + k_2)] \exp(-\Delta G'/kT) \tag{2.54}$$

where $k_1 = \exp[(B/(T - T'_0)]$ and $G'_0 = G_0 \varphi$, as before; $k_2 = 2\langle D \rangle/l$ is the rate at which the amorphous component can be removed from the growth front, l is lamella thickness, $\langle D \rangle = [\varphi p_1 D_1 + (1 - \varphi)p_2 D_2][\varphi/p_1 + (1 - \varphi)/p_2 + 2\varphi(1 - \varphi)\chi]$ is the coefficient of mutual diffusion between the components, and D_i are the corresponding self-diffusion coefficients.

Equation (2.54), combined with reasonable estimates of D_1, D_2, l and $\chi = 0$, provided a self-consistent explanation of the trends observed in an extensive experimental study of the growth rate of spherulites in PEO/PMMA blends as a function of the molecular masses of both components [167].

As might have been expected, the simple Kolmogorov–Avrami equation, eq. (1.50), turned out to be applicable only to the initial stages of bulk crystallization of PMMA/PEO [13, 164, 165, 168] and PMMA/PVDF [13, 164], with non-integer values of the shape parameter n. No clear-cut, systematic trends with blend composition, temperature or pressure could be observed, the values of n randomly fluctuating between 2 and 3. In contrast, the two-stage eq. (1.50a) quantitatively described the same data for PMMA/PEO in the whole range of transformation degree α with integer values of both n_1 and n_2 (Fig. 2.7) [169]. In this case, crystallization at whatever temperature or pressure could be characterized by $n_1 = 3$ and $n_2 = 2$ for pure PEO and sample PMMA/PEO = 10/90, whereas $n_1 = 4$ and $n_2 = 2$ were obtained for sample PMMA/PEO = 20/80. Literally [160, 170], these data suggest three-dimensional growth on athermal (for the former two systems) and thermal (for the latter blend) nuclei in the first crystallization stage, and either one-dimensional growth on thermal nuclei, or two-dimensional growth on athermal nuclei in the second stage of crystallization for all samples studied.

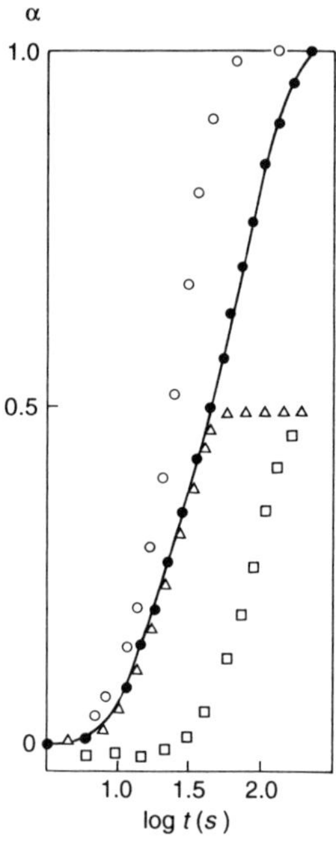

Fig. 2.7. Crystallization isotherm of PEO-2/PMMA = 90.10 at $P = 50.3\,\text{MPa}$, $T = 317.9\,\text{K}$ ($\bullet$) and theoretical plots according to eq. (1.50) ($\bigcirc$), eq. (1.50a) (solid line) and contributions from the first ($\triangle$) and second ($\square$) mechanisms.

In conclusion, the salient features of crystallization of CCPB will be briefly discussed. As a general trend for CCPB comprising polymers with widely different melting temperatures T_m, the crystallization exotherm of the low-melting component is shifted to high temperatures, which is commonly explained by heterogeneous nucleation on the high-melting component already solidified [157]. The occurrence of a small-intensity exotherm in between two major crystallization exotherms of pure components in PE/poly(methylene oxide) (POM) blends was tentatively attributed to the effect of homogeneous nucleation in small volumes of POM dispersed in a continuous PE matrix [171]. However, the idea that the intermediate process is associated with retarded crystallization of low-molecular weight fractions of POM rejected into the BI [172–174] seems physically more realistic.

2.5 CONCLUSIONS

As is clear from the evidence presented in this chapter, all binary polymer blends may be conveniently characterized by a universal structural parameter, i.e. a degree of phase separation α_{sep} which varies, theoretically, from zero (for fully compatible components) to unity (for two-phase blends with a sharp interface). In a strict sense, however, both theoretical limits may never be attained for the following reasons.

(i) According to the experimental evidence available [12, 13, 24–27], strong specific interactions between the components ensuring their compatibility on a 'macroscale' (i.e. on the order of macromolecular dimensions) are, however, responsible for the composition heterogeneity on the 'microscale' (on the order of dimensions of the chain segment). Such a 'nanoheterogeneity' may also arise when energetically unfavorable interactions between the components (interchain repulsion) are counterbalanced by the concomitant non-local entropy gain providing for increased intramolecular mobility [175]. Whatever the physical reason, in either case the 'true' value of α_{sep} should be above zero.

(ii) Even in the case of apparently incompatible pairs in the equilibrium melt state the 'mathematical' (i.e. of zero thickness) interface is smeared out into a 'physical' BI, the thickness of which may be varied from a few nanometers to fractions of a micron by choosing the appropriate conditions of phase separation (as in IPN) and/or blending (as in blends of linear polymers). It is obvious that the upper limit of α_{sep} will be below unity.

Although this latter situation is qualitatively similar to the one encountered with filled polymes (cf. Chapter 1), the fundamental difference between the two is, of course, the composition of the BI. Namely, in filled polymers the BI consists of the same polymer (albeit in a structurally different state), whereas in phase-separated polymer blends the composition of the BI varies with distance from the pure phases of each component. Moreover, for filled polymers the polymer component is always a continuous phase, while the filler remains a disperse phase at whatever filler loadings; on the other hand, in polymer blends each component may form either a continuous or disperse phase depending on its relative content.

This latter morphological difference manifested itself, for example, in samples of PS filled with cellulose (CL) [176]. Initially, the heats of the interactions with solvent of equilibrium PS melt H'_1 increased with filler w, as was the case with PS filled with silica powder (cf. section 1.3.2); however, while the saturation of interfacial interactions for the latter system led to a levelling-off of H'_1 above a characteristic filler content w^* (Fig. 1.9), the values of H'_1 for PS/CL, on the contrary, passed through a minimum and then went up again (Fig. 2.8). Presumably, the minimum of H'_1 was a result of increased interaction density in the region of 'phase inversion' where both components existed as continuous, interpenetrating phases, whereas loss of interfacial area due to 'coagulation' of deformable CL particles at higher w led to a subsequent increase in H'_1 [1976].

One more fundamental difference between filled polymers and polymer blends is the decisive role of kinetic effects in the final morphology (and hence the BI content) of the latter. In fact, a whole spectrum of different morphologies may by obtained from a homogeneous melt of fixed overall composition by changing either the depth of penetration below binodal (as in a liquid–liquid type phase separation for AAPB) or the degree of undercooling below melting temperature of a crystallizable component (as in a liquid–crystal type phase separation for CAPB). Even more spectacular consequences may be expected [113] from the

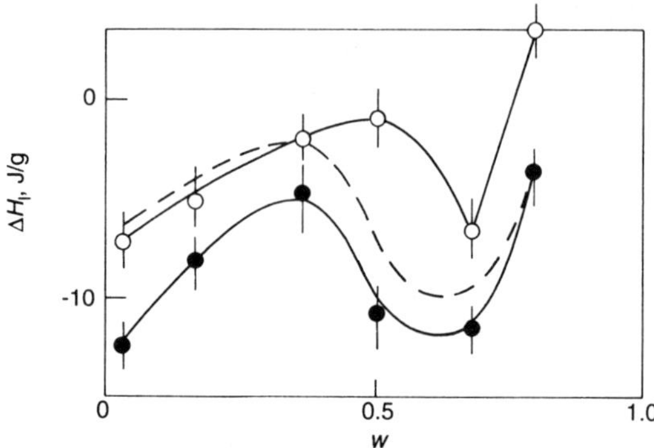

Fig. 2.8. Dependence of the heats of dissolution of equilibrium PS melt on CL content for specimens prepared at normal (○) and elevated (●) pressures

interference of eventual solidification with the phase separation event (e.g. when the binodal is crossed either by the glass transition line for AAPB, or by the melting point line for CAPB). The morphological implications of these cases were treated in [177–181].

REFERENCES

1. Faminskaya L. A. (1983) Thesis, Gorky State University.
2. Flory P. J. (1953) *Principles of Polymer Chemistry*, Cornell University Press, Ithaca, New York.
3. Patterson D. (1967) Thermodynamics of non-dilute polymer solutions, *Rubber Chem. Technol.*, **40**, 1–35.
4. Nesterov A. E. and Lipatov Yu. S. (1984) *Thermodynamics of Polymer Solutions and Blends*, Naukova Dumka, Kiev (in Russian).
5. Roland C. M. (1987) Entropically driven miscibility in a blend of high molecular weight polymers, *Macromolecules*, **20**, 2557–2563.
6. Zoller P. and Hoehen H. H. (1982) Pressure–volume–temperature properties of blends of poly(2,6-dimethyl-1,4-phenylene ether) with polystyrene, *J. Polymer Sci.: Polymer Phys. Ed.*, **20**, 1385–1397.
7. Simha R. and Somcynsky T. (1969) On the statistical thermodynamics of spherical and chain molecule fluids, *Macromolecules*, **2**, 342–350.
8. Jain R. K., Simha R. and Zoller P. (1982) Theoretical equation of state of polymer blends: the poly(2,6-dimethyl-1,4-phenylene ether)–polystyrene pair, *J. Polymer Sci.: Polymer Phys. Ed.*, **20**, 1399–1408.
9. Lipatov Yu. S., Petrenko K. D. and Privalko V. P. (1985) Intermolecular interactions in oligo(ethylene glycole)/poly(methyl methacrylate) blend, *Dokl. Acad. Nauk UkrSSR*, Ser. B, **9**, 50–53.
10. Lipatov Yu. S., Petrenko K. D. and Privalko V. P. (1986) Compatibility of polymers in the framework of 'lattice fluid' model, *Dokl. Acad. Nauk UkrSSR*, Ser. B, **6**, 50–53.
11. Privalko V. P., Lipatov Yu. S. and Petrenko K. D. (1987) Thermodynamics of binary polymer blends. Poly(ethylene oxide)/poly(methyl methacrylate) pair, *Vysokomol. Soed.*, Ser. A, **29**, 2062–2069.

12. Privalko V. P., Petrenko K. D. and Lipatov Yu. S. (1990) Miscible polymer blends. I. Thermodynamics of the blend melts poly(methyl methacrylate)–poly(ethylene oxide) and poly(methyl methacrylate)/poly(vinylidene fluoride), *Polymer*, **31**, 1277–1281.

13. Petrenko K. D. (1987) Thesis, Institute of Macromolecular Chemistry, Academy of Sciences of Ukraine, Kiev.

14. Wolf M. and Wendorff J. H. (1989) Studies of the excess volume of mixing in blends of poly(methyl methacrylate) and poly(vinylidene fluoride), *Polymer*, **30**, 1524–1529.

15. Sanchez I. C. and Lacombe R. H. (1977) An elementary equation of state for polymer liquids, *J. Polymer Sci.: Polymer Lett. Ed.*, **15**, 71–75.

16. Sanchez I. C. (1978) Statistical thermodynamics of polymer mixtures, in *Polymer Blends*, Vol. 1, ed. by D. R. Paul and S. Newman, Academic Press, New York–San Francisco–London, pp. 145–171.

17. Paul D. R. (1978) General concepts and prospects, in *Polymer Blends*, Vol. 1, ed. by D. R. Paul and S. Newman, Academic Press, New York–San Francisco–London, pp. 11–25.

18. Ting S. P., Bulkin B. J., Pearce E. M. and Kwei T. K. (1981) Compatibility studies of styrene and hydroxyl containing styrene copolymers with poly(ethylene oxide), *J. Polymer Sci.: Polymer Chem. Ed.*, **19**, 1451–1473.

19. Coleman M. M. and Zarian J. (1979) Fourier transform infrared studies of polymer blends. II. Poly(ε-caprolactone)–poly(vinyl chloride) systems, *J. Polymer Sci.: Polymer Phys. Ed.*, **17**, 837–850.

20. Hara M. and Eisenberg A. (1984) Miscibility enhancement via ion-dipole interactions. I. Polystyrene ionomer/poly(alkylene oxide) systems, *Macromolecules*, **17**, 1335–1340.

21. Lu X. and Weiss R. A. (1991) Development of miscible blends of polyamide-6 and manganese sulfonated polystyrene using specific interactions, *Macromolecules*, **24**, 4381–4385.

22. Vasilyevskaya V. V., Starodubtsev S. G. and Khokhlov A. R. (1987) Improvement of compatibility of polymer blends due to charging of one component, *Vysokomol. Soed.*, Ser. B, **29**, 930–933.

23. Sanchez I. C. and Balasz A. C. (1989) Generalization of the lattice-fluid model for specific interactions, *Macromolecules*, **22**, 2325–2331.

24. Petrenko K. D., Privalko V. P., Shilov V. V. and Lipatov Yu. S. (1986) Fluctuations of density and concentration in oligo(ethylene glycole)–poly(methyl methacrylate) blends, *Dokl. Acad. Nauk USSR*, **291**, 645–647.

25. Shimada S., Kashima K., Hori Y. and Kashiwabara H. (1990) Concentration fluctuation of poly(ethylene oxide) in a blend with poly(methyl methacrylate) and phase separation behavior studied by spin-label method, *Macromolecules*, **23**, 3769–3772.

26. Saito H., Matsuura M., Okada T. and Inoue T. (1989) Short-range order in a miscible polymer blend, *Polymer J.*, **21**, 357–360.

27. Murthy N. S. and Aharoni S. M. (1987) X-ray and neutron scattering studies of poly(ester carbonate)/poly(ethylene terephthalate) alloys, *Polymer*, **28**, 2171–2175.

28. Privalko V. P., Lipatov Yu. S., Besklubendo Yu. D. and Yarema G. E. (1985) Thermodynamics of binary polymer blends. Melt blended polystyrene–polycarbonate alloys, *Vysokomol. Soed.*, Ser. A, **27**, 1021–1028.

29. Yarema G. E. (1993) Theses, Institute of Macromolecular Chemistry, Academy of Sciences of Ukraine, Kiev.

30. Lipatov Yu. S. and Sergeeva L. M. (1979) *Interpenetrating Polymer Networks*, Naukova Dumka, Kiev (in Russian).

31. Sperling L. H. (1981) *Interpenetrating Polymer Networks and Related Materials*, Plenum Press, New York.

32. Frisch H. L., Frisch K. C. and Klempner D. (1981) Advances in interpenetrating polymer networks, *Pure Appl. Chem.*, **53**, 1557–1566.
33. Frisch K. C., Klempner D. and Frisch H. L. (1982) Recent advances in interpenetrating polymer networks, *Polymer Eng. Sci.*, **22**, 1143–1152.
34. Karabanova L. V. (1980) Thesis, Institute of Macromolecular Chemistry, Academy of Sciences of Ukraine, Kiev.
35. Sergeeva L. M. and Lipatov Yu. S. (1986) Interpenetrating polymer networks, in *Physical Chemistry of Multi-Component Polymer Systems*, Vol. 2, ed. by Yu. S. Lipatov, Naukova Dumka, Kiev, pp. 137–228 (in Russian).
36. Frisch K. C. and Frisch H. L. (1974) Morphology of polyurethan–polyacrylate interpenetrating polymer network, *Polymer Eng. Sci.*, **14**, 76–78.
37. Shilov V. V., Karabanova L. B., Lipatov Yu. S. and Sergeeva L. M. (1978) Structure of interpenetrating polymer networks based on polyurethane and poly(urethane acrylate), *Vysokomol. Soed.*, Ser. A, **20**, 643–650.
38. Binder K. and Frisch H.L. (1984) Phase stability of weakly cross-linked interpenetrating polymer networks, *J. Chem. Phys.*, **81**, 2126–3136.
39. Frisch H. L. and Grosberg A. Yu. (1993) On microphase segregation of interpenetrating polymer networks, *Macromol. Chem. Theory Simul.*, **2**, 517–522.
40. Singh S., Ghiradella H. and Frisch H. L. (1990) Poly(2,6-dimethyl-1,4-phenylene oxide)–poly(methyl methacrylte) IPN's, *Macromolecules*, **23**, 375–378.
41. Mengnjoh P. C. and Frisch H. L. (1989) Interpenetrating polymer networks of poly(2,6-dimethyl-1,4-phenylene oxide) and poly(urethane acrylate), *J. Polymer Sci.: Polymer Chem.*, **27**, 3363–3370.
42. Frisch H. L., Zhou P., Frisch K. C. *et al.* (1991) Interpenetrating polymer networks of poly(carbonate-urethane) and polymethyl methacrylate, *J. Polymer Sci.: Polymer Chem.*, **29**, 1031–1038.
43. Privalko V. P., Azarenkov V. P., Baibak A. V. *et al.* (1994) Thermodynamic characterization of interpenetrating polymer networks of poly(carbonate-urethane) and polymethyl methacrylate, *J. Polymer Sci.: Polymer Chem.* (submitted).
44. Ichihara S., Komatsu A. and Hata T. (1971) Thermodynamic studies of the glass transition and the glassy state of polymers. II. Etnahlpies and specific heats of polystyrene glasses of different thermal histories, *Polymer J.*, **2**, 644–649.
45. Weitz A. and Wunderlich B. (1974) Thermal analysis and dilatometry of glasses formed under elevated pressures, *J. Polymer Sci.: Polymer. Phys. Ed.*, **12**, 2473–2491.
46. Price C. (1975) Effect of pressure on the formation of poly(methyl methacrylate) glasses, *Polymer*, **16**, 585–589.
47. Binder K. (1983) Collective diffusion, nucleation and spinodal decomposition in polymer mixtures, *J. Chem. Phys.*, **79**, 6387–6409.
48. Lifshits I. M. and Slyozov V. V. (1961) The kinetics of precipitation from supersaturated solid solutions, *J. Phys. Chem. Solids*, **19**, 35–50.
49. Cahn J.W. (1965) Phase separation by spinodal decomposition in isotropic systems, *J. Chem. Phys.*, **62**, 93–99.
50. McMaster L. P. (1975) Aspects of liquid–liquid phase transition phenomena in multicomponent polymeric systems, *Adv. Chem. Ser.*, **142**, 43–65.
51. Nishi T., Wang T. T. and Kwei T. K. (1975) Thermally induced phase separation behavior of compatible polymer mixtures, *Macromolecules*, **8**, 227–231.
52. Hashimoto T., Kurami J. and Kawai H. (1983) Time-resolved light scattering studies on kinetics of phase separation and phase dissolution of polymer blends.: I. Kinetics of phase separation of a binary mixture of polystyrene and poly(vinyl methyl ether), *Macromolecules*, **16**, 641–648.
53. Snyder H. L. and Meakin P. (1983) Phase separation dynamics: comparison of experimental results, *J. Chem. Phys.*, **79**, 5588–5593.

54. Gelles R. and Frank C. W. (1982) Phase separation in polystyrene/poly(vinyl methyl ether) blends as studied by excimer fluorescence, *Macromolecules*, **15**, 1486–1491.
55. Gelles R. and Frank C. W. (1983) Effect of molecular weight on polymer blend phase separation kinetics, *Macromolecules*, **16**, 1448–1456.
56. Bogdanovich V. A. (1985) Thesis, Institute of Macromolecular Chemistry, Academy of Sciences of Ukraine, Kiev.
57. Shilov V. V. and Lipatov Yu. S. (1986) Phase separation in multi-component polymer systems, in *Physical Chemistry of Multi-Component Polymer Systems*, Vol. 2, ed. by Yu. S. Lipatov, Naukova Dumka, Kiev, pp. 25–73.
58. Lipatov Yu. S., Shilov V. V., Bogdanovich V. A. *et al.* (1983) Microphase structure of IPN's based on oligo(isoprene dihydrazide) epoxy oligomer and copolymer of styrene and divinyl benzene, *Dokl. Acad. Nauk UkrSSR*, Ser. B, **2**, 34–39.
59. Gilmer J., Goldstein N. and Stein R. S. (1982) Light-scattering studies of phase-demixing of polystyrene/poly(0-chlorostyrene) blends, *J. Polymer Sci.: Polymer Phys. Ed.*, **20**, 2219–2227.
60. Nojima S., Tsutsumi K. and Nose T. (1982) Phase separation process in polymer systems. I. Light scattering studies on a polystyrene and poly(methylphenylsiloxane) mixture, *Polymer J.*, **14**, 255–232.
61. Nojima S., Ohyama Y., Yamaguchi M. and Nose T. (1982) Phase separation process in polymer systems. III. Spinodal decomposition in the critical mixture of polystyrene and poly(methylphenyl siloxane) and scaling analysis, *Polymer J.*, **14**, 907–912.
62. Cook H. E. (1970) Brownian motion in spinodal decomposition, *Acta Met.*, **18**, 297–306.
63. Siggia E. D. (1979) Later stages of spinodal decomposition in binary mixtures, *Phys. Rev.*, Ser. A. **20**, 595–605.
64. Kretschmer R., Binder K. and Stauffer D. (1979) Linear and nonlinear relaxations and cluster dynamics near critical points, *J. Phys. Lett.*, **40**, 69–72.
65. Bonart R. and Mueller E. H. (1974) Phase separation in urethane elastomers as judged by low-angle X-ray scattering. I. Fundamentals, *J. Macromol. Sci.*, Ser. B, **10**, 345–357.
66. Lipatov Yu. S., Shilov V. V., Gomza Yu. P. and Kruglyak N. E. (1982) *X-ray Methods of Characterization of Polymer Systems*, Naukova Dumka, Kiev (in Russian).
67. Shilov V. V. (1983) Thesis, Institute of Macromolcular Chemistry, Academy of Sciences of Ukraine, Kiev.
68. Shilov V. V., Lipatov Yu. S. and Tsukruk V. V. (1986) Structure of interfacial layers in multi-component polymer systems, in *Physical Chemistry of Multi-Component Polymer Systems*, Vol. 2, ed. by Yu. S. Lipatov, Naukova Dumka, Kiev, Vol. 1, pp. 101–120 (in Russian).
69. Gibbs J. W. (1948) Collected Works, Vol. 1, Yale University Press, New Haven
70. Kammer H. W. (1975) Thermodynamik der Grenzschicht zwischen fluessigen Polymeren, *Wiss. Z. Tech. Univ. Dresden*, **24**, 35–39.
71. Kammer H. W. (1978) Grenzflaechenerscheinungen in Polymerschmelzen. 4. Theorie der Grenzschicht zwischen Polymeren, *Faserforsch. Textiltechn.*, **29**, 459–466.
72. Helfand E. and Sapse A. M. (1975) Theory of unsymmetric polymer–polymer interfaces, *J. Chem. Phys.*, **62**, 1327–1333.
73. Helfand E. (1975) Block compolymers, polymer–polymer interfaces, and the theory of inhomogeneous polymers, *Acc. Chem. Res.*, **8**, 295–299.
74. Poser C. I. and Sanchez I. C. (1981) Interfacial tension theory of low and high molecular weight liquid mixtures, *Macromolecules*, **14**, 361–370.
75. Vrij A. and Robeson G. J. (1977) Inhomogeneous polymer solutions: theory of the interfacial free energy and the composition profile near the consolute point, *J. Polymer Sci.: Polymer Phys. Ed.*, **15**, 109–125.

76. Nose T. (1976) Theory of liquid–liquid interface of polymer solutions, *Polymer J.*, **8**, 96–113.
77. Lipatov Yu. S. (1980) *Interphase Phenomena in Polymers*, Naukova Dumka, Kiev (in Russian).
78. Lebedev E. V. (1986) Interfacial region in polymer–polymer systems: in *Physical Chemistry of Multi-Component Polymer Systems*, Vol. 2, ed. by Yu.S. Lipatov, Naukova Dumka, Kiev, pp. 74–100 (in Russian).
79. Chalykh A. E. (1987) *Diffusion in Polymer Systems*, Khimia, Moscow (in Russian).
80. Ruland W. (1971) SAXS of two-phase systems: determination of systematic deviations from Porod law, *J. Appl. Crystallogr.*, **4**, 70–73.
81. Koberstein J. T., Morra B. and Stein R. S. (1980) The determination of diffuse interphase boundaries thickness in polymers by SAXS method, *J. Appl. Crystallogr.*, **13**, 34–45.
82. Lipatov Yu. S., Shilov V. V. and Gomza Yu. P. (1980) Structural features of boundary layers in polymer systems based on polystyrene, poly(methyl methacrylate) and epoxy resin, *Dokl. Acad. Nauk USSR*, **252**, 1393–1397.
83. Fernandez M. L., Higgins J. S., Penfold J. *et al.* (1988) Neutron reflection investigation of the interface between an immiscible polymer pair, *Polymer*, **29**, 1923–1928.
84. Fernandez M. L., Higgins J. S., Penfold J. and Shackleton C. (1991) Interfacial structure for a compatible polymer blend for short and intermediate annealing times, *J. Chem. Soc. Faraday Trans.*, **87**, 2055–2061.
85. Cowie J. M. G. and Ferguson R. (1989) Physical aging studies in polymer blends. 2. Enthalpy relaxation as a function of aging temperature in a poly(vinyl methyl ether)/polystyrene blend, *Macromolecules*, **22**, 2312–2317.
86. Mijovic J., Ho T. and Kwei T. K. (1989) Physical aging in poly(methyl methacrylate)/poly(styrene-co-acrylonitrile) blends, Part II: enthalpy relaxation, *Polymer Eng. Sci.*, **29**, 1604–1610.
87. Ho. T. and Mijovic J. (1990) Physical aging in poly(methyl methacrylate)/poly(styrene-co-acrylonitrile) blends. 3. Simulation of enthalpy relaxation using the Moynihan model, *Macromolecules*, **23**, 1411–1419.
88. Naito K. Johnson G. E., Allara D. L. and Kwei T. K. (1978) Compatibility in blends of poly(methyl methacrylate) and poly(styrene-co-acrylonitrile). 1. Physical properties, *Macromolecules*, **11**, 1260–1265.
89. Jelenic J., Kirste R., Oberthuer R. C. *et al.* (1984) Investigation of exothermic polymer blends by neutron scattering, *Makromol. Chem.*, **185**, 129–156.
90. McBrierty V. J., Douglass D. C. and Kwei T. K. (1978) Compatibility in blends of poly(methyl methacrylate) and poly(styrene-co-acrylonitrile). 2. An NMR study, *Macromolecules*, **11**, 1265–1267.
91. Bosma M., ten Brinke G. and Ellis T. S. (1988) Polymer–polymer miscibility and enthalpy relaxations, *Macromolecules*, **21**, 1465–1470.
92. Grooten R. and ten Brinke G. (1989) Enthalpy relaxations in blends of polystyrene and poly(2-vinylpyridine), Macromolecules, **22**, 1761–1766.
93. Privalko V. P., Demchenko S. S., Besklubenko Yu. D. *et al.* (1980) Thermodynamic and morphological properties of polystyrene/low molecular weight poly(methyl methacrylate) prepared from a common solvent, *Komposits. Polim. Mater.*, **5**, 10–16.
94. Schultz A. R. and Young A. L. (1980) DSC on freeze-dried poly(methyl methacrylate)–polystyrene blends, *Macromolecules*, **13**, 663–668.
95. Lipatov Yu. S., Shilov V. V. and Gomza Yu. P. (1982) Microheterogeneous structure of polystyrene/poly(methyl methacrylate) blends, *Dokl. Acad. Nauk UkrSSR*, Ser. B, **5**, 59–63.
96. Burns C. M. and Kim Woo Nyon (1988) Solution blending of polystyrene and poly(methyl methacrylate), *Polymer Eng. Sci.*, **28**, 1362–1372.

97. MacKnight W. J., Karasz F. E. and Fried J. R. (1978) Phase transitions and relaxations in solid polymer blends in *Polymer Blends*, Vol. 1, ed. by D. R. Paul and S. Newman, Academic Press, New York–San Francisco–London, pp. 219–281.

98. Olabisi O., Robeson L. M. and Shaw M. T. (1979) *Polymer–Polymer Miscibility*, Academic Press, New York–San Francisco–London.

99. Kelley F. N. and Bueche F. (1961) Viscosity and glass temperature relations for polymer–diluent systems, *J. Polymer Sci.*, **50**, 549–556.

100. Couchman P. R. (1980) Prediction of glass-transition temperatures for compatible blends formed from homopolymers of arbitrary degree of polymerization. Compositional variation of glass-transition temperatures. 5, *Macromolecules*, **13**, 1272–1276.

101. Gibbs J. H. and DiMarzio E. (1958) Nature of the glass transition and the glassy state, *J. Chem. Phys.*, **28**, 373–383.

102. DiMarzio E. A. (1990) The glass temperature of polymer blends, *Polymer*, **31**, 2294–2298.

103. Lu X. and Weiss R. A. (1992) Relationship between the glass transition temperature and the interaction parameter of miscible binary polymer blend, *Macromolecules*, **25**, 3242–3246.

104. Breckner M.-J., Schneider H. A. and Cantow H.-J. (1988) Approach to the composition dependence of the glass transition temperature of compatible polymer blends: 1, *Polymer*, **29**, 78–85.

105. Schneider H. A. and DiMarzio E. A. (1992) The glass temperature of polymer blends: comparison of both the free volume and the entropy predictions with data, *Polymer*, **33**, 3453–3461.

106. Shen M. C. and Eisenberg A. (1970) Glass transitions in polymers, *Rubber Chem. Technol.*, **43**, 95–170.

107. Lin A. A., Kwei T. K. and Reiser A. (1989) On the physical meaning of the Kwei equation for the glass transition temperature of polymer blends, *Macromolecules*, **22**, 4112–4119.

108. Schneider H. S. (1989) Glass transition behavior of compatible polymer blends, *Polymer*, **30**, 771–779.

109. Couchman P. R. (1984) Composition-dependent glass transition: relations between temperature, pressure, and composition, *Polymer Eng. Sci.*, **24**, 135–143.

110. Aukett P. N. and Brown C. S. (1988) The measurements of heats of mixing of polymer alloys by a heat of solution method, *J. Thermal Anal.*, **33**, 1079–1084.

111. Allard D. and Prud'homme R. E. (1982) Miscibility of poly(caprolactone)/chlorinated polypropylene and poly(caprolactone)/poly(chlorostyrene) blends, *J. Appl. Polymer Sci.*, **27**, 559–568.

112. Wolf M. and Wendorff J. H. (1990) Modified equation of state treatment for volume changes of mixing, *Polymer*, **31**, 226–228.

113. Mijovic J., Luo H.-L. and Han C. D. (1982) Property–morphology relationships of polymethylmethacrylate/polyvinylidenefluoride blends, *Polymer Eng. Sci.*, **22**, 234–240.

114. Kim W. N. and Burns C. M. (1989) Thermal behavior, morphology and the determination of the polymer–polymer interaction parameter of polycarbonate–poly(butylene terephthalate) blends, *Makromol. Chem.*, **190**, 661–676.

115. Godovsky Yu. K. and Bessonova N. P. (1979) Calorimetric study of two-component polymer blends, *Vysokomol. Soed.*, Ser. A, **21**, 2293–2304.

116. Razinskaya I. N., Shtarkman B. P. Batuyeva L. I. *et al.* (1979) Phase structure and properties of polymer blends illustrated by poly(methyl methacrylate)–poly(vinyl chloride) pair, *Vysokomol. Soed.*, Ser. A, **21**, 1860–1872.

117. Lipatov Yu. S., Rosovitsky V. F. and Maslak Yu. B. (1984) On the glass transition phenomenon in heterogeneous polymer systems with high degree of phase separation, *Vysokomol. Soed.*, Ser. A, **26**, 1029–1032.

118. Wand B. and Krause S. (1987) Properties of dimethylsiloxane microphases in phase-separated dimethylsiloxane block copolymers, *Macromolecules*, **20**, 2201–2208.
119. Privalko V. P., Laskovenko N. N., Korskanov V. V. and Vorona V. V. (1990) Phase morphology and thermoelastic properties of perchlorovinyl–polyurethane systems, *Vysokomol. Soed.*, Ser. A, **32**, 1646–1654.
120. Korskanov V. V. (1992) Thesis, Institute of Macromolecular Chemistry, Academy of Sciences of Ukraine, Kiev.
121. Privalko V. P. (1984) Properties of Bulk Polymers in *Handbook on Physical Chemistry of Polymers*, Vol. 2, ed. by Yu. S. Lipatov, Naukova Dumka, Kiev (in Russian).
122. Couchman P. R. and Karasz F. E. (1978) Some domain-size effects in polymer transitions, *J. Polymer Sci.: Polymer Symp.*, **63**, 271–280.
123. Wu S. (1979) Interfacial energy, surface structure and adhesion between polymers in *Polymer Blends*, Vol. 1, ed. by D. R. Paul and S. Newman, Academic Press, New York–San Francisco– London, pp. 282–336.
124. Romankevich O. V. and Frenkel S. Ya. (1980) On the possibility of existence of thermodynamically stable disperse polymer systems, *Vysokomol. Soed.*, Ser. A, **22**, 1779–1787.
125. Martuschelli E. and Denma G. R. (1980) Morphology, crystallization and melting behavior in poly(ethylene oxide)/poly(methyl methacrylate) blends in *Polymer Blends: Processing, Morphology and Properties*, ed. by E. Martuscelli, R. Palumbo and M. Kryszewski, Plenum Press, New York–London, pp. 101–121.
126. Martuscelli E., Silvestre C., Addonizio M. L. and Amelino L. (1986) Phase structure and compatibility studies of poly(ethylene oxide)/poly(methyl methacrylate) blends, *Makromol. Chem.*, **187**, 1557–1571.
127. Li H. and Hsu S. L. (1984) An analysis of the crystallization behavior of poly(ethylene oxide)/poly(methyl methacrylate) blends by spectroscopic and calorimetric techniques, *J. Polymer Sci.:* Polymer *Phys. Ed.*, **22**, 1331–1342.
128. Nishi T. and Wang T. T. (1975) Melting point depression and kinetic effects of cooling on crystallization in poly(vinylidene fluoride)/poly(methyl methacrylate) mixtures, *Macromolecules*, **9**, 909–915.
129. Hirata Y. and Kotaka T. (1981) Phase separation and viscoelastic behavior of semicompatible polymer blends: poly(vinylidene fluoride)/poly(methyl methacrylate) system, *Polymer J.*, **13**, 273–281.
130. Morra B. and Stein R. (1982) Melting studies of poly(vinylidene fluoride) and its blends with poly(methyl methacrylate), *J. Polymer Sci.: Polymer Phys. Ed.*, **20**, 2443–2459.
131. Khambatta F. B., Warner F., Russell T. and Stein R. (1976) Small-angle X-ray and light scattering studies of the morphology of blends of poly(ε-caprolactone) with poly(vinyl chloride), *J. Polymer Sci.: Polymer Phys. Ed.*, **14**, 1391–1424.
132. Hubbell D. S. and Cooper S.L. (1977) The physical properties and morphology of poly(ε-caprolactone) polymer blends, *J. Appl. Polymer. Sci.*, **21**, 3035–3061.
133. Paul D. R. and Barlow J. W. (1980) Crystallization from miscible polymer blends in *Polymer Alloys. II.*, ed. by D. Klempner and K. C. Frisch, Plenum Press, New York, pp. 239–253.
134. Cousin P. et Prud'homme R. E. (1982) Les melanges polyester/poly(bromure de vinyle), *Eur. Polymer. J.*, **18**, 957–962.
135. Privalko V. P., Lipatov Yu. S. Titov G. V. *et al.* (1984) Phase state of components in a binary blend, polycarbonate/oligoester, *Komposits. Polim. Mater.*, **21**, 3–7.
136. Elyashevich G. K. and Frenkel S. Ya. (1973) Kinetic model of polymer crystallization in conditions of molecular orientation, *Vysokomol. Soed.*, Ser, A, **15**, 2752–2757.

137. Privalko V. P. (1986) *Molecular Structure and Properties of Polymers*, Khimia, Leningrad (in Russian).
138. Ellis T. S. (1988) Miscibility in blends of aliphatic polyamides and an aromtic polyamide, nylon 3Me6T, *Polymer*, **29**, 2015–2026.
139. Hahn B. R., Herrman-Schoenhert O. and Wendorff J. H. (1987) Evidence for a crystal–amorphous interphase in PVDF and PVDF/PMMA blends, *Polymer*, **28**, 201–208.
140. Morra B. S. and Stein R. S. (1984) The crystalline morphology of poly(vinylidene fluoride)/ poly(methyl methacrylate) blends, *Polymer Eng. Sci.*, **24**, 311–318.
141. Russell T. P. and Stein R. S. (1983) An investigation of the compatibility and morphology of semi-crystalline poly(ε-caprolactone)–poly(vinyl chloride) blends, *J. Polymer Sci.: Polymer Phys. Ed.*, **21**, 999–1010.
142. Bliznyuk V. N. (1985) Thesis, Institute of Macromolecular Chemistry, Academy of Sciences of Ukraine, Kiev.
143. Myers M. E., Wims A. M., Ellis T. S. and Barnes J. (1990) An investigation of the morphology of amorphous/semicrystalline nylon blends using small-angle X-ray scattering, *Macromolecules*, **23**, 2807–2814.
144. Kumar S. K. and Yoon D. Y. (1989) Lattice model for interphases in binary semicrystalline/amorphous polymer blends, *Macromolecules*, **22**, 4098–4101.
145. Kwei T. K. and Wang T. T. (1978) Phase separation in polymer–polymer blends in *Polymer Blends*, Vol. 1, ed. by D. R. Paul and S. Newman, Academic Press, New York–San Francisco– London, pp. 172–218.
146. Cortazar M. M., Calahorra M. E. and Guzman G. M. (1982) Melting point depression in poly(ethylene oxide)–poly(methyl methacrylate) blends, *Eur. Polymer. J.*, **18**, 165–166.
147. Aubin M. and Prud'homme R. E. (1980) Miscibility in blends of poly(vinyl chloride) and polylactones, *Macromolecules*, **13**, 365–369.
148. Margaritis A. G. and Kalfoglou N. K. (1988) Compatibility of poly(vinyl chloride) with polyalkylene oxides. I. Poly(methylene oxide) and poly(ethylene oxide), *J. Polymer Sci.: Polymer Phys. Ed.*, **26**, 1595 1612.
149. Kwei T. K. and Frisch H. L. (1978) Interaction parameter in polymer mixtures, *Macromolecules*, **11**, 1267–1271.
150. Chow T. S. (1990) Miscible blends and block copolymers. Crystallization, melting and interaction, *Macromolecules*, **23**, 333–337.
151. Rim P. B. and Runt J. P. (1984) Melting point depression in crystalline/compatible polymer blends, *Macromolecules*, **17**, 1520–1526.
152. Skorodumov V. F. and Godovsky Yu. K. (1987) Melting and crystallization of poly(ethylene oxide) blended with poly(methyl methacrylate) under elevated pressures, *Komposits. Polim. Mater.*, **32**, 15–20.
153. Skorodumov V. F. (1988) Thesis, Karpov Institute of Physical Chemistry, Moscow, Russia.
154. Harrison I. R. and Runt J. (1980) Incompatible blends: thermal effects in a model system, *J. Polymer Sci.: Polymer Phys. Ed.*, **18**, 2257–2261.
155. Runt J. (1981) Polymer blends containing isotactic polystyrene: thermal behavior of model systems, *Macromolecules*, **14**, 420–423.
156. Brusentsova V. G., Gerasimov V. I., Vasilyevskaya L. P. and Bakeev N. F. (1972) Studies of the phase transitions and structure of oriented polyethylene, *Vysokomol. Soed., Ser. A*, **14**, 1109–1115.
157. Nadkarni V. M. and Jog J. P. (1989) Crystallization of polymers in thermoplastic blends and alloys in *Handbook of Polymer Sciences and Technology. Vol. 4. Composites and Specialty Applications*, ed. by N. P. Cheremissinoff, Marcel Dekker, New York and Basel, pp. 81–120.

158. Wunderlich B. (1973) *Macromolecular Physics. Vol. 1: Crystal Structure, Morphology, Defects*, Academic Press, New York and London.
159. Sato T. and Takahashi M. (1969) Study of polyethylene blends by differential scanning calorimetry, *J. Appl. Polymer Sci.*, **13**, 2665–2676.
160. Mandelkern L. (1964) *Crystallization of Polymers*, McGraw-Hill Co., New York–San Francisco–Toronto–London.
161. Hoffman J. D., Davis G. T. and Lauritzen J. I. (1976) The rate of crystallization of linear polymers with chain folding in *Treatise on Solid State Chemistry*, Vol. 3, ed. by N.B. Hannay, Plenum Press, New York, pp. 497–614.
162. Calahorra E., Cortazar M. and Guzman G. M. (1982) Spherulitic crystallization in blends of poly(ethylene oxide) and poly(methyl methacrylate), *Polymer*, **23**, 1322–1324.
163. Wang T. T. and Nishi T. (1977) Spherulitic crystallization in compatible blends of poly(vinylidene fluoride) and poly(methyl methacrylate), *Macromolecules*, **10**, 421–425.
164. Lipatov Yu. S., Privalko V. P. and Petrenko K. D. (1986) Phase state and kinetic of isothermal crystallization of poly(ethylene glycol) in blends with poly(methyl methacrylate) under elevated pressures, *Komposits. Polim. Mater.*, **30**, 49–56.
165. Petrenko K. D., Privalko V. P. and Lipatov Yu. S. (1990) Miscible polymer blends: 2. Kinetics of the high pressure crystallization of poly(ethylene oxide) and poly(vinylidene fluoride) in blends with poly(methyl methacrylate), *Polymer*, **31**, 1283–1287.
166. Bliznyuk V. N., Shilov V. V. and Lipatov Yu. S. (1985) Crystal-liquid phase transition in the poly(ethylene oxide)–poly(methyl methacrylate) system, *J. Mater. Sci. (Lett.)*, **4**, 284–288.
167. Alfonso G. C. and Russell T. P. (1986) Kinetics of crystallization in semicrystalline/amorphous polymer mixtures, *Macromolecules*, 1.9, 1143–1152.
168. Calahorra E., Cortazar M. and Guzman G. M. (1983) Effect of poly(methyl methacrylate) on the crystallization kinetics of poly(ethylene oxide), *Polymer Commun.*, **24**, 211–213.
169. Rymarenko N. L., Petrenko K. D., Voitenko A. I. and Privalko V. P. (1989) Effect of poly(methyl methacrylate) on melt crystallization of poly(ethylene oxide) under elevated pressures, *Komposits. Polim. Mater.*, **42**, 54–57.
170. Wunderlich B. (1976) *Macromolecular Physics. Vol 2: Crystal Nucleation, Growth, Annealing*, Academic Press, New York–San Francisco–London.
171. Romankevich O. V., Grzhimalovskaya I. V., Zabello S. E. and Yudin V. A. (1975) Some characteristics of the crystallization process of the formaldehyde copolymer-polyethylene system, *Sintez Fiz. Khim. Polim.*, **17**, 22–26.
172. Klemmer N. and Jungnickel B.-J. (1984) Ueber eine kristallisationskinetische Anomalie in Mischungen aus Polyethylen und Polyoxymethylen, *Colloid. Polym. Sci.*, **262**, 381–386.
173. Lebedev E. V., Lipatov Yu. S. and Privalko V. P. (1975) Morphological evaluation of interactions between polyethylene and poly(methylene oxide) in extruded blends, *Vysokomol. Soed.*, Ser. A, **17**, 148–153.
174. Lipatov Yu. S., Komoto T., Lebedev E. V. *et al.* (1977) Structural features of low density polyethylene blended with poly(methylene oxide), *Ukr. Khim. Zhurn.*, **43**, 532–536.
175. Khokhlov A. R. and Erukhimovich I. Ya. (1993) A new class of systems exhibiting microphase separation: polymer blends with a nonlocal entropy of mixing, *Macromolecules*, **26**, 7195–7202.
176. Titov G. V., Demchenko S. S., Lipatov Yu. S. and Privalko V. P. (1986) Interfacial interactions in polystyrene–cellulose pair, *Komposits. Polim. Mater.*, **31**, 17–21.
177. Shibanov Yu. D. and Godovsky Yu. K. (1985) Phase separation and glass transition point verus composition diagrams of binary mixtures with covalent bond between components, *Colloid. Polymer Sci.*, **263**, 202–216.

178. Tanaka H. and Nishi T. (1985) New types of phase separation behavior during the crystallization process in polymer blends with phase diagram, *Phys. Rev. Lett.*, **55**, 1102–1105.
179. Komaricheva L. I. (1990) Thesis, Karpov Institute of Physical Chemistry, Moscow.
180. Radzhbov T. M. (1990) Thesis, Karpov Institute of Physical Chemistry, Moscow.
181. Shibanov Yu. D. and Godovsky Yu. K. (1991) Coupling of the phase and relaxation transitions in polymer mixtures, *Makromol. Chem., Macromol. Symp.*, **44**, 61–74.

3 Copolymers

3.1 MOLECULAR PROPERTIES

Binary copolymers of the general formula $[A_F B_{(1-F)}]$ are intrinsically heterogeneous by composition as soon as chemically different species A (component 1) and B (component 2) are linked together to form a common chain. Keeping the overall composition (i.e. the molar content F of component 1) constant, the molecular properties of a copolymer will depend on the pattern of intramolecular distribution of like and unlike units, which may be characterized by the 'distribution index' Q, i.e. [1]

$$Q = Z_{11} Z_{22}/(Z_{12} Z_{21}) \tag{3.1}$$

where Z_{ij} is the conditional probability of formation of a dyad ij by linking of an arbitrarily chosen i-unit to a j-unit (where i and j may be either 1 or 2).

Depending on the value of Q, we may identify the following cases:

(i) alternating copolymers, ACP ($Q < 1$);
(ii) random copolymers, RCP ($Q = 1$); and
(iii) block copolymers, BCP ($Q > 1$).

The simple additivity rule demands that any molecular property P of a binary copolymer should depend on the relative contents of like (f_{11}, f_{22}) and unlike ($f_{12} = f_{21}$) dyads as

$$P = f_{11} P_{11} + f_{22} P_{22} + 2 f_{12} P_{12} \tag{3.2}$$

where P_{11}, P_{22} and $P_{12} = P_{21}$ are the relevant 'partial' properties of the corresponding dyads.

According to eq. (3.2), the contribution of molecular junctions of type 12 between unlike units will change from a maximum for ACP ($f_{11} \Rightarrow 0$, $f_{22} \Rightarrow 0$, $f_{12} \Rightarrow 1$) to a minimum for BCP ($f_{12} \Rightarrow 0$).

In the case of RCP, the relative content of each dyad (at least, in the initial stage of the copolymerization reaction) may be estimated by the following classical formulas [2, 3]:

$$f_{11} = r_1 F_i/K; \quad f_{12} = 1/K; \quad f_{22} = r_2/F_i K \tag{3.3}$$

where $K = r_1 F_i + 2 + r_2/F_i$; F_i is the initial molar ratio of monomers 1 and 2 in the reaction mixture, and r_1 and r_2 are the corresponding monomer reactivities.

A feeling for the change of intramolecular interactions when either of like dyads 11 or 22 is replaced by a heterodyad 12 may be obtained by comparison of corresponding partial properties P_{ij}. For example, a higher equilibrium flexibility of an equimolar ACP of acrylonitrile (AN) and butadiene (B) compared with a RCP of the same overall composition [4] suggests decreased steric hindrances to the internal rotation in a heterodyad 12. Essentially similar information may be derived from dilute solution studies of ACP and RCP of ethylene (E) and α-methylstyrene (MS) [5].

In practice, such data are not always available; therefore, the values of P_{12} may be extracted from treatment of the relevant experimental data for RCP by eq. (3.2). A fairly typical case is that of AN/S (styrene) RCP for which an extremum at equimolar composition was observed for both unperturbed dimensions and their temperature coefficients [6]. Similar patterns of composition dependence of the coil expansion coefficient, unperturbed chain dimensions, etc. from which estimates of P_{12} could be extracted by eq. (3.2), were reported for other RCP [7–11].

Apparently, this procedure may be valid in principle only on condition that uninterrupted sequences of each component in the copolymer chain are shorter than the statistical segment length of either of the homopolymers [12] [this implies, in effect, the validity of eq. (3.3) for f_{12}]. Otherwise [i.e. in the BCP case when the fraction of heterodyads is considerably smaller than that required by eq. (3.3)], the overall size of the copolymer chain will depend primarily not on P_{12} but rather on the degree of swelling of the long uninterrupted sequences (blocks) of each component of the copolymer chain in a common solvent. This is apparently the case with dilute solutions of S/I (isoprene) diblock copolymers [13, 14] for which anomalous temperature dependence of the limiting viscosity number may be explained by the different response of S and I microblocks to a temperature-induced solvent quality change.

In a broad sense, perfectly isotactic vinyl polymers may also be regarded as a special case of BCP consisting of sequences of right-handed and left-handed helices, the length of which may be estimated from eq. (3.4) [15]

$$\zeta_{iso} = (1 + P_{iso})/P_{iso} \qquad (3.4)$$

where $P_{iso} = \exp(-E_{iso}/kT)$, $E_{iso} = [E_{rr} + E_{ll} - (E_{rl} + E_{lr})]/2$ is the energy of chain 'reversal', E_{rr} and E_{ll} are the conformational energies of two adjacent monomeric units in right-handed and left-handed helices, respectively, and E_{rl} and E_{lr} are those for junctions between different helices.

Enrichment of the isotactic chains with 'heterotactic' dyads (i.e. transition to atactic polymer) leads to a drastic reduction in unperturbed dimensions; however, much more elaborate schemes [16] than the simple eq. (3.2) should be used for a quantitative account of this effect.

3.2 THERMODYNAMICS OF THE MELT STATE

Apparently, the S/B copolymers were the first to be characterized by parameters of the Tait equation, eq. (2.9), derived from PVT data [17, 18]. According to eq. (3.2), higher density ρ and lower compressibility β of BCP compared with RCP of the same overall composition F may be explained, qualitatively, by a local packing density deficit associated with 1–2 heterodyads [18]. The same conclusion applies to RCP of N-vinyl carbazole (VC) and n-octyl methacrylate (OMA) [19, 20], on the one hand, and to E/P (propylene) RCP [21, 22], on the other hand, for which the composition dependences of zero-pressure values of the melt-specific volume v_l, thermal expansivity α_l and compressibility β_l could be fitted to eq. (3.2) with $P_{12} > (P_{11} + P_{22})/2$. A still deeper understanding of the effect of chain microstructure on the melt state properties, however, may provide the treatment of relevant PVT data for random copolymer within the framework of an appropriately chosen equation of state.

The extensive PVT data for RCP of VC and 3-iodine-substituted VC (IVC) with various alkyl methacrylates (AMA) [19, 23–25] and alkylstyrenes (AS) [26], as well as RCP of ethylene (E) and 1-olefins (O) [21, 22, 27–29] and E/VA (vinyl acetate) [30] were analyzed by the Simha–Somcynsky reduced equation of state [31, 32], namely

$$\tilde{P}\tilde{V}/\tilde{T} = [1 - 2^{-1/6}y(y\tilde{V})^{1/3}]^{-1} + (2y/\tilde{T})(y\tilde{V})^{-2}[1.011(y\tilde{V})^{-2} - 1.2045] \quad (3.5)$$

which is valid under the condition

$$(3C/s)^{-1}[1 + y^{-1}\ln(1 - y)] = (y/6\tilde{T})(y\tilde{V})^{-2}[2.409 - 3.033(y\tilde{V})^{-2}]$$
$$+ [2^{-1/6}y(y\tilde{V})^{-1/3} - 1/3][1 - 2^{-1/6}y(y\tilde{V})^{-1/3}]^{-1} \quad (3.6)$$

In eqs. (3.5) and (3.6) $\tilde{P} = P/P^*$, $\tilde{V} = V/V^*$ and $\tilde{T} = T^*$ are the reduced pressure, volume and temperature, respectively, $P^* = (z - 2)s^*/v^*$, $V^* = Nav^* = (N_A/m)v^*$ and $T^* = (z - 2)as^*/Ck$ are the corresponding characteristic quantities, s^* and v^* are the energy and size parameters of a potential of non-bonded interactions, N is the number of model chains each comprising s segments of mass m and volume v^* located in one of the N_0 possible sites of a lattice with coordination number z (in usual practice, $z = 12$ is assumed), $y = Na/N_0 < 1$ is the fraction of occupied lattice sites, $3C$ is the number of external (i.e. volume-dependent) degrees of freedom per chain (in calculations, $3C/s = 1$ is recommended), N_A is the Avogadro number and k is Boltzmann's constant.

As can be seen from Table 3.1, for non-polar E/O copolymer series the characteristic parameters P^*, V^* and T^* randomly fluctuate around those for individual homopolymers. On the other hand, the composition dependence of the number of external degrees of freedom per chain repeating unit, $3C/p = (P^*V^*/T^*)(3m_0/R)$ (where p is the number of such units each of mass m_0 in a real macromolecule), for these series (Fig. 3.1), reveals the following features.

(i) For the E/P and E/B1 (butene-1) series the data are reasonably accounted for by eq. (3.2) with $P_{12} = 1.8$ only in the range of low-to-medium E content ($F < 0.8$).

(ii) At $F > 0.8$ significant negative deviations from additivity are observed not only for the E/P and E/B1 series, but also for copolymers of the series E/H1 (hexene-1) in spite of a more than two-fold difference between $3C/p$ for the respective homopolymers (0.88 for PE and 2.02 for polyhexene-1).

Table 3.1 Characteristic parameters of eqs. (3.5) and (3.6) for random copolymers of ethylene

Molar content of ethylene	P^* (MPa)	v^* (cm^3/g)	T^* (10^{-4} K)	$3C/p$
Series E/P				
1.0 (Polyethylene)	850	1.2234	1.2025	0.87
0.98	650	1.2230	1.1450	0.82
0.96	555	1.2350	1.2105	0.69
0.90	535	1.2455	1.2225	0.63
0.78	635	1.1995	1.1270	0.77
0.57	965	1.1190	0.9265	1.50
0.37	750	1.1520	1.0455	1.16
0.32	780	1.1355	0.9560	1.24
0 (Polypropylene)	550	1.2508	1.2774	0.84
Series E/B1				
0.99	605	1.2230	1.1775	0.69
0.97	590	1.2225	1.1770	0.68
0.94	550	1.2170	1.1990	0.69
0.85	575	1.1555	1.0950	0.72
0.73	840	1.1195	1.0550	1.26
0.42	795	1.1160	1.1865	1.31
0.20	480	1.1885	1.2550	0.98
0.11	495	1.1890	1.2480	0.89
0 (Polybutene-1)	435	1.2034	1.2735	0.84
Series E/H1				
0.98	555	1.2305	1.2050	0.66
0.96	550	1.2410	1.2105	0.64
0.92	570	1.2005	1.1990	0.72
0 (Polyhexene-1)	595	1.1926	1.0765	2.02
Series E/VA				
0.93	700	1.1354	1.0636	0.83
0.90	693	1.1043	1.0410	0.88
0.88	719	1.0998	1.0419	0.94
0.82	761	1.0429	1.0276	1.06
0 [Poly(vinyl acetate)]	957	0.8141	0.9419	2.52

Given the physical meaning of $3C/p$ as the measure of thermal mobility of the chain repeating units in the melt state [33], a large excess of that parameter for heterodyads with respect to either of the homodyads [cf. issue (i)] suggests the former act as a sort of 'internal swivel' in copolymer chains at not-too-high E content. On the other hand, the latter observation (ii) may be attributed to a microblock structure of copolymers of all series at high E content, where the chains are presumably composed of sufficiently long, continuous E sequences randomly interrupted by monomeric moieties of a different chemical nature [27, 28].

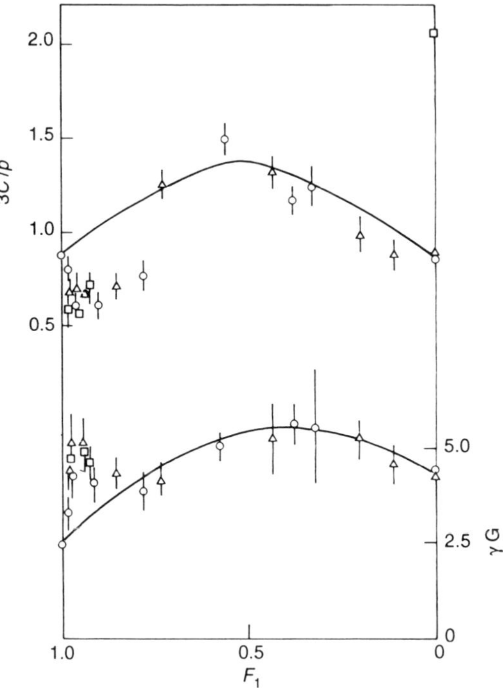

Fig. 3.1. Dependence of the number of external degrees of freedom (upper curve) and of the quasi-lattice Grueneisen parameter (lower curve) on the molar content of ethylene for RCP series E/P ($\bigcirc$), E/B ($\triangle$) and E/H ($\square$)

In the case of polar copolymer series VC/OMA and IVC/OMA the characteristic temperatures T^* tended to increase, whereas the values of the characteristic volume V^* and parameter $3C/p$ decreased with the molar content of stiff, condensed VC and/or IVC cycles (Fig. 3.2) [19, 23, 24]. The patterns of these dependences (except, perhaps, the case of V^*) are, however, not amenable to a quantitative analysis by the simple eq. (3.2) in which dyad additivity is implicitly assumed. It appears that information on contributions from higher-than-dyad microstructural levels (i.e. triads, tetrads, etc.) would be needed for this purpose. This conclusion also applied to similar data for the VC/AMA [20, 23 25] and VC/AS [26] copolymer series.

With regard to the BCP case, the limited PVT data available for S/B block copolymers in the melt state [17, 18] still have not been treated by any modern equation-of-state theory. The results of PVT measurements for a polyurethane elastomer, Solithane 111, could be fitted to eqs. (3.5) and (3.6) [34]; unfortunately, little can be said about the value of the characteristic parameters obtained ($P^* = 1060\,\mathrm{MPa}$, $V^* = 0.889\,\mathrm{cm^3/g}$ and $T^* = 8300\,\mathrm{K}$) except that they are in the range expected for slightly polar polymers.

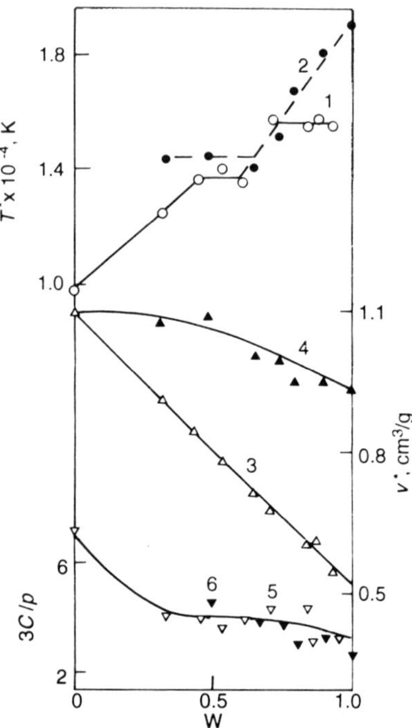

Fig. 3.2. Dependence of the characteristic temperatures (1, 2), volumes (4, 3) and the number of external degrees of freedom per chain-repeating unit (5, 6) on molar content of VC (2, 4, 6) and IVC (1, 3, 5) for their random copolymers with OMA

3.3   PHASE SEPARATION IN BLOCK COPOLYMERS

3.3.1   PHASE MORPHOLOGY

Non-polar Chain Components

Chain connectivity (i.e. the occurrence of intramolecular 1–2 type junctions between components 1 and 2) is the fundamental feature by which BCP may be distinguished from polymer blends of the same nominal composition. While the contribution of excess enthalpy H_{ex} to the excess free energy G_{ex} [cf. eq. (2.2)] may be safely approximated by eq. (2.3) for both systems, the obvious requirement of localization of $(m-1)$ 1–2 type junctions of the m-block copolymer at the interface after phase separation will lead to the following effects [35] (implicitly assuming a sharp interface [36]).

(i) Excess entropy will be decreased by the amount

$$\Delta S_{ex,1}/k = 2(m-1)\Delta S_{dis} \tag{3.7}$$

where

$$\Delta S_{dis}/k \cong \ln[(z-1)e] \qquad (3.8)$$

is the entropy deficit ('disorientation entropy') caused by immobilization of 1–2 type junctions at the interface.

(ii) Excess entropy will be increased by the amount

$$\Delta S_{ex,2}/k = \ln(m-1) \qquad (3.9)$$

which is essentially Boltzmann's formula for the entropy of random placement of $(m-1)$ junctions at the interface.

Substituting H_{ex} from eq. (2.3) and $S_{ex} = \Delta S_{comb} - \Delta S_{ex,1} + \Delta S_{ex,2}$ [where ΔS_{comb} is defined by eq. (2.5)] into eq. (2.2), we obtain from condition $G_{ex} = 0$

$$\chi_{cr} = S_{ex}/k\varphi(1-\varphi) \qquad (3.10)$$

According to numerical estimates [35], the critical values of χ_{cr} calculated by eq. (3.10) for BCP are invariably larger than the values of χ_{cr} from eq. (2.12) for linear blends of the same composition, the more so the larger is m. In other words, chain connectivity makes phase separation more difficult in block copolymers than in the corresponding blends, although in other respects both systems are expected to exhibit similar behavior (e.g. with p_1 and p_2 increasing, the components are predicted to become less compatible due to decreasing χ_{cr}).

These qualitative predictions well agree with the following experimental data.

(i) DTA and dilatometry studies of S/MS diblock copolymers and of the corresponding polymer blends revealed the persistence of a single-phase state in the former to significantly higher molecular weights of both components [34]. A similar result was reported for binary blends and BCP of ethylene oxide (EO) and arylate (AR) [38, 39].

(ii) Equimolar di- and triblock copolymers of S and MS [37, 40], as well as 1, 4- and 1, 2-butadienes [41], become incompatible and separate into two amorphous phases above certain critical molecular weights.

(iii) Transition to a single-phase, amorphous state was observed for E/B [42], EO/AR [38, 39] and EO/C (bis-phenol A carbonate) [43] BCP when the degree of polymerization of a crystallizable component (E in the former and EO in the latter) decreased below a critical value $p_1^* \cong 20$.

An alternative approach to assessing the effect of chain connectivity in phase-separated BCP involves consideration of their unique morphology. In contrast to polymer blends for which phase separation might ultimately result (at least, theoretically) in the formation of two macroscopic phases of pure components, in the case of BCP the finite chain length of each component prohibits macroscopic phase separation; hence, only microphases ('phase domains') of a limited size may eventually form. Assuming the domain shape to be spherical, one may write, approximately [44]

$$R \cong 4\alpha \langle h_0^2 \rangle^{1/2}/3 \qquad (3.11)$$

where α is the 'forced' expansion coefficient accounting for extension of the chain from its unperturbed dimensions, $\langle h_0^2 \rangle^{1/2} \sim p_1^{1/2}$, to domain size R.

This size effect [neglected so far in the derivation of eq. (3.10)] may be accounted for theoretically if the concentration of 1–2 type junctions at the domain–continuous matrix interface is looked at as a source of the following extra contributions to the excess free energy [44].

(i) Placement entropy difference

$$\Delta S_p/k = \ln \{ [3p_1/(p_1 + p_2)](\Delta r/R) \} \tag{3.12}$$

where Δr is the thickness of the boundary layer (BI) composed of 1–2 type junctions at the periphery of component 1 domains of radius R.

(ii) Restricted volume entropy difference

$$\Delta S_v/k = \ln P(p_1:r',r < R) + \ln P(p_2:r',r > R) \tag{3.13}$$

where the first term accounts for the probability that all chain segments of component 1 (including the first one fixed at r') are confined within the domain, while the probability that all chain segments of component 2 are outside the domain is given by the second terms.

(iii) Elasticity entropy difference

$$\Delta S_{el}/k = - 3(\alpha^2 - 1 - 2 \ln \alpha)/2 \tag{3.14}$$

(iv) Surface free energy

$$\Delta G_s = \sigma_{12} S_{12} \tag{3.15}$$

where σ_{12} is the interfacial tension and S_{12} is the total surface area.

Substituting eqs. (2.3) and (3.12)–(3.15) into (2.2) we obtain for $G_{ex} = 0$ [44]

$$\chi_{cr} = [(p_1 + p_2)/p_1] \{ \ln [(1 + p_2/p_1)(R/\Delta r)] - \ln P(p_1:r',r < R)$$
$$- \ln P(p_2:r',r > R) + g(\alpha_m^2 - 1)/2 - 3 \ln \alpha_m \} \tag{3.16}$$

where α_m is the maximum possible chain extension.

Qualitatively, eq. (3.16) agrees with the simpler eq. (3.10) as concerns the predicted chain length dependence of χ_{cr}. Moreover, the validity of the starting assumptions (i)–(iv) was apparently proved by a gratifying quantitative agreement between the theoretical values of the domain's dimensions, R, from eq. (3.11) (assuming $\alpha \cong 2^{1/2}$) and the BI thickness obtained from condition $\partial G_{ex}/\partial(\Delta r) = 0$

$$\Delta r = (6R/\chi) + [(6R/\chi)^2 + (1/6)\pi^2 l_D]^{1/2} \tag{3.17}$$

(where χ is approximated by eq. (2.4) and $l_D \cong 0.8$ nm is the characteristic Debye distance of intermolecular interactions) and the corresponding SAXS data ($R \cong 7$–30 nm; $\Delta r \cong 2$ nm) for 80/20 S/I diblock copolymers of variable molecular masses in which isoprene spheres embedded in a continuous styrene matrix were observed [45]. These data can be also rationalized within the framework of other theoretical approaches [46–49].

The spherical morphology of domains is not unique for BCP. According to the structural information available [50, 51], spherical domains is the characteristic morphological feature of either of components at low (say, below 30%) volume content; as the composition changes, a morphological transformation to a system of regularly packed cylinders (amorphous 'superlattice') and, finally, to interpenetrating, continuous lamellar structures around the 'phase inversion' point is observed (Fig. 3.3 [52]). Simple scaling arguments were invoked [53, 54] to correlate the characteristic interdomain spacings (D) for different morphologies with solvent quality ($v = 1/2$ and $3/5$ for perfect and good solvents, respectively), concentration (C) of solution and molecular dimensions (p_1 and p_2) of di- and triblock copolymers (Table 3.2). Both theoretical [55, 56] and experimental [57, 58] data available suggest a spinodal decomposition mechanism of phase

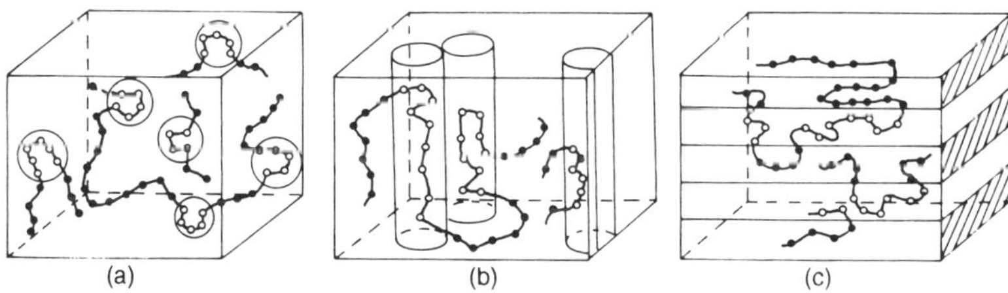

Fig. 3.3 Morphology evolution of a block copolymer as a function of the relative content and/or length of the second component: spherical micells (a); cylindrical micells (b); alternating lamellae (c)

Table 3.2 Scaling relationships for the morphology of diblock BCP

Regime	BCP composition	R	D
		Spherical micelles	
I^*_a	$p_1 < p_2^{v/6}$	$p_2^{2/3}$	p_1^v
I^{**}_b	$p_2^{v/6} < p_1 < p_2^{(1+2v)/6v}$	—	$p_1 p_2^{(v-1)/6v}$
II^{***}	$p_2^{(1+2v)/6v} < p_1 < p_2^{(1+2v)/5v}$	$p_2 p_1^{-2v/(1+2v)}$	$p_1^{3v/(3v+1)}$
III^{****}	$p_1 > p_2^{(1+2v)/5v}$	$p_2^{3/5}$	$p_1^v p_2^{2(1-v)/5}$
		Cylindrical micelles	
I^*_a	$p_1 < p_2^{v/6}$	$p_2^{2/3}$	p_1^v
I^{**}_b	$p_2^{v/6} < p_1 < p_2^{(1+2v)/6v}$	—	$p_1 p_2^{(v-1)/6v}$
II^{***}	$p_2^{(1+2v)/6v} < p_1 < p_2^{(1+2v)/5v}$	$p_2 p_1^{-2v/(1+2v)}$	$p_1^{3v/(3v+1)}$
III^{****}	$p_1 > p_2^{(1+2v)/5v}$	$p_2^{(2+v)/(3+v)} \times p_1^{-v/(3+v)}$	$p_2^{(1+v)/(3+v)} \times p_1^{4v/(3+v)}$

*Isolated coils of component 1 at the periphery of micelle core of component 2.
**Onset of the overlap of coils of component 1.
***Quasi-planar overlap of coils of component 1.
****Onset of the effect of core curvature.

separation (cf. section 2.2.2) in the melts of BCP. Moreover, as can be assessed from molecular dynamics simulations [59–61], the lamellar morphology is the ultimate result of a series of successive structural transformations from initial spheres via transient cylinders.

It appears that the most salient structural features of BCP may be reasonably explained by a model that explicitly accounts for issues (i)–(iv) cited above [46–54, 62–64]. However, the major issue (iii) implying the build-up of internal stresses within the domains due to chain extension by ends fixed at the interface, is not consistent, for example, with the IR dichroism study of S/B/S triblock copolymers in which no evidence for other-than-random chain orientation within cylindrical domains was found [65]. Thus, the physical origin of issue (iii) (if any) would be the external forces from the surrounding medium acting all along the chain [66], rather than the stretching force applied to the other chain end on the domain's surface [44–49, 53, 54].

Chain Components with Specific Interactions

The theoretical and experimental data discussed so far referred essentially to non-polar BCP with randomly oriented (i.e. non-specific) interactions between chain components. 'Segmented' polyurethanes (SPU) are the typical representatives of another large family of multiblock copolymers in which microphase separation is caused by self-association of rigid chain fragments [67, 68]. The main features of SPU may be summarized as follows.

(i) Given the statistical nature of the polyaddition reaction by which rigid fragments (mostly, aromatic diisocyanates) are bonded to soft fragments (oligomeric ethers, esters, dienes, etc.) via chain extenders (short-chain diols and/or diamines), as well as rather wide molecular weight distribution of soft fragments, the size of the rigid domains separated from a continuous soft phase is usually characterized by a fairly broad dispersion [67–69].

(ii) The driving force for microphase separation is essentially the enthalpy gain due to the complete saturation of specific interactions (e.g. interchain H-bonds, ionic interactions, etc.) between rigid fragments [68], rather than the difference in solubility parameters of rigid and soft fragments [cf. eqs. (2.3), (2.4) and (2.8)].

(iii) The conformation adopted by rigid fragments within the domains is not the familiar random coil but either a distorted rod (for short sequences) or an assembly of irregular folds (for sufficiently long sequences) [70–72].

(iv) No evidence for the cylindrical shape of domains in SPU has been obtained so far, while ellipsoids and lamellae are the typical morphological forms observed [67, 73–75].

(v) The contraction of rigid domains in the course of microphase separation leads to an extension of soft fragments with ends anchored to peripheries of adjacent domains by some 20–30% above the size of the corresponding random coil [76, 77].

(vi) The degree of microphase separation α_{sep} is lower (i.e. the thickness of the BI Δr is higher) the higher is the polarity of soft fragments [78–80]. Surprisingly, the values of Δr are typically of the same order of magnitude as observed for non-polar block copolymers (i.e. about 1–2 nm), although the amplitude of Δr fluctuations around the mean value is sometimes changing from very high in the composition range of a continuous soft phase to severalfold lower following the morphological 'inversion' to a continuous rigid phase (Fig. 3.4 [81]).

(vii) Transition to a single-phase melt (at least, for SPU based on oligoethers and 4, 4′-diphenyl methane diisocyanate, MDI) is believed to occur above the softening point of MDI domains ($T_{g,2} \cong 420$ K) [80, 82, 83]. Kinetic data on microphase separation (obtained by monitoring the time dependence of either the intensity of characteristic bands of IR spectra [84, 85] or the glass transition quantities [86, 87] during annealing at $T_{g,1} \ll T < T_{g,2}$ of samples quenched from well above $T_{g,2}$ to below the glass transition temperature of the soft phase $T_{g,1}$) revealed two stages with different relaxation times. The data could be fitted to the simple Kolmogorov–Avrami equation, eq. (1.50), for the first-order phase transitions with $n = 2$ for the initial and $n = 0.6$ for the second stage, respectively [84]; moreover, the temperature dependence of the apparent rate constant is

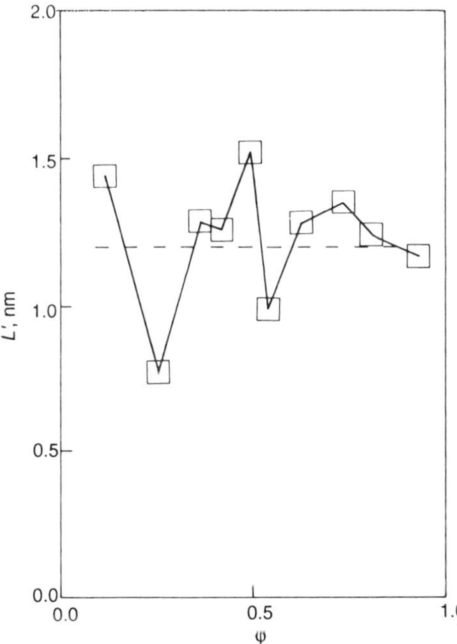

Fig. 3.4. Dependence of the thickness of BI on the volume fraction of the stiff phase in SPU

of a bell-like shape [85] as required by eq. (1.51) for the melt–crystal transition. It is thus likely that phase separation in SPU proceeds by the mechanism of nucleation and growth.

The issue of structure formation in block copolymers with other (e.g. ionic) types of specific interactions between components was reviewed in [88, 89].

Properties of a BI

Although the validity of both the physical concept as well as the experimental evidence for a BI with thickness significantly exceeding the length of 1–2 type junctions may be questioned (at least, in case of non-polar block copolymers with nearly monodisperse blocks) [34, 35], it is difficult to understand why block copolymers should be fundamentally different (especially taking into consideration the chain connectivity effect) from corresponding polymer blends for which the existence of a BI seems now firmly established (cf. section 2.2.3). In the case of block copolymers, the evidence for a BI is provided, e.g. by the following experimental data.

(i) As expected, only two heat capacity jumps at the glass transition temperatures of I blocks ($T_{g,1}$) and S blocks ($T_{g,2}$) were observed for rapidly quenched samples of S/I/S triblock copolymers, whereas annealing at the intermediate temperature, $T_{g,1} < T_a < T_{g,2}$, produced the onset of endothermic enthalpy relaxation near $T_{max} > T_a$, the intensity of which (i.e. the amount of heat absorbed, ΔH_{ex}) increased with annealing time t_a [90]. Since no change was observed in pure PS after similar treatment, the effect observed was attributed to structural relaxation in the BI. The total fraction of the latter assumed to increase with ΔH_{ex} as [90]

$$v_{BI} = \frac{2(T_{g,2} - T_{g,1})\Delta H_{ex}}{k_H(T_{max} - T_a)^2} \tag{3.18}$$

where k_H is the appropriate rate constant. According to numerical estimates by eq. (3.18), the values of v_{BI} tended to increase from about 0.1 to as high as 0.5 with an increase in T_a, t_a and/or a decrease in the copolymer molecular mass [90]. These experimental findings may be adequately reproduced [91] by an appropriately adapted kinetic model [cf. eq. (1.17)] assuming the BI to consist of discrete fractions with $T_{g,i}$ values between those of individual components, each contributing independently to H_{ex}.

(ii) In a pulse broad line NMR study of S/B block copolymers [92] the free induction decay data at $T_{g,1} < 293\,\mathrm{K} < T_{g,2}$ could be resolved into three components: the fast one corresponding to a rigid S microphase, and the remaining slow ones to an unperturbed, soft B phase and to a BI phase of intermediate stiffness, respectively.

Moreover, fairly thick BI is the most natural element of the structural model of polar multiblock copolymers with a low degree of phase separation, although

Table 3.3 Sample coding and physical properties of segmented polyurethanes

N	Sample code	Molar ratio OPG/EG/MDI	φ	ΔH_d (J/g)	ρ (g/cm^3)	S/V (m^2/cm^2)
1	SPU-13	1/0.35/1.35	0.13	2.7	1.113	149
2	SPU-26	1/1.5/2.5	0.26	-3.8	1.125	197
3	SPU-38	1/2.5/3.5	0.38	—	1.145	216
4	SPU-43	1/4/5	0.43	-6.8	1.151	205
5	SPU-51	1/5.5/6.5	0.51	-7.6	1.178	224
6	SPU-55	1/7/8	0.55	—	1.188	372
7	SPU-64	1/11/12	0.64	-25.0	1.225	314
8	SPU-75	1/18/19	0.75	-33.3	1.267	265
9	SPU-82	1/24/25	0.82	—	1.285	234
10	SPU-94	1/99/100	0.94	-41.6	1.321	102
11	SPU-100	0/1/2	1.00	-52.4	1.325	—

no clear-cut correlation between Δr and α_{sep} was eastablished so far [67, 73, 79, 81, 93–95]). In this case, however, the following procedure developed for SPUs with oligomeric propylene glycol (OPG-2000) as a soft-chain component, and 4,4′-diphenylmethane diisocyanate (MDI) extended with ethylene glycol (EG) as a stiff-chain component [96], may be recommended.

As can be seen from Table 3.3 and Fig. 3.5, both heats of dissolution ΔH_d and the bulk density ρ of SPU decreased, the lower the volume fraction φ of the stiff-chain fragments. As soon as the plots of composition dependence of both these properties would be straight lines connecting points at $\varphi - 0$ (i.e., pure OPG-200) and $\varphi = 1$ (pure stiff-chain homopolymer) if SPUs had separated into two microphases of pure components with sharp boundaries (i.e., the case of negligibly small contribution from a 'mathematical' interface of zero thickness), the observed deviations of both plots from linear additivity (broken lines in Fig. 3.5) should be considered as experimental evidence for the three-phase morphology of SPU referred to above. To evaluate the contribution of BI (P_{BI}) to any physical property P of SPU, one needs to consider two alternative situations.

(i) BI is composed exclusively of the soft-phase component sterically 'immobilized' in the vicinity of stiff chain domains. In this case, the relevant expression for the composition dependence of P would be

$$P = \varphi P_2 + (1 - \varphi)[v P_{\text{BI}} + (1 - v)P_1]. \tag{3.19a}$$

(ii) If BI is composed of fragments of both soft-chain and stiff-chain components, one should expect the following expression to hold

$$P - (1 - v)[(1 - \varphi)P_1 + \varphi P_2] + v P_{\text{BI}}. \tag{3.19b}$$

In the above expressions, subscripts 1, 2 and BI refer to respective components, and v is the volume fraction of BI. A look at the ρ vs. φ plot (Fig. 3.5) shows

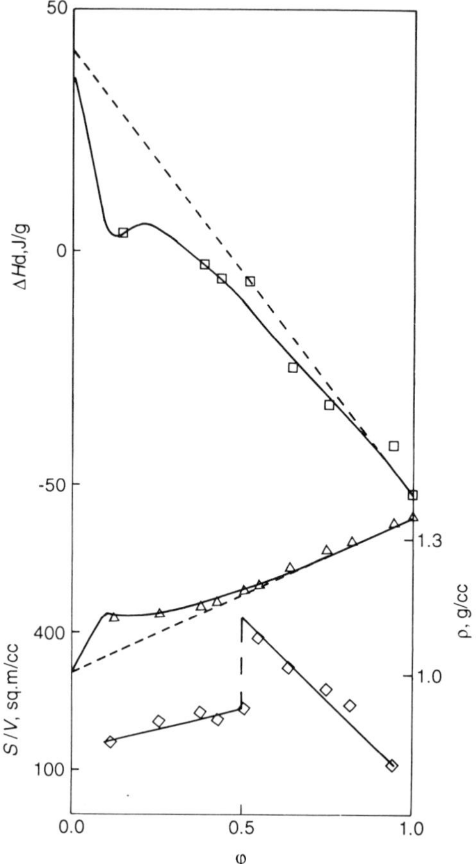

Fig. 3.5. Dependences of the heat of dissolution ($\bigcirc$) density ($\triangle$) and the ratio S/V ($\square$) on the volume fraction of the stiff phase in SPU

that the former expression (3.19a) would not apply as soon as an unreasonably high (well above 10% in excess of $\rho_1 = 1.0\,\text{g/cc}$ for pure OPG-2000 [33]) value of ρ_{BI} would be required to account for a fairly large positive deviations of the majority of experimental data from the additivity line. Thus, in subsequent analysis the latter expression (3.19b) was used throughout.

For numerical estimates, the fraction of BI was approximated by the corresponding linear fraction, $v = \langle L \rangle / [\langle L \rangle + \langle d \rangle]$, where $\langle L \rangle$ and $\langle d \rangle$ are the thickness of BI and the characteristic longitudinal dimension of stiff-chain domains, resp. The values of $\langle L \rangle$ for all SPUs estimated from SAXS data analysis [81] were found to fluctuate randomly around the mean, composition-invariant value $\langle L \rangle = 1.2\,\text{nm}$. No clear-cut dependence on composition was observed for other structural parameters [81] except the surface-to volume ratio of domains, S/V, which exhibited a smooth variation with φ both above and below the apparent "phase inversion" point in the vicinity of $\varphi^* = 0.5$ (Fig. 3.5). Thus, for internal consistency the values of $\langle d \rangle$ were estimated from the following theoretical expression [69, 79],

$$\langle d \rangle = X\varphi/(S/V), \tag{3.20}$$

using the best-fit value $X = 3$ [96].

The composition dependences of ΔH_d and ρ were analyzed by eqs. (3.19b) and (3.20) making use of experimental values of S/V [81] assuming $\langle L \rangle = \text{const} = 1.2$ nm. As can be easily verified, the experimental data are adequately accounted for assuming constant contributions $\Delta H_{d,BI} = -(45.6 \pm 10.3)$ J/g and $\rho_{BI} = 1.27 \pm 0.09$ g/cc (cf. solid lines in Fig. 3.5).

This approach was also applied to the analysis of temperature and composition dependencies of basic thermodynamic functions (i.e., absolute enthalpy H, absolute entropy S and absolute Gibbs free energy G) derived from precise heat capacity measurements of the same SPU in the temperature interval 4.2–420 K [96]. As can be judged by the excess thermodynamic quantities calculated via eqs. (2.1), (2.2), thermodynamic stability of SPU decreased, the higher the relative amount of BI. Structural implications of this purely phenomenological result are uncertain; it may be speculated, however, that soft-chain and/or stiff-chain fragments within BI adopt somewhat extended conformations (compared to those in corresponding pure phases), favoring stronger interchain interactions (hence, $H_{ex} \langle 0 \rangle$; however, the concomitant entropy losses turn out so severe ($S_{ex} > 0$) that the thermodynamic stability of SPU is decreased ($G_{ex} > 0$). Stated otherwise, incomplete microphase separation resulting in formation of BI should be considered as the main source of intrinsic instability of the three microphase-type morphology of SPU.

It can be thus concluded that the three-phase morphological model (i.e. two microphses of pure components and the BI microphase) is consistent with the experimental data available for the majority of block copolymers. Special care should be taken, however, to properly assess the significance of structural data like those mentioned below.

(i) Lower-than-unity values of α_{sep} suggest incomplete phase separation; however, higher-than-unity values of α_{sep} calculated by eq. (2.24) may also be encountered, as was the case with siloxane–urea segmented copolymers [93]. This unusual result was tentatively attributed to the excess contribution of isolated urea fragments in the siloxane matrix to the experimental electron density fluctuations [the numerator of eq. (2.24)].

(ii) As already emphasized in section 2.2.3, values of Δr calculated by eq. (2.31) strongly depend on the procedure used to analyze the 'tails' of experimental SAXS curves. Therefore, consistency checks are needed to extract the most reliable values of Δr from the SAXS data [81, 93–95]. In this respect, eqs. (2.30) and (2.33) seem preferable.

3.4 GLASS TRANSITION IN NON-CRYSTALLINE COPOLYMERS

3.4.1 RANDOM COPOLYMERS

As expected for single-phase, two-component systems, RCP in the amorphous state are characterized by a single glass transition temperature, T_g, located in between those for individual homopolymers, $T_{g,1}$ and $T_{g,2}$. According to the limited evidence available [92,98], the kinetics of structural relaxation at T_g for RCP of S/MMA and S/AN may be quantitatively described by eq. (1.17); unfortunately, the data are too scarce to warrant any meaningful analysis of the kinetic quantities involved by eq. (3.2).

Quasi-equilibrium, composition-dependent values of T_g may be estimated by numerous, more or less elaborated empirical schemes assuming additivity by tabulated increments assigned to individual molecular groups of chain-repeating units [99–101]. With regard to model-based approaches, we may use, pragmatically, eqs. (2.36), (2.37), (2.41), etc. for compatible polymer blends which should presumably apply to any binary system of randomly mixed components, provided appropriately defined free volume fractions [eq. (2.36)], excess entropies [eq. (2.37)], fractions of flexible bonds (eq. (2.41) [102]), etc. are additive at T_g values. Physically more appealing are, however, the following approaches which explicitly account for the chain connectivity effect (i.e. contributions from 1–2 type junctions) through the non-linear additivity eq. (3.2).

(i) Assuming additivity of the glass transition temperatures [i.e. setting $p_{ij} = T_{g,ij}$ in eq. (3.2)], we obtain

$$T_g = T_{g,11} f_{11} + T_{g,22} f_{22} + 2T_{g,12} f_{12} \qquad (3.22)$$

where $T_{g,11} = T_{g,1}$ and $T_{g,22} = T_{g,2}$ are the glass transition temperatures of the corresponding pure components (i.e. homopolymers) and $T_{g,12}$ is that of an alternating copolymer. In the special case of equal reactivities of comonomers [i.e. $r_1 = r_2$ in eq. (3.3)], eq. (3.22) transforms into [103]

$$T_g = T_{g,1} F^2 + T_{g,2}(1 - F)^2 + 2F(1 - F)T_{g,12} \qquad (3.23)$$

where F is the molar fraction of component 1 in a copolymer.

In the derivation of eq. (3.23), identical fractions of flexible bonds [i.e. $B_1 = B_2$ in eqs. (2.39)] in repeating units of both components was tacitly assumed; if, however, $B_1 \neq B_2$, eq. (3.24) should apply [104]

$$T_g = T_{g,1} B_1 + T_{g,2} B_2 + T_{g,12} B_{12} \qquad (3.24)$$

where $B_{12} \cong (B_1 + B_2)/2$ is the fraction of flexible bonds in 1–2 type junctions.

It can be also shown [104] that eq. (3.23) is a special case of eq. (3.25) [105]

$$T_g \cong T_{g,2} - 1.5(F_{v,1}/R)[F_{v,1}(2A_{12}^* - A_{11}^* - A_{22}^*) + (A_{22}^* - A_{12}^*)] \qquad (3.25)$$

(where $F_{v,1} = \Delta\alpha_1 m_1 / [\Delta\alpha_1 m_1 + \Delta\alpha_2 m_2]$ is the fraction of 'vibrational' volume of component 1, $\Delta\alpha_i$ and m_i are the thermal expansivity jump at $T_{g,1}$ and the molar mass of the ith component, respectively, and $A_{ij}^* \sim T_{g,ij}$ are enthalpies of the hole formation in corresponding environments) one the condition $\Delta\alpha_1 m_1 = \Delta\alpha_2 m_2$.

(ii) Substitution of $P_{ij} = w_{ij}/T_{g,ij}$ into eq. (3.2) (where w_{ij} is the weight fraction of the ij dyads) yields [106]

$$1/T_g = w f_{11}/T_{g,1} + (1-w) f_{22}/T_{g,2} + 2 f_{12}/T_{g,12} \qquad (3.26)$$

where w is the weight fraction of component 1.

(iii) Assuming additivity of 'configurational' entropies at the T_g values defined by eq. (1.24) and setting the limits of integration for each contribution in the temperature intervals from $T_{g,ij}$ to T_g, we obtain from eq. (3.2), after some rearrangements [107],

$$\ln T_g = \frac{f_{11}\Delta C_{p,1} \ln T_{g,1} + f_{22}\Delta C_{p,2} \ln T_{g,2} + 2 f_{12}\Delta C_{p,12} \ln T_{g,12}}{f_{11}\Delta C_{p,1} + f_{22}\Delta C_{p,2} + 2 f_{12}\Delta C_{p,12}} \qquad (3.27)$$

where $\Delta C_{p,ij}$ is the contribution of the ij dyads to the heat capacity jump at corresponding $T_{g,ij}$.

The composition-dependent values of T_g for many RCP series [108–119] exhibiting both negative (curve passing through minimum as for INVC/OMA RCP) and positive (curve passing through maximum) deviations from linear additivity [i.e. from $T_{g,12} = (T_{g,1} + T_{g,2})/2$ in eq. (3.23)] could be reasonably well fitted to eqs. (3.23), (3.24), (3.25) and (3.27) with values of $T_{g,12}$ listed in Table 3.4 [104, 106, 110, 114–119]. In terms of the molecular model [120, 121], higher-than-average values of $T_{g,12}$ may be attributed to increased chain stiffness (and hence a deficit of the conformational entropy), whereas the reverse will be true for lower-than-average $T_{g,12}$ values the linear dependence of T_g on fraction of isotactic dyads for a series of poly(α-chloroacrylates) [122] suggests a smooth variation of chain flexibility with composition, which is at variance with data for other stereoregular polymers [14, 15].

The relative merits of different approaches is difficult to assess; it appears, however, that the latter equation, eq. (3.27), may be preferred as concerns its predictive ability since any eventual deviations from linear additivity of ΔC_p values (either negative as for the ω-dodecanelactam/ε-caprolactam system [114], or zero as for randomly chlorinated PVC [123], or positive as for IVC/OMA systems) may be explicitly accounted for [cf. the denominator in eq. (3.27)]. Otherwise, the agreement between theory and experiment may be improved only at the expense of semi-empirical corrections, e.g. [124]

$$T_g = T_g^*(1 - K_{g,12} f_{12}^2) \qquad (3.28)$$

where T_g^* is given by eq. (3.22) and $K_{g,12}$ is the fitting parameter, presumably accounting for the non-linear additivity of free volume fractions at T_g values by eq. (3.2).

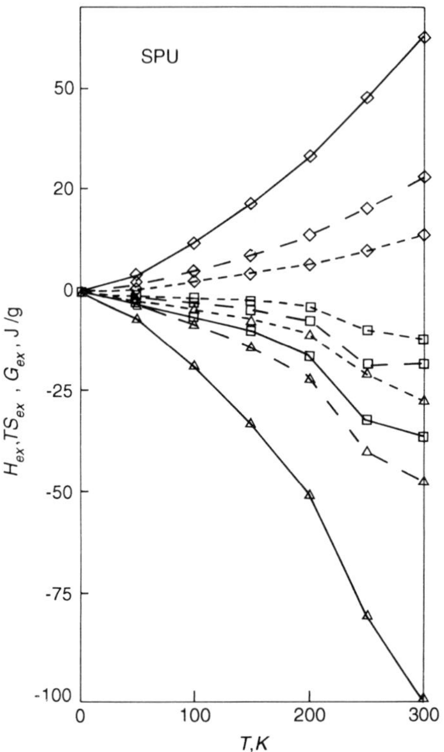

Fig. 3.6. Temperature dependences of H_{ex} (squares), TS_{ex} (triangles) and G_{ex} (diamonds) for SPU-13 (solid lines), SPU-55 (broken lines) and SPU-82 (dotted lines)

3.4.2 BLOCK COPOLYMERS

The occurrence of two glass transition temperatures located near the $T_{g,i}$ values of corresponding homopolymers is a standard test for phase separation in two-component, amorphous polymer systems (BCP included); however, more insight into the phase state of the components may be obtained from a detailed analysis of the other relevant quantities involved (see below).

(i) At low-to-moderate content of stiffer blocks (φ) in BCP of dimethylsiloxane (DMS) and S [125], phenyl-sil-sesquioxane (PSSO) [126], carbonate urethane (CU) [127], hydroxy ether (HE) [128], etc. the glass transition temperature of DMS blocks ($T_{g,2}$) coincided with that of pure PDMS. However, in the composition interval $\varphi > 0.5$ one observed either a decrease in $T_{g,2}$ by 20–30 K (for as-cast samples of S/DMS, PSSO/DMS, CU/DMS and HE/DMS) or an increase by about the same magnitude (for samples of CU/DMS and PSSO/DMS quenched from the melt above the glass transition temperatue of stiff component

$T_{g,1}$); moreover, in both cases the apparent activation energy, $\Delta E_{g,2}/R = -\,\mathrm{d}\ln q/\mathrm{d}(1/T_{g,2})$, dropped severalfold, while the 'partial' heat capacity jump was lower compared with those for pure PDMS (here q is the DSC heating rate).

(ii) In as-cast samples of S/DMS, $T_{g,1}$ is depressed, whereas the corresponding 'partial' $\Delta C_{p,1}$ is higher than that in pure PDMS [125]. A similar effect was observed for CU/DMS; however, the glass transition of stiff blocks (CU) manifests itself as several relaxations, one with highest activation energy at $T_{g,1} = 420\,\mathrm{K}$ (i.e. the glass transition temperature of pure PCU), and the other two with severalfold lower ΔE_i values observed in the temperature interval 400–300 K [127]. Only two low-strength relaxations at 350–370 K and at 320 K were found in samples quenched from the melt.

Apparently, both issues (i) and (ii) could be rationalized by assuming that the highest activation energy of the transitions observed was associated with cooperative segmental motion at T_g, whereas significantly lower ΔE_i values referred to secondary transitions (i.e. the onset of non-cooperative mobility of chain segments either in the glassy state above $T_\beta \cong 0.75\,T_g$, or in the melt state above $T_{11} \cong 1.2\,T_g$) [129]. In this context, the structure of as-cast samples with $\varphi > 0.5$ may be visualized as a continuous matrix of stiff CU with isolated inclusions of DMS which are presumably expanded by negative internal pressure (cf. eq. (2.46) [125]), arising from the difference in thermal expansion coefficients between both phases (and hence depression of $T_{g,2}$). This conclusion is consistent with the WAXS data in which increased interchain distances in DMS domains were observed in samples with high CU content [129].

In these samples the continuous matrix itself is, however, heterogeneous since it must comprise microregions of pure CU (with highest $T_{g,1}$ and $\Delta E_{g,1}$) and two mixed phases of different compositions (with lower $T_{g,i}$ and $\Delta E_{g,i}$ values). It was argued [126, 127, 129] that partial mixing of DMS and CU blocks (hence, depression of $T_{g,1}$ with a concomitant increase in 'partial' $\Delta C_{p,1}$) results in degeneration of the cooperative glass transition into non-cooperative, secondary β-relaxation (since the ratios $\Delta E_{g,1}/\Delta E_{g,i}$ obtained are close to a seemingly universal ratio, $\Delta E_g/\Delta E_\beta \cong 4$ [129]). In a similar fashion, an increased degree of mixing in quenched samples may also be a natural explanation for the degeneration of cooperative relaxation of DMS segments at $T_{g,2}$ to non-cooperative relaxation at $T_{g,i} \cong T_{11} > T_{g,2}$ with $\Delta E_{g,i} < \Delta E_{g,2}$ [127, 129].

In the case of SPU, the glass transition temperature of stiff segments ($T_{g,2}$) usually remained approximately constant independent of their weight content (w), whereas that of soft segments ($T_{g,1}$) increased with w (more precisely, the lower their molecular mass, the higher $T_{g,1}$), presumably due to the enhanced miscibility of stiff and soft segments [68]. This phenomenon of the soft chain length-dependent miscibility of components is responsible for the occurrence of multiple transitions in SPU, namely one composition invariant $T_{g,2}$ in microregions enriched with stiff fragments and two composition-dependent transitions, the first one (at lower temperature) corresponding to microphases

Table 3.4 Dyad contributions to the glass transition temperatures of RCP (in K)

Component 1	Component 2	$T_{g,1}$	$T_{g,2}$	$(T_{g,1} + T_{g,2})/2$	$T_{g,12}$
S	AN	373	378	375.5	384.5
S	B	373	195	284	284
S	MA (methyl acrylate)	364	279	321.5	331
S	BA (butyl acrylate)	363	216	289.5	291
S	AA (acrylic acid)	363	413	388	428
S	MMA	372	368	370	363
S	BS (4-bromo-styrene)	376	417	396.5	392
AN	B	378	195	286.5	260
MS	AN	450	378	414	395
MMA	AN	378	381	379.5	343
MMA	MA	376	278	327	366
MMA	VC (vinyl chloride)	378	350	364	323
MMA	VDC (vinylidene chloride)	379	262	320.5	368
BMA (butyl methacrylate)	VC	293	350	321.5	286
OMA (octyl methacrylate)	IVC	250	470	360	390
FMA (furfuryl methacrylate)	NVP (N-vinyl pyrrolidone)	393	358	375.5	347
MA	VDC	279	254	266.5	375.5
EA (ethyl acrylate)	VDC	249	254	251.5	354
VC	AN	353	378	366	362
VC	VA	353	303	328	307
VC	VDC	346	254	300	280
VDC	VA	254	303	278.5	323
VDC	BA	254	221	237.5	303
VDC	VP (vinyl propionate)	254	281	267.5	296
VDC	IB (isobutylene)	254	203	228.5	230
TFE (tetrafluoroethylene)	CTFE (chloro-trifluoroethylene)	(233)	322	277.5	222
TFE	BTFE (bromotrifluoro-ethylene	(193)	330	261.5	270
VDF (vinylidene fluoride)	HFP (hexafluoropro-pylene)	227.5	433	330.3	273

of essentially nominal composition with $T_{g,1}$ obeying to eqs. (2.36) and (2.37) for compatible polymer blends, and the second one (at higher temperature) associated with the BI of intermediate composition [130].

The above data may be regarded as an additional piece of experimental evidence for the BI in phase-separated BCP; however, we should recognize the hazards of estimating the BI content by a straightforward application of eq. (2.23) in view of the extreme sensitivity of the $\Delta C_{p,i}$ values to the phase morphology of BCP (see above).

3.5 CRYSTALLIZATION AND MELTING OF CRYSTALLIZABLE COMPONENTS

3.5.1 PROPERTIES OF A CRYSTALLIZED COMPONENT

Crystallinity

A unique feature of crystallizable, flexible-chain polymers in the solid state is their lamellar morphology [131–133]. In simple terms, the degree of crystallinity X is essentially the fraction of folded-chain lamellae, while $1 - X$ is the remaining fraction of defect material rejected into the interlamellar space. In the case of homopolymers, the theoretical upper limit $X = 1$ corresponds to a hypothetical defect-free crystalline state of infinitely long, monodisperse chains at their complete extension; therefore, violation of any of these requirements (e.g. broader-than-unity molecular weight distribution, occurrence of chain folding, etc.) should invariably result in $X < 1$, even under the most favorable conditions of crystallization. It becomes obvious, therefore, that the worse the latter becomes (i.e. the higher is the rate of continuous cooling of the melt, and/or the higher is the degree of undercooling, $\Delta T = T_m^0 - T$), the larger will be the fraction $(1 - X)$ of 'kinetic' defects (i.e. chain folds, tie chains, etc.) in the interlamellar space. In a single-chain approximation one may thus intuitively write [134]

$$X = (n - a)/n \tag{3.29}$$

where $n - a$ and a are the numbers of main chain bonds inside and outside (i.e. in folds) the crystalline lamella, respectively.

In the case of copolymers, monomers of a minor component may be regarded as 'equilibrium' defects. When the local conformation and the size of the latter satisfy the criterion of crystallographic compatibility (i.e. 'isomorphism') with the major crystallizable component, the overall crystallinity X remains almost unchanged regardless of copolymer composition, as was the case with copolymers of VF (vinyl fluoride) and VDF (vinylidene fluoride) [135], hydroquinone and substituted hydroquinones [136], as well as with several series of copolyesters and copolyamides [132].

Crystalline isomorphism is, however, a much less frequent event compared with the more general case of a crystallographic misfit between different chain components of copolymers. In this case equilibrium chemical defects (i.e. units of component 2) will be rejected from the crystal lattice of component 1, and it is only sufficiently long (compared with the corresponding dimension ξ^* of a critical nucleus), uninterrupted sequences of the latter which can crystallize. Thus, the theoretical upper limit to crystallinity of RCP will be set up by the molar fraction of such sequences, i.e.

$$X = 0.5(1 - p)p^{\xi^* - 1} \tag{3.30}$$

where p is the probability that an arbitrarily chosen monomeric unit of the crystallizable component 1 is bonded to an identical unit (in the case of RCP, this probability may be identified as the molar fraction F of component 1 [2, 3]). Since $\zeta^* \gg 1$, eq. (3.30) predicts a dramatic loss in crystallinity with an attendant increase of the molar content of component 2, $1 - F$, in RCP.

A typical example of the validity of this prediction are the relevant data for RCP of E with O, VA and AA in which even small quantities $[(1 - F) < 0.1]$ of comonomers caused the lateral expansion of the E crystalline unit cell concomitant with an overall X decrease, until complete loss of crystallizability occurred at $F < 0.8$ [132, 137]. As can be seen from Fig. 3.7, in log–log plots the crystallinity of RCP of E with P, B1 and H1 (expressed as the melting enthalpy, ΔH_m^*) exhibits a linear dependence on F according to the empirical equation, eq. (3.31) [26, 27, 138]

$$\log \Delta H_m^* = A + B \log F \tag{3.31}$$

where $A = -10.8 \pm 2.2$, $B = 6.6 \pm 0.9$ for the series E/P, and $A = -30.4 \pm 3.4$, $B = 16.4 \pm 1.3$ for the series E/B1 and E/H1. Eq. (3.31) is nearly equivalent to eq. (3.30), assuming $A = \log[0.5(1 - F)]$ and $B = \zeta^* - 1$. Larger values of the

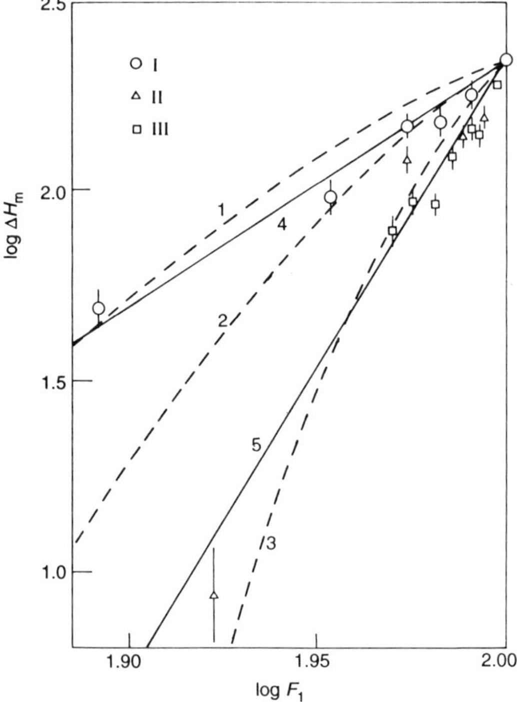

Fig. 3.7. Composition dependence of the melting heat of the crystalline phase of RCP series E/P (I), E/B (II) and E/H (III). Curves 1, 2 and 3 were constructed using eq. (3.28), and the straight lines, 4 and 5, we constructed using eq. (3.27)

slope B for the series E/B1 and E/H1 compared with E/P suggest that at the same nominal composition of RCP (F = const) the hindrances to the formation of a crystallization nucleus are higher the bulkier are the side chains.

The temperature dependence of ζ^* (and hence X) neglected by simple empirical equations, eqs. (3.30) and (3.31), is, however, explicitly accounted for by the following more elaborate treatments.

(i) Assuming comonomer units to be non-crystalline, equilibrium defects, eq. (3.32) was derived [131]:

$$X = (F/p)(1-p)^2 p^{\zeta^*}\{p(1-p)^{-2} - e^{-\theta}(1-e^{-\theta})^{-2} + \zeta^*[(1-p)^{-1}(1-e^{-\theta})^{-1}]\}$$

(3.32)

where

$$\zeta^* = \frac{\ln(DF/p) + 2\ln[(1-p)/(1-e^{-\theta})]}{\theta + \ln p},$$

$\theta = (\Delta H^0_m/R)(1/T - 1/T^0_m)$ is the reduced undercooling, T^0_m is the equilibrium melting temperature and ΔH^0_m is the corresponding melting enthalpy of the extended-chain, defect-free crystal of component 1. $D = \exp(-2\sigma_e/RT)$ is the contribution of interfacial energy at the crystal end face/melt boundary, σ_e, to the thermodynamic barrier of crystallization.

(ii) Allowing for mixing effects in the solid state (i.e. incorporation of 2 units into the end faces of extended-chain lamellae of component 1) and in the melt, eq. (3.33) was obtained [139]:

$$X = (1 - A/3)F^{\zeta^*-1}[(\zeta^* - \zeta')(1 - F) + F]$$

(3.33)

where $\zeta' = (B/3 - 1/2)/(1 - A/3) > 0$, and the coefficients $A > 0$ and $B > 0$ define the empirical parameter of solid-state mixing, $M = A\zeta + B$.

It may be easily verified that eq. (3.33) predicts a stronger effect of chain composition on crystallinity of RCP compared with eq. (3.30). By an appropriate choice of 'material parameters' both eqs. (3.32) and (3.33) adequately describe the experimental data on composition dependence of crystallinity of various RCP series [29, 30, 132, 138, 139].

With regard to diblock BCP with one crystallizable component, the crystallinity of the latter is relatively little affected provided its length, ζ, is sufficiently higher than ζ^*; otherwise, crystallizability is lost [38, 39, 42, 43, 125–128]. On the other hand, in triblock BCP with one or both components intrinsically crystallizable, the degree of crystallinity depends both on the placement within the chain as well as on the length of each component. For example, no evidence for crystallization of even fairly long ($MM_1 = 50 \times 10^3$) central EO blocks in triblock MMA/EO/MMA copolymers was observed above a 'critical' chain length of MMA blocks ($MM_2 - 5 \times 10^3$) [140]. The absence of such an effect both in diblock BCP and in physical mixtures prepared from the same blocks suggests [139] a fundamental influence of the mobility of sufficiently long MMA sequences linked to both ends of the central EO block on its ability to crystallize from the presumably homogeneous, single-phase melt.

Somewhat different behavior was observed for triblock ε-caprolactone (CL)/EO/CL copolymers, both crystallizable components having nearly identical T_m values and densities in the amorphous and crystalline states, but different crystalline unit cells, thus prohibiting cocrystallization [141, 142]. The degree of crystallinity of each of the crystallizable components proved crucially dependent on the relationship between the variable chain length of CL blocks (p_2) and $p_1 = $ const of the central EO block (see below).

(i) $p_1 > p_2$. The temperature dependences of both crystallinity and SAXS long spacings of EO blocks (X_1 and l_1, respectively) were essentially those observed for a pure PEO homopolymer, although the values of the chain-folding index became non-integer. The lower was the degree of undercooling ΔT, the larger l_1 became; therefore, CL lamellae were also 'forced' to thicken.

(ii) $p_1 < p_2$. Similar to pure PCL homopolymers, no change with temperature of X_2 and l_2 of CL lamellae was detected; EO blocks located between CL lamellae remained amorphous after cooling to room temperature, and it was only isothermal storage at small ΔT which promoted crystallization of EO blocks.

(iii) $p_1 \cong p_2$. The crystallinities of both components were strongly dependent on the degree of undercooling; namely, X_1 and X_2 were lower than and close to those for corresponding homopolymers at high ΔT, while the reverse was true at low ΔT.

Summarizing these data, it may be concluded that it is the major (in the case of $p_1 \neq p_2$) or faster-crystallizing (in the case of $p_1 \cong p_2$) component which controls the ultimate morphology of BCP with crystallizable blocks [141, 142].

Melting Temperature

As expressed by eq. (1.44), the melting temperature T_m of chain-folded polymer lamellae of thickness l is depressed below that for a defect-free, extended-chain crystal T_m^0, the more so the higher is the contribution of 'kinetic' defects (folds) to the interfacial free energy σ_e (i.e. the thinner is the lamella [131–133], or, equivalently, the more chain units are incorporated in the folds [134]).

It seems obvious that the melting temperature T_m of RCP will also be depressed below that for a crystalline homopolymer 1 (T_m^0) by equilibrium chain defects (comonomers 2); however, the expected magnitude of such a depression will depend on the model of crystallization assumed.

(i) Assuming that units 2 are forbidden to be incorporated into the crystal lattice of component 1 and remain in the amorphous phase means, formally, a corresponding change in the melting entropy of component 1, $\Delta S_m^0 = \Delta H_m^0 / T_m^0$, which may be accounted for in several alternative ways.

(a) In terms of an equilibrium approach, the molar content of the crystallizable component F becomes a single factor controlling T_m, i.e. [131, 132]

$$1/T_m - 1/T_m^0 = -(R/\Delta H_m^0)\ln F \tag{3.34}$$

(b) Alternately, the 'kinetic' exclusion of 2 units from the crystal lattice of component 1 is assumed equivalent to the change of ΔS_m^0 by $\Delta S_k = - R \ln F$. The resulting depression of non-equilibrium T_m is expressed by eq. (3.35) [143]:

$$T_m = T_m^0(1 + \Delta S_k/\Delta S_m^0) \tag{3.35}$$

(c) Assuming that, besides ΔS_k, the ideal mixing of RCP chain components in the melt makes an additional contribution to ΔS_m^0, i.e. $\Delta S_{mix} = R/\langle \zeta \rangle$ (where $\langle \zeta \rangle = 0.5/(1 - F)F$ is the mean sequence length of 1 and 2 units), the expected depression of T_m will be given by eq. (3.36) [144]

$$T_m = \Delta H_m^0/(\Delta S_m^0 + \Delta S_{seq} + \Delta S_{mix}) \tag{3.36}$$

(ii) Alternatively, assuming a (partial) incorporation of units 2 into crystals of component 1 implies a composition-invariant melting entropy ΔS_m^0; therefore, the depression of T_m may be considered to be a result of the change in the melting enthalpy ΔH_m^0 due to this mixing effect. The latter may be formally accounted for by the following approaches.

(a) Eq. (3.37) should apply on the condition of identical compositions of RCP in both the solid and the melt state [145, 146]

$$T_m = T_m^0[1 - (\Delta H_m^d/\Delta H_m^0)(1 - F)] \tag{3.37}$$

where ΔH_m^d is the enthalpy deficit associated with the mixing effect.

(b) Assuming F and $Y = (1 - F)\exp(- \varepsilon/kT)$, respectively, to be the statistical weights of units 1 and 2 in RCP crystals (here ε is the excess free energy of incorporation of units 2 into crystals of component 1), eq. (3.38) was derived [143]:

$$1/T - 1/T_m^0 = - (R/\Delta H_m^0)[\ln(F + Y)] \tag{3.38}$$

(c) Taking into consideration the difference between the actual (i.e. 'kinetic') concentration of 2 units in crystals of component 1, $(1 - F)_c$, and that in the equilibrium melt, $(1 - F)$, eq. (3.39) was obtained [147]:

$$1/T - 1/T_m^0 = (R/\Delta H_m^0)\{(1 - F)_c\varepsilon/RT_m + [1 - (1 - F)_c]$$
$$\times \ln([(1 - (1 - F)_c]/F) + (1 - F)_c \ln[(1 - F)_c/(1 - F)]\} \tag{3.39}$$

It can be easily verified that eq. (3.39) reduces to eq. (3.38) on substitution of $(1 - F)_c = Y/(F + Y)$.

Extensive tests of eqs. (3.34)–(3.39) for a large number of RCP [29, 30, 130, 131, 142–147] proved that each was capable of adequately representing the composition dependence of T_m (at least, at the expense of treating T_m^0 and ΔH_m^0 as fitting variables). An appropriately modified eq. (3.34) was also applied to the analysis of the effect of chain microstructure on T_m of stereoregular polymers [148, 149].

As was the case with crystallinities, the melting temperature of a crystallizable component in BCP is close to that for a pure homopolymer, provided the sequence length ζ is sufficiently high to ensure phase separation in the melt phase. Marked depression of T_m with decreasing F was observed, however, for multiblock copolymers of the type $B_y(A_xB_y)_n$, the more so the lower was the ratio $y/(n+1)$ [131]. In the case of multiblock copolyether-esters [150] and SPUs [151] the observed depression of T_m of a crystallizable component could be quantitatively fitted to eq. (3.34); however, the values of ΔH_m^0 derived from such fits turned out to be much too low.

3.5.2 KINETICS OF MELT CRYSTALLIZATION

Random Copolymers

As was already observed for homopolymers (cf. section 1.4.2), the simple eq. (1.50) failed to describe the melt crystallization kinetics of non-isomorphous RCP of E/VA [152] and E/B [153] as well (except perhaps in the initial stage, $\alpha < 0.1$–0.2). In case of homopolymers, this failure was recognized to be an indication of the two-stage crystallization mechanism described by eq. (1.50a); however, for RCP violation of the starting assumptions inherent in the derivation of eq. (1.50) seemed more likely. In fact, the continuous enrichment of the melt phase by non-crystallizable units of component 2 in the course of deposition of longer-than-critical sequences of component 1 on the crystal growth face may be considered responsible for variation with time of both the radial growth rate of spherulites, G_R, and their crystallinity (and hence the rate constant K_n in eq. (1.50) [152, 153]).

In the case of a series of aliphatic copolyesters obtained by melt polycondensation of 1,16-hexanediol with the acid chlorides [154], both equilibrium melting temperatures T_m^0 estimated by eq. (1.46) from apparent T_m, apparent melting enthalpies ΔH_m from DSC measurements and interchain spacings b_0 from the WAXS data exhibited a linear dependence on composition, which is typical for isomorphous RCP. The isothermal kinetics of crystallization from the melt for this copolyesters series could be reasonably well fitted to eq. (1.50) over a much wider range of transformation degree (albeit with non-integer values of the shape parameter n randomly fluctuating between 3 and 4). The temperature dependence of the bulk crystallization rate constant, K_n, also obeyed the general eq. (1.51) assuming the onset of crystallization by the mechanism of secondary nucleation [i.e. $m = 1$ in eq. (1.52)]. The values of the nucleation parameters σ and σ_e (i.e. surface-free energies at the melt interface with lateral and basal, or fold-containing, nucleus faces, respectively) obtained also exhibited a linear dependence on composition [namely, with b_0 increasing from 0.369 to 0.383 nm, σ decreased from $10.4 \times 10^{-3}\,\mathrm{J/m^2}$ to $8.8 \times 10^{-3}\,\mathrm{J/m^2}$, and σ_e from $224 \times 10^{-3}\,\mathrm{J/m^2}$ to $110 \times 10^{-3}\,\mathrm{J/m^2}$]. The relatively weak composition dependence of σ observed was explained by mutually compensating contributions from b_0 and ΔH_m^0 to the standard empirical equation, eq. (3.37) [133]

$$\sigma \cong 0.1 b_0 \Delta H_m^0 \tag{3.40}$$

whereas a more than twofold decrease in σ_e was loosely attributed to the higher chain flexibility of more unsaturated chains [154]. However, since the σ_e for flexible-chain polymers proved an extremely strong function of the packing coefficient in the crystalline state K_c, namely [133]

$$\sigma_e[J/m^2] = 5.3K_c^{12} \tag{3.41}$$

the observed change in σ_e may be associated with an about 5% decrease in K_c (from 0.76 to 0.72) which is consistent with an about 4% increase in b_0 [154].

Therefore, the notion of either an increase or a decrease in the nucleation parameter σ_e, which was invoked to explain the effects of retardation or acceleration, respectively, in crystallization from the melt studies of other (presumably isomorphous) series of copolyesters [155, 156] and copolyamides [157], may be rationalized by eq. (3.41) as the effect of chain composition on the molecular packing density in the crystallization nucleus. A similar approach also seems feasible to account for the influence of chain microstructure on the melt crystallization rates of stereoregular polymers [158].

Block Copolymers

So far, the problem of crystal nucleation and growth kinetics in BCP has remained essentially unexplored. In the case of triblock MMA/EO/MMA copolymers the apparent temperature of crystal nucleation in EO blocks (T_N) defined by the onset of a steep upswing in the melting heat-vs.-annealing temperature plots, tended to decrease the longer were (i.e. the higher was the content of) the end MMA blocks [140]. In the case of incompatible polymer blends a similar effect is usually regarded as moving closer to the temperature of homogeneous nucleation of a crystallizable component as the size of its microregions decreases [159]; this argument would apply to MMA/EO/MMA copolymers as well, provided both blocks were immiscible in the melt state. Although physical blends of PMMA and PEO in the melt state are in fact single-phase systems on the 'macroscale' (i.e. on the scale of dimensions of macromolecular coils), there is evidence (cf. section 2.1) for melt microheterogeneity on the 'microscale' (of the order of chain segment length). This effect manifests itself even stronger in case of MMA/EO/MMA copolymers with longer-than critical MMA blocks ($MM_2 = 10^4$); namely, although no crystallization event associated with the central EO block was found in DSC experiments, the SAXS data suggest the possibility of the onset of microphase separation which developed to the scale of chain segments and then was arrested due to the concomitant elevation of the glass transition temperature of the MMA-rich microphase [140].

Different crystallization behavior of both crystallizable blocks was observed in multiblock copolymers prepared by polycondensation of EO ($M_1 = 3 \times 10^3$) with random copolymers (AA) of polyamides 6 and 66 ($M_2 = 6 \times 10^3$); 4,4'-diphenylmethane diisocyanate served as the chain extender [160]. The apparent overlap of the liquid–liquid and liquid–crystal phase separation mechanisms was assumed to be responsible for the following results.

(i) Acceleration of the bulk crystallization of AA blocks observed in isothermal DSC experiments for samples containing 10% and 60% (weight) of EO was explained by a partial miscibility of EO and AA in the melt concomitant with a slight decrease in the nucleation parameter σ_e, whereas an unexpected decrease in the AA crystallization rate in the sample containing 25% of EO was attributed to the elevation of nucleation barrier height in eq. (2.53) resulting from an increase in both the miscibility of components in the melt and the value of the nucleation parameter σ_e.

(ii) On further cooling to below T_m of EO blocks of specimens with AA blocks crystallized to completion, we could observe on DSC traces two exothermic processes with peaks at $T' = 300\,\mathrm{K}$ and $T'' = 255\,\mathrm{K}$, respectively, the relative intensity of the latter process tending to increase the lower was EO content. These findings, together with electron microscopy data, suggested that EO blocks were separated from crystallized AA blocks into microphases of at least two different sizes (and, presumably, compositions) [160]. Namely, the composition of the larger-size EO domains was assumed essentially identical to that of a pure PEO homopolymer. This conclusion is consistent with coincidence of T' in BCP with the temperature of non-isothermal crystallization of pure PEO, as well as with the identical values of σ_e derived from isothermal crystallization experiments of EO in BCP and of PEO. On the other hand, the process at T'' was attributed to crystallization of the remaining EO blocks on homogeneous nuclei in smaller-size microregions.

Thus, the above data provide the additional documentation to the effect of overlapping temperature intervals of liquid–liquid and liquid–crystal phase separation mechanisms on the final morphology of BCP with both crystallizable components.

3.6 CONCLUSIONS

As demonstrated in this chapter, by their general behavior RCP and BCP resemble compatible and incompatible polymer blends, respectively; however, the influence of the chain-connectivity effect becomes evident from a comparison of the properties (especially kinetics-dependent ones) of both systems at the same nominal composition. For example, 'intrachain plasticization' of the stiff-chain polymer with the comonomer in RCP is an efficient means to avoid the problem of eventual phase separation and migration of the minor component to the surface, which is notorious for interchain plasticization in physical blends. On the other hand, although the equilibrium thickness of the BI, Δr, is predicted to be roughly the same for both incompatible blends and the corresponding BCP, the probabilty of kinetics-controlled freezing-in of the thicker BI should be higher for the latter due to the intrinsically slower phase separation processes. Thus, the chain-connectivity effect in copolymers offers unique possibilities to monitor the phase state of components and the ultimate morphology (and hence the physical properties) of binary polymer systems.

REFERENCES

1. Theil M. H. (1969) The effect of interachain order on solid solutions of copolymers, *Macromolecules*, **2**, 351–354.
2. Alfrey T., Jr, Bohrer J. and Mark H. (1952) *Copolymerization*, Interscience, New York.
3. Bamford C. H., Barb W. G., Jenkins A. D. and Onyon P. F. (1958) *The Kinetics of Vinyl Polymerization by Radical Mechanisms*, Butterworths, London.
4. Furukawa J., Nishioka A. and Kotani T. (1970) Statistical segment size and ultimate elongation of acrylonitrile and butadiene copolymers, *J. Polymer Sci.: Polymer Lett. Ed.*, **8**, 25–28.
5. Tanzawa H., Tanaka T. and Soda A. (1969) Dilute solution properties of poly(α-methylstyrene/ethylene) 'regular' sequence copolymers, *J. Polymer Sci.*, Part A-2, **7**, 929–946.
6. Shimura-Kambe Y. (1968) Temperature dependence of the limiting viscosity number of the solutions of acrylonitrile–styrene copolymers in dimethylformamide, *J. Phys. Chem.*, **72**, 4104–4110.
7. Dondos A. and Benoit H. (1968) Influence de la nature du solvant et de la temperature sue les dimensions non perturbees des copolymeres statistiques, *Makromol. Chem.*, **118**, 165–176.
8. Danon J. and Derar B. (1969) Etude du comportement en solution diluee des copolymeres statistiques poly(styrene)–poly(methacrylate de methyle), *Europ. Polymer J.*, **5**, 656–674.
9. Fischer H. and Maechtle W. (1969) Untersuchungen an hochverduennten Loesungen von statistischen Styrol–Acrylsaeuremethylester–Copolymeren, *Kolloid-Z. Z. Polymere*, **230**, 221–229.
10. Wunderlich W. (1970) Flexibilitaet und thermodynamische Eigenschaften von Polyacrylaten, Polymethacrylaten und statistischen Acrylat/Methacrylat–Copolymeren, *Angew. Makromol. Chem.*, **11**, 201–210.
11. Lange H. and Baumann H. (1970) Struktur unde Eigenschaften von ABS-Polymeren. Y. Grenzviscositaet-Molekulargewichtsbeziehungen fuer unterschiedlich zusammengesetzte Styrol/Acrylnitril–Copolymere, *Angew. Makromol. Chem.*, **14**, 25–42.
12. Ehrenburg E. G., Kartashova G. G., Eryomina M. A. and Poddunbyi I. Ya. (1967) Conformational features of copolymer chains in solutions, *Vysokomol. Soed.*, Ser. A, **9**, 2709–2717.
13. Cramond D. N. and Urwin J. R. (1969) Solution properties of block copolymers of poly(isoprene/styrene). I. Molecular dimensions by viscosity and light scattering methods, *Europ. Polymer J.*, **5**, 35–44.
14. Girolamo M. and Urwin J. R. (1971) Anomalous viscosity behavior of block copolymers as a function of the temperature for two sequence or AB type poly(isoprene/styrene), *Europ. Polymer J.*, **7**, 693–698.
15. Birshtein T. M. and Ptitsyn O. B. (1964) *Conformations of Macromolecules*, Nauka, Moscow (in Russian).
16. Flory P. J. (1969) *Statistical Mechanics of Chain Molecules*, Interscience, New York–London–Sydney–Toronto.
17. Sasuga T. and Takehisa M. (1977) Pressure–volume–temperature behavior of several synthetic rubbers, *J. Macromol. Sci.*, Ser. B, **13**, 215–229.
18. Renuncio J. A. R. and Prausnitz J. M. (1977) Volumetric properties of block and random copolymers of butadiene and styrene at pressures to 1 kilobar, *J. Appl Polymer Sci.*, **21**, 2867–2872.
19. Arbuzova A. P., Besklubenko Yu. D., Lipatov Yu. S., Privalko V. P. *et al.* (1983) Thermodynamics of random copolymers of *n*-octyl methacrylate and *N*-vinyl carbazole, *Vysokomol. Soed.*, Ser. A, **25**, 914–920.

20. Arbuzova A. P., Lipatov Yu. S., Pas'ko S. P. and Privalko V. P. (1986) Thermodynamics of random copolymers of *N*-vinyl carbazole and alkyl methacrylates, *Vysokomol. Soed.*, Ser. A, **28**, 2157–2162.
21. Privalko V. P., Severova N. N. and Shmorgun A. V. (1986) Thermal diffusivity of ethylene/propylene copolymers in the melt state under elevated pressures, *Prom. Teplotechn.*, **8**, 39–41.
22. Privalko V. P. and Shmorgun A. V. (1991) Thermodynamic properties of random copolymers of ethylene and 1-olefins in the solid state and in the melt, *Vysokomol. Soed.*, Ser. A, **33**, 1698–1707.
23. Privalko V. P., Arbuzova A. P., Zagdanskaya N. E. *et al.* (1987) Thermodynamics of random copolymers of 3-iodo-9-*N*-vinyl carbazole and *n*-octyl methacrylate, *Vysokomol. Soed.*, Ser. A, **29**, 1892–1896.
24. Privalko V. P., Pas'ko S. P., Arbuzova A. P. and Zagdanskaya N. E. (1987) Thermodynamics of random copolymers of 3-iodo-9-N-vinyl carbazole and alkyl methacrylates in the melt state under elevated pressures, *Ukr. Khim. Zhurn.*, **53**, 550–553.
25. Arbuzova A. P. (1987) Thesis, Institute of Macromolecular Chemistry, Academy of Sciences of Ukraine, Kiev.
26. Privalko V. P., Arbuzova A. P., Korskanov V. V. and Zagdanskaya N. E. (1994) Thermodynamics of random copolymers of *N*-vinyl carbazole and 4-alkylstyrenes, *Polymer Int.* (submitted).
27. Privalko V. P., Shmorgun A. V., Rafailovich G. M. and Severova N. N. (1988) Thermodynamic properties and heat transfer in the melt state of random copolymers of ethylene and butene-1, *Komposits. Polim. Mater.*, **36**, 9–12.
28. Privalko V. P. and Shmorgun A. V. (1992) Comprehensive thermal analysis of random copolymers of ethylene and 1-olefins, *J. Thermal Analysis*, **38**, 1257–1270.
29. Shmorgun A. V. (1993) Thesis, Shevchenko State University, Kiev.
30. Zoller P., Jain R. K. and Simha R. (1986) Equation of state of copolymer melts: the poly(vinyl acetate)–polyethylene pair, *J. Polymer Sci.: Polymer Physics*, **24**, 687–696.
31. Simha R. and Somcynsky T. (1969) On the statistical thermodynamics of spherical and chain molecule fluids, *Macromolecules*, **2**, 342–350.
32. Olabisi O. and Simha R. (1975) Configurational thermodynamic properties of amorphous polymers and polymer melts. II. Theoretical considerations, *Macromolecules*, **8**, 211–218.
33. Privalko V. P. (1986) *Molecular Structure and Properties of Polymers*, Khimia, Leningrad (in Russian).
34. Questad D. L., Pae K. D., Scheinbeim J. I. and Newman B. A. (1982) Pressure–volume–temperature studies of a polyurethane elastomer, *J. Appl. Phys.*, **53**, 6578–6580.
35. Krause S. (1970) Microphase separation in block copolymers. Zeroth approximation including surface free energies, *Macromolecules*, **3**, 84–86.
36. Krause S. (1978) Why the phase boundaries should be sharp in most phase-separated block copolymers, *Macromolecules*, **11**, 1288–1290.
37. Dunn D. J. and Krause S. (1974) Phase separation in diblock copolymers of styrene and α-methyl styrene and in mixtures of the corresponding homopolymers, *J. Polymer Sci.: Polymer. Lett. Ed.*, **12**, 591–596.
38. Shibanov Yu. D. and Godovsky Yu. K. (1983) Features of liquid–liquid and amorphous–crystalline separation in block copolymers and in mixtures with variable molecular masses, *Vysokomol. Soed.*, Ser. A, **25**, 339–345.
39. Shibanov Yu. D. (1985) Thesis, Karpov Institute of Physical Chemistry, Moscow.
40. Gaur U. and Wunderlich B. (1980) Study of microphase separation in block copolymers of styrene and α-methyl styrene in the glass transition region using quantitative thermal analysis, *Macromolecules*, **13**, 1618–1625.

41. Bates F. S., Bair H. E. and Hartney M. A. (1984) Block copolymers near the microphase separation transition. I. Preparation and physical characterization of a model system, *Macromolecules*, **17**, 1987–1993.

42. Hay J. N. and Wiles M. (1979) The crystallization characteristics of ethylene block copolymers, *J. Polymer Sci.*: Polymer *Chem. Ed.*, **17**, 2223–2231.

43. Suzuki T. and Kotaka T. (1980) Dielectric and mechanical relaxations in randomly coupled multiblock copolymers with varying block length: bisphenol-A polycarbonate–poly(oxyethylene) systems, *Macromolecules*, **13**, 1495–1501.

44. Meier D. J. (1969) Theory of block copolymers. I. Domain formation in A–B block copolymers, *J. Polymer, Sci.*, Part C, **26**, 81–98.

45. Todo A., Uno H., Miyoshi K. *et al.* (1977) Domain–boundary structure of styrene–isoprene block copolymer films cast from solutions. III. Preliminary results on spherical microdomains, *Polymer Eng. Sci.*, **17**, 587–597.

46. Bianchi U., Pedemonte E. and Turturro A. (1969) Statistical thermodynamics of styrene–butadiene block copolymers, *J. Polymer Sci.*, Ser. B, **7**, 785–788.

47. Bianchi U., Pedemonte E. and Turturro A. (1970) Morphology of styrene–butadiene –styrene block copolymers, *Polymer*, **11**, 268–276.

48. Helfand E. (1975) Block copolymers, polymer–polymer interfaces, and the theory of inhomogeneous polymers, *Acc. Chem. Res.*, **8**, 295–299.

49. Helfand E. (1975) Block copolymer theory. III. Statistical mechanics of the microdomain structure, *Macromolecules*, **8**, 552–556.

50. Dlugosz J., Keller A. and Pedemonte E. (1970) Electron microscopy evidence of a macroscopic 'single crystal' from a three block copolymer, *Kolloi-Z. Z. Polymere*, **242**, 1125–1131.

51. Sadron Ch. and Gallot B. (1973) Heterophases in block-copolymer systems in the liquid and in the solid state, *Makromol. Chem.*, **164**, 301–332.

52. Grosberg A. Yu. and Khokhlov A. R. (1989) *Statistical Physics of Macromolecules*, Nauka, Moscow, Ch. 5 (in Russian).

53. Zhulina E. B. and Birshtein T. M. (1985) Conformation of molecules of block copolymers in selective solvents (micellar structures), *Vysokomol. Soed.*, Ser. A, **27**, 511–517.

54. Birshtein T. M. and Zhulina E. B. (1985) Geometry of lamellar superstructures in block copolymers, *Vysokomol. Soed.*, Ser. A, **27**, 1613–1620.

55. Erukhimovich I. Ya. (1982) Fluctuations and domain structure formation in heteropolymers, *Vysokomol. Soed.*, Ser. A, **24**, 1942–1949.

56. Erukhimovich I. Ya. (1982) Influence of chemical composition of two-component melts of heteropolymers on domain formation therein, *Vysokomol. Soed.*, Ser. A, **24**, 1950–1957.

57. Roe R.-J., Fishkis M. and Chang J. C. (1981) Small-angle X-ray diffraction study of thermal transition in styrene–butadiene block copolymers, *Macromolecules*, **14**, 1091–1103.

58. Vilesov A. L. and Frenkel S. Ya. (1983) Dynamics of superlattices formed by homodisperse block copolymers in *Synthesis and Properties of Block Copolymers*, ed. by Yu. S. Lipatov, Naukova Dumka, Kiev, pp. 98–102 (in Russian).

59. Lipatov Yu. S., Elyashevich A. M., Pletneva S. G. *et al.* (1986) Initial stages of a liquid–liquid phase separation in mixtures of short chains studied by computer simulation technique, *Dokl. Acad. Nauk USSR*, **279**, 386–389.

60. Tkach A. I., Shilov V. V., Elyashevich A. M. *et al.* (1987) Structural analysis of phase separation in binary blends of short chains by molecular dynamics (computer) experiments, in *Methods of Calculations in Physical Chemistry*, ed. by Yu. G. Papulov, Kalinin University, pp. 57–62 (in Russian).

61. Tkach A. I. (1988) Thesis, Institute of Macromolecular Chemistry, Academy of Sciences of Ukraine, Kiev.

62. Skoulios A., Helffer P., Gallot Y. and Selb J. (1971) Solubilization and chain conformation in a block copolymer system, *Makromol. Chem.*, **148**, 305–310.
63. Hashimoto T., Todo A., Itoi H. and Kawai H. (1977) Domain–boundary structure of styrene–isoprene block copolymer films cast from solutions. 2. Quantitative estimation of the interfacial thickness of lamellar microphase systems, *Macromolecules*, **10**, 377–384.
64. Richards R. W. and Thomason J. L. (1983) Small-angle neutron scattering measurement of block copolymer interphase structure, *Polymer*, **24**, 1089–1096.
65. Folkes M. J., Keller A. and Scalisi F. P. (1971) A test for molecular orientation in a 'single crystal' of SBS three-block copolymer by infrared spectroscopy, *Polymer*, **12**, 793–796.
66. Witten T. A. (1990) Heterogeneous polymers and self-organization, *J. Phys.: Condens. Matter*, **2**, SA1–SA8.
67. Bonart R. (1977) Segmentierte Polyurethane, *Angew. Makromol. Chem.*, **58/59**, 259–297.
68. Kercha Yu. Yu. (1979) *Physical Chemistry of Polyurethanes*, Naukova Dumka, Kiev (in Russian).
69. Miller J. A., Lin S. B., Hwang K. K. S. *et al.* (1985) Properties of polyether–polyurethane block copolymers: effect of hard segment length distribution, *Macromolecules*, **18**, 32–44.
70. Blackwell J. and Gardner K. H. (1979) Structure of the hard segments in polyurethane elastomers, *Polymer*, **20**, 13–17.
71. Koberstein J. T. and Stein R. S. (1983) Small-angle X-ray scattering studies of microdomain structure in segmented polyurethane elastomers, *J. Polymer Sci.: Polymer Phys. Ed.*, **21**, 1439–1472.
72. Leung L. M. and Koberstein J. T. (1985) Small-angle scattering analysis of hard microdomain structure and microphase mixing in polyurethane elastomers, *J. Polymer Sci.: Polymer Phys. Ed.*, **23**, 1883–1913.
73. Abouzahr S., Wilkes G. L. and Ophir Z. (1982) Structure–property behavior of segmented polyether–MDI–butanediol based urethanes: effect of composition ratio, *Polymer*, **23**, 1077–1086.
74. Xu M., MacKnight W. J., Chem-Tsai C. H. Y. and Thomas E. L. (1987) Structure and morphology of segmented polyurethanes. 4. Domain structure of different scales and the composition heterogeneity of the polymers, *Polymer*, **28**, 2183–2189.
75. Fok J., Michler G. and Naumann J. (1985) Determination of lamellae in segmented polyurethanes by electron microscopy, *Polymer*, **26**, 2195–2199.
76. Miller J. A., Cooper S. L., Han C. C. and Pruckmayr G. (1984) Small-angle neutron scattering from a polyurethane block copolymer, *Macromolecules*, **17**, 1063–1071.
77. Miller J. A., Pruckmayr G., Epperson E. and Cooper S. L. (1985) The thermal response of the polyether soft segment chain conformation in a polyurethane block copolymer measured by small-angle neutron scattering, *Polymer*, **26**, 1915–1920.
78. Hu C. B., Ward R. S. and Schneider N. S. (1982) A new criterion of phase separation: the effect of diamine chain extender on the properties of polyurethaneureas, *J. Appl. Polymer Sci.*, **27**, 2167–2177.
79. Lipatov Yu. S., Dmitruk N. V., Tsukruk V. V. *et al.* (1984) Microphase state of oligobutadienediol-based polyurethaneureas, *J. Appl. Polymer Sci.*, **29**, 1919–1927.
80. Comargo R. E., Macosco C. W., Tirrell M. and Wellighoff S. T. (1985) Phase separation studies in RIM polyurethanes. Catalyst and hard segment crystallinity effects, *Polymer*, **26**, 1145–1154.
81. Privalko V. P., Usenko A. A., Vorona V. V. and Letunovsky M. P. (1994) Boundary interphase in segmented polyurethanes. I. Structural characterization, *J. Polymer Eng.* **13**, 203–222.

82. Leung L. M. and Koberstein J. T. (1986) DSC annealing study of microphase separation and multiple endothermic behavior in polyether-based polyurethane block copolymers, *Macromolecules*, **19**, 706–713.

83. Gibson P. E., Van Bogart J. W. C. and Cooper S. L. (1986) Small-angle X-ray scattering studies of thermally-induced morphological changes in segmented polyurethane elastomers, *J. Polymer Sci.: Polymer Physics*, **24**, 885–907.

84. Lee H. S., Wang Y. K., MacKnight W. J. and Hsu S. L. (1988) Spectroscopic analysis of phase-separation kinetics in model polyurethanes, *Macromolecules*, **21**, 270–273.

85. Lee H. S. and Hsu S. L. (1989) An analysis of phase separation kinetics of model polyurethanes, *Macromolecules*, **22**, 1100–1105.

86. Camberlin Y. and Pascault J. P. (1984) Phase segregation kinetics in segmented linear polyurethanes: relations between equilibrium time and chain mobility and between equilibrium degree of segregation and interaction parameter, *J. Polymer Sci.: Polymer Phys. Ed.*, **22**, 1835–1844.

87. Chee K. K. and Farris R. J. (1984) Kinetics of phase separation in segmented polyurethanes, *J. Appl. Polymer Sci.*, **29**, 2529–2535.

88. Mauritz K. (1988) Review and critical analyses of theories of aggregation in ionomers, *J. Macromol. Sci.—Rev. Macromol. Chem. Phys.*, **C28**, 65–98.

89. Eisenberg A., Hird B. and Moore R. B. (1990) A new multiplet-cluster model for the morphology of random ionomers, *Macromolecules*, **23**, 4098–4107.

90. Quan X., Bair H. E. and Johnson G. E. (1989) Thermal characterization of block copolymer interfaces, *Macromolecules*, **22**, 4631–4635.

91. ten Brinke G. (1990) Mathematical modelling of the effects of annealing on enthalpy relaxations of polymer interfaces, *Macromolecules*, **23**, 1225–1227.

92. Segre A. L., Capitani D., Fiordiponti P. *et al.* (1992) Interface in styrene–butadiene copolymers and blends: a solid state ^{1}H–NMR study, *Eur. Polymer J.*, **10**, 1165–1172.

93. Tyagi D., McGrath J. E. and Wilkes G. L. (1986) Small angle X-ray studies of siloxane–urea segmented copolymers, *Polymer Eng. Sci.*, **26**, 1371–1398.

94. Tyagi D., Hedrick J. L., Webster D. C. *et al.* (1988) Structure–property relationships in perfectly alternating segmented polysulphone/poly(dimethylsiloxane) copolymers, *Polymer*, **29**, 833–844.

95. Koberstein J. T., Morra B. and Stein R. S. (1980) The determination of diffuse interphase boundary thickness in polymers by SAXS method, *J. Appl. Crystallogr.*, **13**, 34–45.

96. Privalko V. P., Shapoval R. L., Shtompel V. I., Usenko A. A. and Vorona V. V. (1994) Boundary interphase in segmented polyurethanes. II. Characterization by heats of solution measurements, *Colloid & Polymer Sci.* (submitted for publication).

97. Privalko V. P., Demchenko S. S. and Lipatov Yu. S. (1986) Kinetics of enthalpy relaxation at the glass transition of flexible-chain polymers, *Vysokomol. Soed.*, Ser. A, **28**, 1296–1303.

98. Ho T. and Mijovic J. (1990) Physical aging in poly(methyl methacrylate)/poly(styrene-co-acrylonitrile) blends. 3. Simulation of enthalpy relaxation using the Moynihan model, *Macromolecules*, **23**, 1411–1419.

99. Van Krevelen D. W. (with the collaboration of P. J. Hoftyzer) (1972) *Properties of Polymers: Correlations With Chemical Structure*, Elsevier Publ. Co., Amsterdam–London–New York.

100. Becker R. (1978) Beziehungen zwischen der Glastemperatur und der Chemischen Struktur von Polymeren, *Faserforsch. Textiltech.*, **29**, 361–385.

101. Askadsky A. A. and Matveev Yu. M. (1983) *Chemical Structure and Physical Properties of Polymers*, Khimia, Moscow (in Russian).

102. DiMarzio E. A. and Gibbs J. H. (1959) Glass temperature of copolymers, *J. Polymer Sci.*, **40**, 121–131.

103. Ellerstein S. M. (1963) The glass temperature of random addition copolymers, *J. Polymer Sci.*, Part B, **1**, 223.

104. Barton J. M. (1970) Relation of glass transition temperature to molecular structure of addition copolymers, *J. Polymer Sci.*, Part C, **30**, 573–597.

105. Kanig G. (1963) Zur Theorie der Glastemperatur von Polymerhomologen, Copolymeren und weichgemachten Polymeren, *Kollod. Z. Z. Polymere*, **190**, 1–16.

106. Johnston N. W. (1976) Sequence distribution–glass transition effects, *J. Macromol. Sci.—Rev. Macromol. Chem.*, **C14**, 215–250.

107. Couchman P. R. (1982) Compositional variation of glass-transition temperatures. 7. Copolymers, *Macromolecules*, **15**, 770–773.

108. Beevers R. B. and White E. F. T. (1960) Physical properties of vinyl polymers, Part 2: The glass-transition temperature of block and random acrylonitrile + methyl methacrylate copolymers, *Trans. Farad. Soc.*, **56**, 1529–1534.

109. Beevers R. B. (1962) Physical properties of vinyl polymers, Part 4: Glass-transition temperatures of methyl methacrylate + styrene copolymers, *Trans. Farad. Soc.*, **58**, 1465–1472.

110. Illers. K.-H. (1963) Die Glastemperatur von Copolymeren, *Kolloid-Z. Z. Polymere*, **190**, 16–34.

111. Pavlinov L. I., Rabinovich I. B., Okladnov N. A. and Arzhakov S. A. (1967) Heat capacity of methyl methacrylate/methacrylic acid copolymers in the interval 25–190 C, *Vysokomol. Soed.*, Ser. A, **9**, 483–487.

112. Slonimsky G. L., Askadsky A. A., Mzhelsky A. I. *et al.* (1969) On the glass transition temperature of amorphous copolymers, *Vysokomol. Soed.*, Ser. A, **11**, 2265–2272.

113. Godovsky Yu. K., Dubovik I. I., Ivanova S. L. *et al.* (1977) Glass transition, melting and crystallization behavior of ω-dodecanelactam/ε-caprolactam copolymers, *Vysokomol. Soed.*, Ser. A, **19**, 392–398.

114. Wilhelm T., Hoffman R. and Fuhrmann J. (1982) Glass transition temperatures of styrene/4-bromostyrene copolymers, *Makromol. Chem. Rapid. Commun.*, **4**, 81–85.

115. Moggi G., Bonardelli P., Monti C. and Bart J. C. J. (1985) Copolymers of tetrafluoroethylene with chlorotrifluoroethylene and with bromotrifluoroethylene, *J. Polymer Sci.: Polymer Phys. Ed.*, **23**, 1099–1108.

116. Bonardelli P., Moggi G. and Turturro A. (1986) Glass transition temperatures of copolymer and terpolymer fluoroelastomers, *Polymer*, **27**, 905–909.

117. Cowie J. M. G. and Harris J. H. (1992) The influence of the microstructure of some chlorine containing polymers on their miscibility with poly(butadiene-*stat*-acrylonitrile), *Polymer*, **33**, 4592–4596.

118. Wessling R. A., Dicken D. F., Kurowsky S. R. and Gibbs D. S. (1974) Glass transition temperatures of vinylidene chloride copolymers, *Appl. Polymer Symp.*, **24**, 83–105.

119. Zaldivar D., Peniche C., Bulay A. and San Roman J. (1992) Free radical copolymerization of furfuryl methacrylate and *N*-vinylpyrrolidone, *Polymer*, **33**, 4625–4629.

120. Tonelli A. E. (1974) Possible molecular origin of sequence distribution–glass transition effects in copolymers, *Macromolecules*, **7**, 632–634.

121. Tonelli A. E. (1975) Sequence distribution–glass transition effects in copolymers of vinyl chloride and vinylidene chloride with methyl acrylate, *Macromolecules*, **8**, 544–547.

122. Lehr M., Parker R. G. and Komoroski R. A. (1985) Thermal property–structure relationships of solution-chlorinated poly(vinyl chlorides), *Macromolecules*, **18**, 1265–1272.

123. Braun D., Kohl P. R. and Hellmann G. P. (1988) The glass transition temperatures

of homogeneous blends of polystyrene, poly(methyl methacrylate), and copolymers of styrene and methyl methacrylate, *Makromol. Chim.*, **189**, 1671–1679.

124. Dever G. R., Karasz F. E., MacKnight W. J. and Lenz R. W. (1975) Poly(alkyl-α-chloroacrylates). Y. Preparation and properties of methyl, ethyl and isopropyl polymers of varied tacticity, *J. Polymer Sci.: Polymer Chem. Ed.*, **13**, 2151–2180.

125. Wang B. and Krause S. (1987) Properties of dimethylsiloxane microphases in phase-separated dimethylsiloxane block copolymers, *Macromolecules*, **20**, 2201–2208.

126. Bershtein V. A., Levin V. Yu., Egorova L. M. *et al.* (1987) Studies of relaxation transitions in block copolymers of polydimethylsiloxane and poly(phenyl–silsesquioxane) by differential scanning calorimetry technique, *Vysokomol. Soed.*, Ser. A, **29**, 2360–2366.

127. Bershtein V. A., Levin V. Yu., Egorova L. M. *et al.* (1987) Heat capacity anomalies and relaxation transitions in poly(carbonate urethane)/siloxane block copolymers, *Vysokomol. Soed.*, Ser. A, **29**, 2553–2559.

128. Hedrick J. L., Haidar B., Russell T. P. and Hofer D. C. (1988) Synthesis and properties of segmented and block poly(hydroxy ether-siloxane) copolymers, *Macromolecules*, **21**, 1967–1977.

129. Bershtein V. A. and Egorov V. (1990) *Differential Scanning Calorimetry in Physical Chemistry of Polymers*, Khimia, Leningrad (in Russian).

130. Privalko V. P., Khaenko E. S., Khmelenko G. I. *et al.* (1990) Heat capacity of crown ether-containing poly(etherurethaneureas), *Vysokomol. Soed.*, Ser. A, **32**, 1600–1605.

131. Mandelkern L. (1964) *Crystallization of Polymers*, McGraw-Hill Co., New York–San Francisco–Toronto–London.

132. Wunderlich B. (1973) *Macromolecular Physics, Vol. 1: Crystal Structure, Morphology, Defects*, Academic Press, New York and London.

133. Hoffman J. D., Davis G. T. and Lauritzen J. I. (1976) The rate of crystallization of linear polymers with chain folding in *Treatise on Solid State Chemistry*, Vol. 3, ed. by N. B. Hannay, Plenum Press, New York, pp. 497–614.

134. Knox J. R. (1967) The nature of the lamellar crystalline structure in branched polyethylenes as deduced from melting point and branching determinations, *J. Polymer Sci.*, Part C, **18**, 69–77.

135. Natta G., Allegra G., Bassi I. *et al.* (1965) Isomorphism in systems containing fluorinated polymers and in new fluorinated copolymers, *J. Polymer Sci.*, Part A, **3**, 4263–4278.

136. Rubin I. D. (1963) Isomorphism in polycarbonates. Copolymers of hydroquinone and substituted hydroquinones, *J. Polymer Sci.*, Part A, **1**, 1645–1650.

137. Kortleve G., Tuijnman C. A. F. and Vonk C. G. (1972) Crystallization of branched polymers. I. Ethylene–vinyl acetate and ethylene–acrylic acid copolymers, *J. Polymer Sci.*, Part A-2, **10**, 123–131.

138. Burfield D. R. (1987) Correlation between crystallinity and ethylene content in LLDPE and related ethylene copolymers. Demonstration of the applicability of a simple empirical relationship, *Macromolecules*, **20**, 3020–3023.

139. Glenz W., Kilian H. G., Klattenhoff D. and Stracke Fr. (1977) Thermodynamics of the melting of pseudoeutectic linear copolymer systems, *Polymer*, **18**, 685–696.

140. Donth E., Kretzschmar H., Schulze G. *et al.* (1987) Influence of the chain-end mobility on the melt crystallization of the ethylene oxide (B) sequences in systems containing diblock AB and triblock ABA copolymers with methyl methacrylate (A), *Acta Polymer*, **38**, 260–270.

141. Perret R. et Skoulios A. (1972) Etude de la cristallisation des copolymeres trisequences poly-ε–caprolactone/polyoxyethylene. I. Copolymeres dont les sequences ont des longeurs tres inegales, *Makromol. Chem.*, **162**, 147–162.

142. Perret R. et Skoulios A. (1972) Etude de la cristallisation des copolymeres trisequences

poly–ε–caprolactone/polyoxyethylene/poly–ε–caprolactone. II. Copolymeres dont les sequences ont des longeurs voisines, *Makromol. Chem.*, **162**, 163–177.

143. Helfand E. and Lauritzen J. I. (1973) Theory of copolymer crystallization, *Macromolecules*, **6**, 631–638.

144. Baur H. (1966) Einfluss der Sequenzlaengenverteilung auf das Schmelz-Ende von Copolymeren, *Makromol. Chem.*, **98**, 297–301.

145. Colson J. P. and Eby R. K. (1966) Melting temperatures of copolymers, *J. Appl. Phys.*, **37**, 3511–3514.

146. Sanchez I. C. and Eby R. K. (1973) Crystallization of random copolymers, *J. Res. NBS*, Ser. A, **77**, 353–360.

147. Sanchez I. C. and Eby R. K. (1975) Thermodynamics and crystallization of random copolymers, *Macromolecules*, **8**, 638–641.

148. Miller R. L. (1962) On the characterization of stereoregular polymers. IY. Application to polypropylene, *J. Polymer Sci.*, **57**, 975–991.

149. Aggarwal S. L., Marker L., Kollar W. L. and Geroch R. (1965) Characterizing stereosequence length of propylene oxide polymers from different catalysts, *Adv. Chem. Ser.*, **52**, 88–104.

150. Castles J. L., Vallance M. A., McKenna J. M. and Cooper S. L. (1985) Thermal and mechanical properties of short-segment block copolyesters and copolyether-esters, *J. Polymer Sci.: Polymer Phys. Ed.*, **23**, 2119–2147.

151. Vallance M. A., Castles J. L. and Cooper S. L. (1984) Microstructure of as-polymerized thermoplastic polyurethane elastomers, *Polymer*, **25**, 1734–1746.

152. Johnsen U., Nachtrab G. und Zachmann H. G. (1970) Untersuchung der isothermen Kristallisation von Aethylen–Vinylacetat–Copolymeren mir Hilfe der Differential–Scanning–Calorimeter, *Kolloid-Z. Z. Polymere*, **240**, 756–761.

153. Amelino L. and Martuscelli E. (1975) Effect of intra-chain double bonds on the crystallization of polyethylene from the melt, *Polymer*, **16**, 864–868.

154. Di Meo A., Maglio G., Martuscelli E. and Palumbo R. (1976) Effect of intra-chain double bonds on the melt crystallization of aliphatic polyesters, *Polymer*, **17**, 802–806.

155. Ueberreiter K. und Felber W. (1970) Kristallisationskinetik von Polymeren. X. Copolyester aus Polyaethylensuccinat mit Sebacinsaure und aus Polyhexamethylensebacat mit verschiedenen Diolen, *Kolloid-Z. Z. Polymere*, **242**, 1173–1179.

156. Bier P., Binsack R., Vernaleken H. und Rempel (1977) Einfluss verzweigter Codiole auf das Kristallisationsverhalten von aromatischen Polyestern, *Angew. Makromol. Chem.*, **65**, 1–21.

157. Maglio G., Martuscelli E., Palumbo R. and Soldati I. (1976) Influence of intra-chain *trans* double bonds on the melt crystallization of polyamides, *Polymer*, **17**, 185–191.

158. Martuscelli E., Pracella M. and Crispino L. (1983) Crystallization behavior of fractions of isotactic polypropylene with different degrees of stereoregularity, *Polymer*, **24**, 693–699.

159. Romankevich O. V. (1984) Thesis, Institute of Macromolecular Chemistry, Academy of Sciences of Ukraine, Kiev.

160. Godovsky Yu. K., Yanul N. A. and Bessonova N. P. (1991) Crystallization of microphase-separated block copolymers with two crystallizable blocks, *Colloid. Polymer Sci.*, **269**, 901–915.

Part II

THERMOPHYSICAL CHARACTERIZATION OF HETEROGENEOUS POLYMERS

4 Heat Conductivity

4.1 THEORETICAL ASPECTS: PRAGMATIC APPROACHES

4.1.1 CLOSURE OF EQUATIONS FOR EFFECTIVE CONDUCTIVITY

An understanding of the effective properties of microheterogeneous materials (MHM) implies knowledge of the pattern of distribution of physical fields in all components of MHM from which the properties of interest may be derived using the approximation of a quasi-homogeneous medium [1–4]. The first step involves specification of the 'bulk representative elements' (BRE) which completely fill the total volume V and which have the same properties as MHM. It is assumed that the volume occupied by the BRE is statistically homogeneous (i.e. the volume content of the components, two-point correlations and other properties of the BRE are invariant with respect to their location in the given volume V) since this would conform to the ergodicity criterion (i.e. equality of the averages over the statistical ensemble and over the volume).

The statistically homogeneous field in MHM may be generated by a proper choice of boundary conditions; in the case of the heat conductivity problem, the latter may be defined as

$$t(s) = \langle \nabla t \rangle r, \qquad q_n(s) = \langle q \rangle n \tag{4.1}$$

where $t(s)$ is the temperature on the surface s surrounding the volume V, $q_n(s)$ is the normal heat flux to the surface, r is the vector-radius, n is the vector-normal to the surface, and $\langle \nabla t \rangle$ and $\langle q \rangle$ are the temperature gradient and the heat flux averaged over the volume V, respectively.

The effective heat conductivity λ and resistivity ρ are defined by eqs. (4.2);

$$\langle q \rangle = -\lambda \langle \nabla t \rangle \tag{4.2a}$$

$$\langle \nabla t \rangle = -\rho \langle q \rangle \tag{4.2b}$$

where $\lambda \rho = 1$. In a similar fashion, for the local regions of MHM (e.g. components) we can write

$$q(r) = -\lambda(r) \nabla t(r); \qquad \nabla t(r) = -\rho(r) q(r) \tag{4.3}$$

where $q(r)$, $\nabla t(r)$, $\lambda(r)$ and $\rho(r)$ are the random functions of coordinates.

Equations (4.2) and (4.3) also apply to several other properties (e.g. electrical conductivity, dielectric permittivity, magnetic susceptibility, diffusivity, etc.) all of which may be united under a common generic name, i.e. generalized conductivity.

Since the properties of a homogeneous material are usually assumed to be invariant with respect to its dimensions, down to the infinitesimal differential volumes, the field equations may be written in derivatives. However, as far as any real material has an intrinsic microstructure, the appropriate quantity in this case would be the 'bulk differential element' (BDE), which is assumed to consist of a sufficiently large number of small (compared to both the total volume V and the BRE) microelements ('crystals').

In the case of MHM, the usual differential field equations of generalized conductivity are assumed to apply, i.e. the mean value of the function $f(r)$ at the point r is defined as

$$\langle f(r) \rangle = V^{-1} \int_V f(r,r')\,dr' \tag{4.4}$$

where r is the vector-radius of the point in the volume V and r' is the local system of coordinates with r as the origin.

The structure of MHM is characterized at each of the following scales [5].

(i) Microscale: l_0 (corresponds to the dimensions of microheterogeneities like crystals, dispersed particles, etc.).

(ii) Miniscale: l (corresponds to the dimensions of the BRE).

(iii) Macroscale: L (corresponds to the specimen's dimensions).

The necessary and sufficient conditions for the validity of the concept of 'effective properties' defined by the following inequality:

$$l_0 \ll l \ll L \tag{4.5}$$

permits us to determine the effective conductivity λ with the aid of eqs. (4.2) and (4.3) from the following system:

$$\lambda = \lambda_1 \varphi A_1 + \lambda_2 (1 - \varphi) A_2 \tag{4.6a}$$

$$\varphi A_1 + (1 - \varphi) A_2 = 1 \tag{4.6b}$$

where A_i are defined by

$$\langle \nabla t_i \rangle = A_i \langle \nabla t \rangle \tag{4.7}$$

$$\langle \nabla t_i \rangle = V_i^{-1} \iiint_{V_i} \nabla t(r)\,dV, \quad i = 1, 2 \tag{4.8}$$

V_i is the volume occupied by ith component and φ is the volume concentration of component 1.

A similar procedure may be applied to determine ρ, i.e.

$$\rho = \rho_1 \varphi B_1 + \rho_2 (1 - \varphi) B_2 \tag{4.9a}$$

$$\varphi B_1 + (1 - \varphi) B_2 = 1 \tag{4.9b}$$

where the B_i values are defined as

$$\langle \boldsymbol{q}_i \rangle = B_i \langle \boldsymbol{q} \rangle, \qquad \langle \boldsymbol{q}_i \rangle = V_i^{-1} \iiint_{V_i} \boldsymbol{q}(\boldsymbol{r}) \, \mathrm{d}V, \quad i = 1, 2 \tag{4.10}$$

Equations (4.6) and (4.9) cannot be used for a straightforward calculation of λ and ρ, respectively, because the number of unknowns (λ, A_1 and A_2 in the former case; ρ, B_1 and B_2 in the latter case) exceeds the number of equations available (two in each case). Thus, additional information on the structure of MHM becomes necessary.

Consider the simplest structure, the stratified MHM. When the heat flux q is directed along the strata, we can write

$$\langle \nabla t_1 \rangle = \langle \nabla t_2 \rangle = \langle \nabla t \rangle \tag{4.11}$$

therefore, since $A_1 = A_2 = 1$, we obtain from (4.6)

$$\lambda_\| = \lambda_1 \varphi + \lambda_2 (1 - \varphi) \tag{4.12}$$

where $\lambda_\|$ is the conductivity along the strata.

When the heat flux q is directed normal to the strata (i.e. $\langle \boldsymbol{q}_1 \rangle = \langle \boldsymbol{q}_2 \rangle = \langle \boldsymbol{q} \rangle$, and hence $B_1 = B_2 = 1$), from eq. (4.9) we obtain $\rho_\perp = \rho_1 \varphi + \rho_2 (1 - \varphi)$; finally, the conductivity in the direction normal to the strata is derived as

$$\lambda_\perp = [\varphi / \lambda_1 + (1 - \varphi) / \lambda_2)^{-1} \tag{4.13}$$

Combining eqs. (4.12) and (4.13), we obtain

$$\lambda_\| - \lambda_\perp = \frac{(\lambda_1 - \lambda_2)^2 \varphi (1 - \varphi)}{\lambda_2 \varphi + \lambda_1 (1 - \varphi)} \tag{4.14}$$

As soon as, generally, $\lambda_\perp \leqslant \lambda \leqslant \lambda_\|$, we can write

$$\lambda = \langle \lambda \rangle - K \frac{(\lambda_1 - \lambda_2)^2 \varphi (1 - \varphi)}{\lambda_2 \varphi + \lambda_1 (1 - \varphi)} \tag{4.15}$$

where $0 \leqslant K \leqslant 1$ is the structure-dependent coefficient.

Thus, structural characterization of MHM by the corresponding A_i and B_i values should be the first step in any theoretical analysis of its effective conductivity.

4.1.2 METHODS OF DIRECT CALCULATIONS

In this approach the unknown parameters A_i and/or B_i are calculated directly from the differential equations in partial derivatives for an appropriate model. In the simplest case of a dilute suspension of spherical inclusions of component 1 in a continuous matrix of component 2, the temperature field around any particle presumably is unperturbed by its neighbors (i.e. the external field gradient is equal to the mean gradient in MHM); hence we can write

$$A_1 = 3\lambda_2/(\lambda_1 + 2\lambda_2) \tag{4.16}$$

Substitution of eq. (4.16) into (4.6) yields [5]

$$\lambda = \lambda_2 + \frac{3\lambda_2\varphi(\lambda_1 - \lambda_2)}{\lambda_1 + 2\lambda_2} \tag{4.17}$$

Assuming, further, equality of the mean gradients within the external field and the matrix means replacing eq. (4.16) by eq. (4.18):

$$A_1 = \frac{3\lambda_2}{\lambda_1 + 2\lambda_2}\left[\frac{3\lambda_2}{\lambda_1 + 2\lambda_2}\varphi + (1 - \varphi)\right]^{-1} = \frac{3\lambda_2}{3\lambda_2\varphi + (\lambda_1 + 2\lambda_2)(1 - \varphi)} \tag{4.18}$$

Finally, substituting eq. (4.18) into (4.6), Maxwell's equation, eq. (4.19), is obtained [6]

$$\lambda = \lambda_2 + \frac{\lambda_1 + 2\lambda_2 - 2\varphi(\lambda_2 - \lambda_1)}{\lambda_1 + 2\lambda_2 + \varphi(\lambda_2 - \lambda_1)} \tag{4.19}$$

Lorenz [7] and Lorentz [8] treated the case of an inclusion separated from a continuous medium by an interlayer with the intermediate conductivity to derive eq. (4.20):

$$\frac{\lambda_2 - \lambda}{\lambda_2 + 2\lambda} = \varphi\frac{\lambda_2 - \lambda_1}{\lambda_2 + 2\lambda_1} \tag{4.20}$$

Recently, Malyshev and Malyshev [9] derived eq. (4.21) for the effective conductivity of MHM modelled by spherical inclusions located at the cites of the regular cubical lattice:

$$\lambda = \lambda_2\left\{1 + 3\varphi\left[\frac{\lambda_1 + 2\lambda_2}{\lambda_1 - \lambda_2} - \varphi + \frac{A}{B} + C\right]^{-1}\right\} \tag{4.21}$$

where

$$A = 1.3091\frac{\lambda_2 - \lambda_1}{\lambda_1 + 4\lambda_2/3}\varphi^{10/3}\left(1 - 0.1173\frac{\lambda_2 - \lambda_1}{\lambda_1 + 6\lambda_2/5}\varphi^{11/3}\right)^2$$

$$B = 1 + 0.4054\frac{\lambda_2 - \lambda_1}{\lambda_1 + 4\lambda_2/3}\varphi^{7/3} - 6.6568\frac{(\lambda_2 - \lambda_1)^2\varphi^6}{(\lambda_1 + 4\lambda_2/3)(\lambda_1 + 5\lambda_2/6)}$$

$$C = 0.0723\frac{\lambda_2 - \lambda_1}{\lambda_1 + 6\lambda_2/5}\varphi^{14/3} + 0.15256\frac{\lambda_2 - \lambda_1}{\lambda_1 + 8\lambda_2/7}\varphi^6$$

Truncation of eq. (4.21) after the first two terms yields eq. (4.17); the Rayleigh equation, eq. (4.22) [10], is recovered by truncation after the first four terms:

$$\lambda = \lambda_1\left(1 + \frac{3\varphi}{\dfrac{\lambda_1 + 2\lambda_2}{\lambda_1 - \lambda_2} - \varphi - 1.3091\dfrac{\lambda_1 - \lambda_2}{\lambda_1 + 4\lambda_2/3}\varphi^{10/3}}\right) \tag{4.22}$$

Finally, eq. (4.23) of Meredith and Tobias [11] is obtained by truncation after the first six terms:

$$\lambda = \lambda_2\left\{1 + 3\varphi\left[\frac{\lambda_1 + 2\lambda_2}{\lambda_1 - \lambda_2} - \varphi\frac{1.3091\dfrac{\lambda_1 - \lambda_2}{\lambda_1 + 4\lambda_2/3}\varphi^{10/3}}{1 - 0.4054\dfrac{\lambda_1 - \lambda_2}{\lambda_1 + 4\lambda_2/3}\varphi} - 0.0723\frac{\lambda_1 - \lambda_2}{\lambda_1 + 6\lambda_2/5}\varphi^{14/3}\right]^{-1}\right\}$$

$$(4.23)$$

A common feature of all the approaches cited is the neglect of contact effects at the interface between components (i.e. ideal thermal contact is assumed).

4.1.3 SELF-CONSISTENT FIELD APPROACH

The basic model is an isolated spherical inclusion embedded into an infinite medium with the effective properties sought for. Making use of arguments similar to those used in the derivation of eq. (4.16), we can write

$$A_1 = 3\lambda/(\lambda_1 + 2\lambda_2) \tag{4.24}$$

to obtain, after substitution of eq. (4.24) into eq. (4.6) [12–14]

$$\lambda = \lambda_1\{[(3\varphi - 1) + (2 - 3\varphi)a]/4 + ([(3\varphi - 1) + (2 - 3\varphi)a]^2/16 + a/2)^{1/2}\} \tag{4.25}$$

where $a = \lambda_2/\lambda_1$.

Extension of this approach to the case of inclusions of ellipsoidal shape yielded [15, 16]

$$\lambda = \lambda_2 + \frac{\varphi}{3(1 - \varphi)}\cdot\frac{\lambda_1 - \lambda_2}{1 + (b/2)(\lambda_1/\lambda_2 - 1)} \tag{4.26}$$

where b is the numerical shape factor equal to zero, 2/3 and unity for sheets, spheres and rods, respectively. In the case of disc-like and rod-like shapes (i.e. compressed and stretched spheroids, respectively) this factor is defined by eqs. (4.27) and (4.28), respectively

$$b = \frac{\psi - 0.5\sin 2\psi}{\sin 2\psi}\cos\psi \tag{4.27}$$

$$b = \frac{1}{\sin^2\psi} - \frac{\cos^2\psi}{\sin^3\psi}\ln\frac{1 + \sin\psi}{1 - \sin\psi} \tag{4.28}$$

where $\psi = \arccos{(d/l)}$, d is the thickness and l is either the disc diameter or the rod length.

This approach fails at $a < 10^{-2}$ [17, 18]; moreover, $\lambda < 0$ is predicted at $a = 0$ and $\varphi < 0.3$.

4.1.4 METHOD OF INTEGRATION

The incremental change in the effective conductivity of MHM on addition of a small amount of inclusions may be defined as [cf. eq. (4.16)]

$$\Delta\lambda = \lambda - \lambda_2 = \frac{3(\lambda_1 - \lambda_2)\lambda_2}{\lambda_1 + 2\lambda_2}\Delta\varphi \tag{4.29}$$

where $\Delta\varphi = \Delta V_1/V$, V_1 is the volume occupied by component 1 and ΔV_1 is the amount of component 1 added. Hence, the volume content of the latter will change from an initial φ to φ^*, i.e.

$$\varphi^* = (V_1 + \Delta V_1)/(V_1 + V_2 + \Delta V_1) = (\varphi + \Delta\varphi)/(1 + \Delta\varphi)$$

therefore the incremental change in the volume content is

$$\Delta\varphi = \varphi^* - \varphi = (1 - \varphi)\Delta\varphi \Rightarrow d\varphi = (1 - \varphi)\,d\varphi \tag{4.30}$$

Substitution of eq. (4.30) into (4.29) yields

$$d\lambda = \frac{3(\lambda_1 - \lambda_2)\lambda_2}{\lambda_1 + 2\lambda_2} \cdot \frac{d\varphi}{1 - \varphi} \tag{4.31}$$

which, after substitution of λ for λ_2, transforms into

$$d\lambda = \frac{3(\lambda_1 - \lambda)\lambda}{\lambda_1 + 2\lambda} \cdot \frac{d\varphi}{1 - \varphi} \tag{4.32}$$

Integration of eq. (4.32) assuming $\lambda_{\varphi=0} = \lambda_2$ gives, finally [12],

$$\frac{\lambda - \lambda_1}{\lambda_2 - \lambda_1}(\lambda_2/\lambda)^{1/3} = 1 - \varphi \tag{4.33}$$

Equation (4.33) is intended to apply only for the case of isolated inclusions.

4.1.5 METHOD OF AVERAGING OVER THE ENSEMBLE

This method combines the technique of averaging over the ensemble of permissible configurations with the self-consistent field (i.e. effective medium) approach [19–22].

Essentially, the effective properties of MHM are derived from an analysis of the pattern of physical fields around a selected spherical inclusion embedded into the medium with effective properties. However, in contrast to the usual 'effective medium model', the inclusion is assumed to be surrounded by a concentric shell with properties dependent on the distance to the surface of the sample inclusion. In this case, the mean volume content of the dispersed component 1 in the immediate vicinity of the sample inclusion, $\varphi(r - r')$, is expected to differ from that in more remote regions, φ; moreover, sample inclusions are assumed to have no effect on the pattern of distribution of other inclusions outside the spherical shell of thickness R around the sample inclusion. Thus, the local concentration is defined as

$$\varphi(x) = \varphi\sigma(x/R); \qquad (x/R) \in (1.3)$$

$$\sigma(x/R) = \frac{27 - 56(x/R) + 30(x/R)^2 - (x/R)^4}{16(x/R)}$$

and $\sigma(x/R) = 1$ if $(x/R) > 3$.

The final result is

$$\begin{aligned}
\lambda = \lambda_1 [7a(1-\varphi) &+ 17 + 7\varphi]^{-1}\{a(1+11\varphi) + 5 - 11\varphi \\
&+ [a(1+11\varphi) + 5 - 11\varphi]^2 + [7a(1-\varphi) + 17 + 7\varphi] \\
&\times [a(5+7\varphi) + 7(1-\varphi)]^{1/2}\}
\end{aligned} \qquad (4.34)$$

Since this rigorous result was obtained from the first principles, it may be recommended as a test for the validity of other model approaches.

A closely related problem of the rate of heat transfer from a heated body embedded into a two-component material was solved (assuming that the undisturbed temperature gradient varies inversely with the square of the distance from the heated body) to yield [23]

$$\lambda/\lambda_2 = 1 + 3(\alpha - 1)[\varphi + f(\alpha)\varphi^2 + 0(\varphi^3)]/[\alpha + 2 - (\alpha - 1)] \qquad (4.35)$$

where $f(\alpha) = \sum_{p=6}^{\infty}[(B_p - 3A_p)/(p-3)2^{p-3}]$ is the function decreasing from $f(\alpha) = \text{const} = 0.108$ (at $\log \alpha < -2$) to zero (at $\log \alpha = 0$) and increasing again up to $f(\alpha) = \text{const} = 0.504$ (at $\log \alpha > 2$), and B_p and A_p are the known functions of $\alpha = \lambda_2/\lambda_1$. It can be shown [23] that eq. (4.35) reduces to eq. (4.19) by neglecting the term in square brackets in the numerator.

4.1.6 LICHTENECKER'S APPROACH

Assuming similar functional dependences of both resistivity ρ and conductivity λ, i.e.

$$\lambda = \Phi(\lambda_1, \lambda_2, \varphi), \qquad \rho = \Phi(\rho_1, \rho_2, \varphi)$$

provided the following conditions hold:

$$\Phi(1/\rho_1, 1/\rho_2, \varphi) = 1/\Phi(\rho_1, \rho_2, \varphi)$$

$$\Phi(\rho_1, \rho_2, 0) = \rho_2, \qquad \Phi(\rho_1, \rho_2, 1) = \rho_1$$

Lichtenecker and Rother [24] derived eqs. (4.36):

$$\rho = \rho_1^{\varphi}\rho_2^{(1-\varphi)} \qquad (4.36a)$$

$$\lambda = \lambda_1^{\varphi}\lambda_2^{(1-\varphi)} \qquad (4.36b)$$

The simplicity of eqs. (4.36) is attractive; however, the limits of their validity are unclear since the vanishing effective conductivity of MHM predicted for the case when the conductivity of either of the components is zero contradicts, of course, the experimental data.

4.1.7 HERRING'S APPROACH

Using the expansion of all fluctuating parameters into the Fourier series inside volume V, i.e.

$$\lambda(r) = \langle\lambda\rangle + {\sum_e}' \lambda_i \exp(ier) \tag{4.37a}$$

$$q(r) = \langle q\rangle + {\sum_k}' q_k \exp(ikr) \tag{4.37b}$$

$$\nabla t(r) = \langle\lambda t\rangle + {\sum_m}' \nabla t_m \exp(imr) \tag{4.37c}$$

and eliminating terms the $k = 0$, $l = 0$ and $m = 0$ in the summations $\sum'$, eq. (4.38) was derived [25, 26]:

$$\lambda = \langle\lambda\rangle \left[1 - (1/3)\frac{\langle(\lambda - \langle\lambda\rangle)^2\rangle}{\langle\lambda\rangle^2} + (1/3)^2 \frac{\langle(\lambda - \langle\lambda\rangle)^3\rangle}{\langle\lambda\rangle^3} - \cdots \right.$$
$$\left. + (-1/3)^{n-2} \frac{\langle(\lambda - \langle\lambda\rangle)^n\rangle}{\langle\lambda\rangle^n} + \cdots \right] \quad (n > 2) \tag{4.38}$$

This method does not differentiate between different shapes (spheres, cylinders, etc.); moreover, eq. (4.38) converges more slowly the more different are the properties of both components.

4.1.8 CORRELATION APPROXIMATION. METHOD OF CONDITIONAL MOMENTS

The local values of the flux, temperature gradient and heat conductivity may be defined as follows:

$$q(r) = \langle q\rangle + q^0(r), \qquad \nabla t(r) = \langle\nabla t\rangle + \nabla t^0(r)$$
$$\lambda(r) = \langle\lambda\rangle + \lambda^0(r), \qquad t(r) = \langle t\rangle + t^0(r)$$

where the quantities with a superscript 0 are the random functions of coordinates for which the following condition holds:

$$\langle q^0\rangle = \langle\nabla t^0\rangle = \langle\lambda^0\rangle = \langle t^0\rangle = 0$$

The conservation equation for a stationary heat flux ($\operatorname{div} q(r) = 0$) may be written as (assuming $\operatorname{div}(\langle\lambda\rangle\langle\nabla t\rangle) = 0$)

$$\langle\lambda\rangle \operatorname{div} \nabla t^0 + \operatorname{div}(\lambda^0 \nabla t) = 0 \tag{4.39}$$

which can be transformed into

$$\langle\lambda\rangle \operatorname{div} \nabla t^0(r) = -f(r), \qquad f(r) = -\operatorname{div}(\lambda^0 \nabla t) \tag{4.40}$$

Equation (4.40) may be solved using Green's function to yield linked chaining with respect to the local temperature $t^0(r)$; hence, we need to split that chaining which is characteristics of non-linear systems.

Using this approach and neglecting higher-than-second moments of $\langle \lambda \nabla t \rangle$ (the so-called 'correlation approximation'), the effective heat conductivity of isotropic, homogeneous MHM with isometric components may be expressed as [4]

$$\lambda = \langle \lambda \rangle - \varphi(1 - \varphi)(\lambda_1 - \lambda_2)^2/3\langle \lambda \rangle \tag{4.41}$$

Equation (4.41) may be regarded as a special case of eq. (4.38) truncated after the first two terms; presumably, it is best suited in the case of either small differences in the properties of the components and/or low concentrations $(\varphi, (1 - \varphi) < 0.2)$. As a natural step to eliminate these limitations, the higher moments were considered to yield [4]

$$\lambda = \langle \lambda \rangle - \varphi(1 - \varphi)(\lambda_1 - \lambda_2)^2/[3\langle \lambda \rangle - (1 - 2\varphi)(\lambda_1 - \lambda_2)] \tag{4.42}$$

It turned out, however, that the agreement between eq. (4.42) and the experimental data was worse the larger was the difference between the properties of the components.

4.1.9 VARIATIONAL APPROACH

This method was applied to estimate the higher and the lower bounds of the effective conductivity of a MHM [27–32]. In one approach, these estimates were derived from the principle of minimum entropy production, i.e. [30]

$$dS/d\tau = - \iiint_{(V)} T^{-1}(\nabla t \cdot q)\, dV \geqslant 0 \tag{4.43}$$

In the case of a stationary temperature field, eq. (4.43) may be transformed into

$$\delta \iiint_{(V)} \lambda[\nabla t(r)]^2\, dV = 0 \tag{4.44}$$

to arrive at the upper bound, eq. (4.45), after substitution of $\nabla t(r) = \langle \nabla t \rangle + \langle \nabla t^0(r) \rangle$ for the local $\nabla t(r)$ into eq. (4.44), i.e.

$$\lambda \leqslant \langle \lambda(n + \nabla t^0)^2 \rangle \tag{4.45}$$

where n is the unit vector parallel to $\langle \nabla t \rangle$ and $t^0(r)$ is the arbitrary limit function.

On the other hand, recalling that $1/\lambda = \langle q^2/\lambda \rangle/\langle q \rangle$ and having rewritten eq. (4.43) as

$$\delta \left(\iiint_{(V)} \lambda^{-1} q^2(r)\, dV \right) = 0 \tag{4.46}$$

the lower bound is derived:

$$\lambda \geqslant \lambda^{-1}\langle[\boldsymbol{n}+(\nabla\times\boldsymbol{A})]^2\rangle^{-1} \qquad (4.47)$$

where $\boldsymbol{A}$ is the arbitrary limit vector. Setting $\nabla t^0 = 0$ and $\nabla\times\boldsymbol{A}=0$, we obtain finally from eqs. (4.45) and (4.47)

$$\langle\lambda^{-1}\rangle^{-1} \leqslant \lambda \leqslant \langle\lambda\rangle \qquad (4.48)$$

In another approach [27–29], introducing a reference body with the shape and dimensions identical to those of a real MHM and with the conductivity tensor λ^c, we may define the vector of polarization as

$$\boldsymbol{P}_i = \boldsymbol{q}_i - \boldsymbol{q}_i^c, \qquad \boldsymbol{q}_i = -\lambda_{ij}\nabla t_j, \qquad \boldsymbol{q}^c = -\lambda_{ij}\nabla t_j^c \qquad (4.49)$$

where $\boldsymbol{q}_i$ and ∇t_j are the heat flux and the temperature gradient in a MHM, respectively.

Denoting by one prime (') the deviations from the mean of the properties in a reference body and by two primes (") the central arbitrary quantities, i.e.

$$\lambda_{ij}' = \lambda_{ij} - \lambda_{ij}^c, \qquad \boldsymbol{P}_i'' = \boldsymbol{P}_i - \langle\boldsymbol{P}_i\rangle \qquad (4.50)$$

and making use of Hashin and Shtrikman's variational principles [27], we may define the functional

$$U = 0.5 \iiint_{(V)} (\lambda_{ij}^c\langle\nabla t_i\rangle\langle\nabla t_j\rangle - \boldsymbol{P}_i t_{ij}\boldsymbol{P}_j + \boldsymbol{P}_i''\nabla_i t_j'' + 2\boldsymbol{P}_i\langle\nabla t_i\rangle)\,\mathrm{d}V \qquad (4.51)$$

for which the stationary state is ensured provided the following main conditions

$$\nabla t_i = \tau_{ij}\boldsymbol{P}_j, \qquad \tau_{ij}\lambda_{jk} \equiv \delta_{ik} \qquad (4.52)$$

and additional conditions

$$\nabla_i[\lambda_{ij}^c\nabla t_j''(\boldsymbol{r}) + \boldsymbol{P}_i''] = 0, \qquad t''(\boldsymbol{r})|_s = 0 \qquad (4.53)$$

are satisfied.

The stationary values of the functional (4.51) are the absolute minimum and the absolute maximum in the cases of $\lambda_{ij}^c - \lambda_{ij} \geqslant 0$ and $\lambda_{ij}^c - \lambda_{ij} < 0$, respectively. The final result for MHM composed of isotropic and homogeneous components is

$$\lambda = \left[\sum_\alpha \varphi_\alpha(\lambda_\alpha + 2\lambda^c)^{-1}\right]^{-1} - 2\lambda^c \qquad (4.54)$$

In the case when the unknown λ^c is estimated from the following relationships [31]:

$$\lambda_+^c = \varphi\lambda_2 + (1 - \varphi)\lambda_1 = \lambda_1\lambda_2\langle\lambda^{-1}\rangle$$
$$\lambda_-^c = [\varphi/\lambda_2 + (1 - \varphi)/\lambda_1]^{-1} = \lambda_1\lambda_2/\langle\lambda\rangle$$

the interval between the lower and upper bounds may be defined as

$$\left[\sum_{\alpha=1}^{N}\varphi_\alpha(\lambda_k+2\lambda_-^c)^{-1}\right]^{-1}-2\lambda_-^c\leqslant\lambda\leqslant\left[\sum_{\alpha=1}^{N}\varphi_k(\lambda_\alpha+2\lambda_+^c)^{-1}\right]^{-1}-2\lambda_+^c \quad (4.55)$$

On substitution of $\lambda_+^c=\lambda_1$ and $\lambda_-^c=\lambda_2$ $(\lambda_2>\lambda_1)$ into eq. (4.55), we obtain Hashin and Shtrikman's eq. (4.56) for two-component MHM [27]:

$$\langle\lambda\rangle-\frac{\varphi(1-\varphi)(\lambda_1-\lambda_2)^2}{\varphi\lambda_2+(1-\varphi)\lambda_1+\lambda_1}\leqslant\lambda\leqslant\langle\lambda\rangle-\frac{\varphi(1-\varphi)(\lambda_1-\lambda_2)^2}{\varphi\lambda_2+(1-\varphi)\lambda_1+\lambda_2} \quad (4.56)$$

Equations. (4.55) and (4.56) should be compared with eq. (4.57) [33, 34]:

$$\left[\langle1/\lambda\rangle-\frac{2(1/\lambda_1-1/\lambda_2)\varphi(1-\varphi)}{\langle1/\lambda\rangle_\chi+2\langle1/\lambda\rangle_\varphi}\right]^{-1}\leqslant\lambda\leqslant\langle\lambda\rangle-\frac{(\lambda_1-\lambda_2)^2\varphi(1-\varphi)}{\langle\lambda\rangle_\varphi+2\langle\lambda\rangle_\chi} \quad (4.57)$$

(where $\langle\lambda\rangle_\chi=\lambda_1\chi+\lambda_2(1-\chi)$, $\langle\lambda\rangle_\varphi=\lambda_1(1-\varphi)+\lambda_2\varphi$ and $0<\chi<1$ is the geometrical factor), and with eq. (4.58) [28, 35]:

$$\lambda_1+\varphi/[1/(\lambda_2-\lambda_1)+\varphi/3\lambda_1]\leqslant\lambda\leqslant\lambda_2+(1-\varphi)/[1/(\lambda_2-\lambda_1)+(1-\varphi)/3\lambda_2] \tag{4.58}$$

which was derived for the case of a single spherical inclusion with either higher (upper bound) or lower (lower bound) conductivity, respectively, compared with the continuous medium.

It can be shown that the gap between the upper and lower bounds predicted by eqs. (4.55)–(4.58) increases and becomes comparable with the magnitude of λ_1 with $\lambda_2/\lambda_1\Rightarrow0$. We should also recognize the crucial influence of the structural model (in fact, the method of closure of equations based on Green's function) chosen in each of the above variational approaches. Since Green's function may be defined only for the simplest shapes like spheres and cylinders, the theoretical predictions are also limited to MHM with heterogeneities of this kind. Thus, it seems pertinent to emphasize once again that the theoretical predictions of the effective properties of MHM depend more on the structural model chosen than on details of the mathematical formalism.

4.1.10 METHOD OF FUNCTIONAL INEQUALITIES

The starting point are eqs. (4.2) and (4.3), which may be rewritten as [36]

$$\lambda\langle\nabla t(r)\rangle^2=\langle\lambda(r)\nabla t'(r)\rangle \tag{4.59a}$$

$$\rho\langle q(r)\rangle^2=\langle\rho(r)q^2(r)\rangle \tag{4.59b}$$

Next, the problem of effective conductivity will be solved in the following steps.

(i) Assuming that the properties of Chebyshev's inequalities for the reciprocally ordered functions $f(r)$ and $g(r)$, i.e. [37]

$$[f(r_1) - f(r_2)][g(r_1) - g(r_2)] \leqslant 0$$
$$\langle f(r)g(r) \rangle \leqslant \langle f(r) \rangle \langle g(r) \rangle$$

also apply to the functions $\lambda(r)$, $\nabla t^2(r)$, $\rho(r)$ and $q^2(r)$, one may write

$$\langle \lambda \rangle \langle \nabla t^2 \rangle \geqslant \langle \lambda(r)\nabla t^2(r) \rangle = \lambda \langle \nabla t \rangle^2 \tag{4.60a}$$

$$\langle \rho \rangle \langle q^2 \rangle \geqslant \langle \rho(r)q^2(r) \rangle = \rho \langle q \rangle^2 \tag{4.60b}$$

(where $1/\lambda = \rho$) to obtain

$$\langle 1/\lambda \rangle^{-1} \leqslant \lambda \leqslant \langle \lambda \rangle \tag{4.61}$$

(ii) Assuming, further, the applicability of Jensen's inequalities for a convex function $f(x)$ [38]

$$f(\langle x \rangle) < \langle f(x) \rangle$$
$$1/\langle f(x)^{-1} \rangle \leqslant \exp\langle \ln f(x) \rangle \leqslant \langle f(x) \rangle$$

(where $1/\langle f(x)^{-1} \rangle$ and $\exp\langle \ln f(x) \rangle$ are harmonic and geometrical averages, respectively), to the effective conductivity of MHM (at least, at sufficient distance from the phase transition point) the following inequality should hold:

$$1/\langle 1/\lambda \rangle \leqslant \exp\langle \ln \lambda \rangle \leqslant \langle \lambda \rangle \tag{4.62}$$

Comparing eqs. (4.61) and (4.62), one may derive, approximately

$$\lambda \cong \exp\langle \ln \lambda \rangle \cong \lambda_1^{\varphi_1} \lambda_2^{\varphi_2} \cdots \lambda_n^{\varphi_n} \tag{4.63}$$

which is essentially Lichtenecker's equation, eq. (4.36b).

(iii) Assuming the applicability of the reverse Gelder's inequality for the functions $f(r)$ and $g(r)$ [39]

$$\langle f(r)g(r) \rangle > [\langle f^p(r) \rangle]^{1/p}[\langle g^l(r) \rangle]^{1/l}$$

(where $0 < p < 1$ or $p < 0$, and p and l are the conjugated numbers, i.e. $1/p + 1/l = 1$), to the functions $\lambda(r), \nabla t^2(r), \rho(r)$ and $q^2(r)$, we may write

$$[\langle \lambda(r)\nabla t^2(r) \rangle] > [\langle \lambda^p(r) \rangle]^{1/p}[\nabla t^{2l}(r) \rangle]^{1/l} \tag{4.64a}$$

$$[\rho(r)q^2(r) \rangle] > [\rho^l(r) \rangle]^{1/l}[\langle q^{2p}(r) \rangle]^{1/p} \tag{4.64b}$$

Substitution of eqs. (4.49) into (4.64) yields

$$\lambda \geqslant (\langle \lambda^p \rangle^{1/p})C_1 \tag{4.65a}$$

$$\lambda \geqslant (\langle \lambda^{-l} \rangle^{1/l})C_2 \tag{4.65b}$$

where $C_1 = (\langle \nabla t^{2l} \rangle^{1/l})/(\nabla t^2 \rangle)$ and $C_2 = (\langle q^{2p} \rangle)^{1/p}/(\langle q^2 \rangle)$. It can be shown that $C_1 < 1$ if $l < 1$, and $C_2 < 1$ if $p < 1$; moreover, having derived $\lambda \leqslant \langle \lambda \rangle$ and $1/\lambda \leqslant \langle 1/\lambda \rangle$ from eqs. (4.61), and $\lambda \geqslant \langle \lambda^p \rangle^{1/p}$ and $1/\lambda \geqslant \langle \lambda^{-l} \rangle^{1/l}$ from eqs. (4.65), we may write

$$\langle \lambda \rangle \geqslant \langle \lambda^p \rangle^{1/p} \tag{4.66a}$$

$$\langle 1/\lambda \rangle \geqslant \langle \lambda^l \rangle^{1/l} \tag{4.66b}$$

provided the following conditions hold:

$$\langle \lambda^p \rangle^{1/p}\big|_{p=-1} = 1/\langle 1/\lambda \rangle, \qquad \langle \lambda^{-l} \rangle^{-1/l}\big|_{l=1} = \langle \lambda \rangle \tag{4.67}$$

Thus, the limits of p may now be established as

$$-1 \leqslant p < 0\,(0 < l \leqslant 1/2) \tag{4.68a}$$

$$0 < p \leqslant 1/2\,(-1 \leqslant l < 0) \tag{4.68b}$$

Combining (4.66a) and (4.66b) we obtain, finally

$$\langle \lambda^p \rangle^{1/p} \leqslant \lambda \leqslant \langle \lambda^{-l} \rangle^{-1/l} \tag{4.69}$$

The gap between the upper and lower bounds of eq. (4.69) at constant conductivity ratio and volume content of both components is not fixed but depends on p; moreover, the validity of eq. (4.69) is restricted to a certain subensemble of the total variety of possible values, $1/\langle 1/\lambda \rangle \leqslant \lambda \leqslant \langle \lambda \rangle$. This latter aspect will be analyzed below for the following special cases.

(a) $p = -1, \quad l = 1/2$:

$$1/\langle 1/\lambda \rangle < \lambda < 1/\langle \lambda^{-1/2} \rangle^2 \tag{4.69a}$$

(b) $p = -1/2, \quad l = 1/3$:

$$1/\langle \lambda^{-1/2} \rangle^2 < \lambda < 1/\langle 1/\lambda^{1/3} \rangle^3 \tag{4.69b}$$

(c) $p = 1/3, \quad l = -1/2$:

$$\langle \lambda^{1/3} \rangle^3 < \lambda < \langle \lambda^{1/2} \rangle^2 \tag{4.69c}$$

(d) $p = 1/2, \quad l = -1$:

$$\langle \lambda^{1/2} \rangle^2 < \lambda < \langle \lambda \rangle \tag{4.69d}$$

All the above inequalities conform to eq. (4.61), however, it seems obvious that different values of p and l refer to different MHM structures. In fact, eq. (4.69a) is expected to apply in the case of highly conductive inclusions embedded into a poorly conductive medium ($\lambda_1 \gg \lambda_2$), while eq. (4.69d) would be better suited in the reverse case; the intermediate situations would be described by eqs. (4.69b) and (4.69c).

The relevance of arbitrary numbers p and l to the structural characteristics of random MHM may be assessed from eq. (4.70):

$$p = 1/(1 + \delta), \qquad l = -1/\delta \tag{4.70}$$

where $\delta = \langle \delta \rangle / d$, in which $\langle \delta \rangle = 1.19 d/\varphi^{1/3}$ is the mean distance between highly

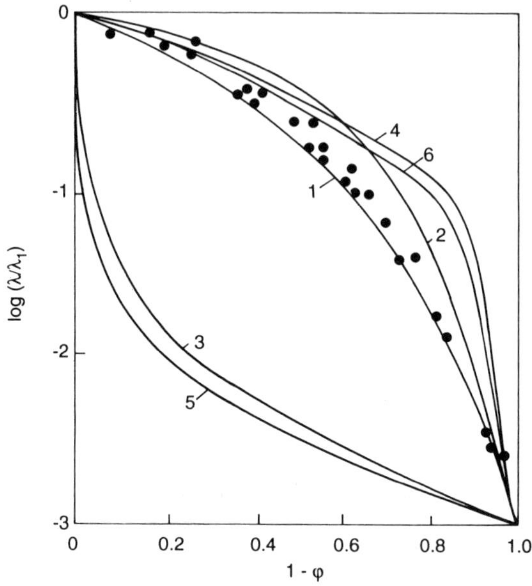

Fig. 4.1. Composition dependence of the reduced electrical conductivity of a NH_3–Li solution (assuming $a = 1.2 \times 10^{-3}$) and theoretical calculations by eqs. (4.72) (curves 1 and 2), eqs. (4.56) (curves 3 and 4) and eqs. (4.58) (curves 5 and 6)

conducting inclusions and d is the mean linear dimension of the latter [39]. Substitution of eq. (4.70) into (4.69) yields

$$\langle \lambda^{1/(1+\delta)} \rangle^{1+\delta} \leqslant \lambda \leqslant \langle \lambda^{1/\delta} \rangle^{\delta} \tag{4.71}$$

which can be easily transformed into eq. (4.72) for a two-component MHM:

$$[\varphi + a^{1/(1+\delta)}(1-\varphi)]^{1+\delta} \leqslant \lambda/\lambda \leqslant [\varphi + a^{1/\delta}(1-\varphi)]^{\delta} \tag{4.72}$$

Equation (4.72) predicts a much narrower gap between the upper and lower bounds of the effective conductivity of two-component MHM compared with other approaches [e.g. eqs. (4.56)–(4.58)] and therefore provides a better agreement with the experimental data available for certain systems (Fig. 4.1). However, special analysis proved that the quantitative applicability of eq. (4.72) was limited both by φ and by a (cases with $\varphi > 0.2$ and $a \geqslant 10^{-4}$ may be recommended).

4.1.11 METHOD OF INTEGRAL SECTIONING

Estimates of the upper and lower bounds within the framework of this method were obtained from general physical arguments, rather than from a rigorous mathematical treatment [1, 10, 32, 40, 41]. According to eq. (4.43), the additional constraint, i.e. div $\boldsymbol{q}(\boldsymbol{r}) = 0$, ensures the stationary state of the integral (i.e. energy dissipation during the heat flux flow):

$$I = (2V)^{-1} \iiint_{(V)} q(r)\nabla t(r)\,dx_1\,dx_2\,dx_3$$

Moreover, the minimum value (i.e. $\delta I = 0$) corresponds to the series of legitimate functions $[q(r), \nabla t(r)]$ that satisfy the conditions

$$\mathrm{div}\,[\lambda(r)\nabla t(r)] = 0$$

$$q(r) = -\lambda(r)\nabla t(r)$$

Therefore, any other choice of pair functions $[q'(r), \nabla t(r)]$ or $[q(r), \nabla t'(r)]$ from the series of legitimate functions that satisfy the same boundary conditions as do the true functions $[q(r), \nabla t(r)]$ would yield the solution I' such that $I' \geqslant I$, where

$$I' = \langle q'\nabla t \rangle, \qquad \langle q' \rangle = -\lambda'\langle \nabla t \rangle$$

or

$$I' = \langle q\nabla t' \rangle, \qquad \langle \nabla t' \rangle = -\rho\langle q \rangle$$

Here λ' and ρ' are the effective properties of the fictive (or reference) bodies characterized by the pair functions $[q'(r), \nabla t(r)]$ or $[q(r), \nabla t'(r)]$, respectively.

The following useful relationship was also proved to hold for quasi-homogeneous bodies [42]:

$$\langle q(r)\nabla t(r) \rangle = \langle q \rangle\langle \nabla t \rangle$$

Two methods of selection of the test functions $q'(r)$ and $\nabla t'(r)$ to estimate the upper and lower bounds of the effective conductivity of MHM are available. For convenience, operators of averaging by coordinates will be introduced:

$$\{f(r)\}_L = L^{-1} \int_0^L f(r)\,dx_k$$

$$\{f(r)\}_S = S^{-1} \iint_{(S)} f(r)\,dx_i\,dx_j$$

assuming the validity of the following equalities

$$\{\{f(r)\}_S\}_L = \{\{f(r)\}_L\}_S = \langle f(r) \rangle$$

The two-bounds estimates by the integral sectioning approach are based essentially on two methods of conditional sectioning of the representative volume V. In the first case (i) the latter (Fig. 4.2(a)) is sectioned in the direction selected along the external field (say, along the axis Ox_3) into the prisms of $dx_1\,dx_2$ for the basal area and L for the height (Fig. 4.2(b)), whereas in the second case (ii) the volume V is sectioned into layers of dx_3 for the thickness and $L \times L$ for the basal area (Fig. 4.2(c)).

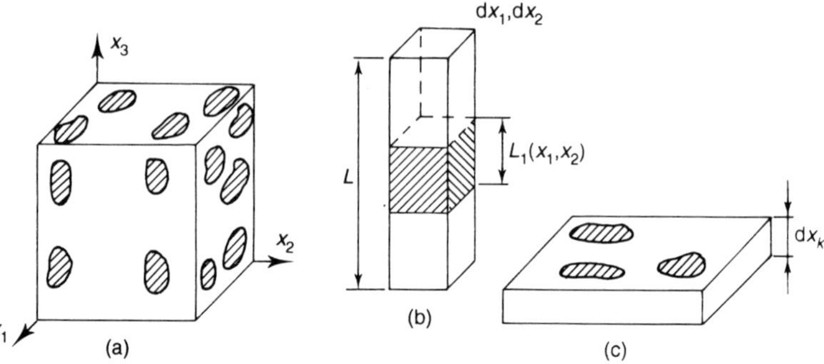

Fig. 4.2. Structural models of MHM: bulk representative element (a); prism-shaped (b) and plate-shaped (c) bulk differential elements.

(i) The heat flux is expressed as

$$\langle q \rangle = - \{\{\lambda(r)\nabla t(r)\}_S\}_L \tag{4.73}$$

and the test functions $[q'(r), \nabla t(r)]$ are chosen such that the following condition will hold:

$$\{q'(r)\}_S = \langle q' \rangle \tag{4.74}$$

Given eq. (4.74) the mean heat flux through the specimen's cross-section [i.e. the term $\{\cdots\}_S$ in eq. (4.73)] may be defined as

$$\{q(r)\}_S = - \{\lambda(r)\nabla t(r)\}_S = - [\lambda]_S[\nabla t(r)]_S \tag{4.75}$$

where $[\lambda]_S = \lambda_1 \langle S \rangle_1(x_3) + \lambda_2 \langle S \rangle_2(x_3)$ is the conductivity of the layer of thickness dx_3, $S = S_1(x_3) + S_2(x_3)$ is the cross-sectional area of the BRE in the normal direction to the Ox_3 axis, $S_i(x_3)$ is that occupied by the ith component ($i = 1, 2$) and $\langle S \rangle_i(x_3) = S_i(x_3)/S$.

Combining $\langle \nabla t \rangle = \{[\lambda]_S^{-1}\}_L \langle q' \rangle$ derived from eq. (4.75) with $(\langle q \rangle \langle \nabla t \rangle) \leqslant (\langle q' \rangle \langle \nabla t \rangle)$, we obtain the upper bound

$$\lambda \leqslant \{\{\lambda\}_S^{-1}\}_L^{-1} \tag{4.76}$$

(ii) The mean gradient is expressed as

$$\langle \nabla t \rangle = \{\{\nabla t(r)\}_L\}_S \tag{4.77}$$

and the test functions $[\nabla t'(r), q(r)]$ are chosen to satisfy the condition

$$\{\nabla t'(r)\}_L = \langle \nabla t' \rangle \tag{4.78}$$

Thus, the term $\{\cdots\}_L$ in eq. (4.77) may be written as

$$\{\nabla t'(r)\}_L = \{\lambda^{-1}\}_L\{q(r)\}_L \tag{4.79}$$

where $\{\lambda^{-1}\}_L = \langle L \rangle_1(x_1, x_2)/\lambda_1 + \langle L \rangle_2(x_1, x_2)/\lambda_2$ is the heat resistance of a prism of height L and of basal area $dx_1\,dx_2$, $\langle L\rangle_i(x_1,x_2)=L_i(x_1,x_2)/L$ and $L_i(x_1,x_2)$ is the specimen's length along the Ox_3 axis occupied by the ith component ($i=1,2$).

Using eq. (4.79) to derive

$$\langle q\rangle = -\{\{\lambda^{-1}\}_L^{-1}\}_S\langle\nabla t'\rangle \tag{4.80}$$

we can write

$$I' = 0.5\langle q\rangle\langle\nabla t'\rangle = 0.5\{\{\lambda^{-1}\}_L^{-1}\}_S^{-1}\langle q\rangle^2 \tag{4.81}$$

Comparing this result with

$$I = 0.5\langle q\rangle\langle\nabla t\rangle = -0.5\lambda^{-1}\langle q\rangle^2 \tag{4.82}$$

and recalling that $I' \geqslant I$ (see above), we can write for the lower bound

$$\lambda \geqslant \{\{\lambda^{-1}\}_L^{-1}\}_S \tag{4.83}$$

Combining eqs. (4.76) and (4.83), we define finally the upper and lower bounds of the effective heat conductivity of MHM, i.e.

$$\{\{\lambda^{-1}\}_L^{-1}\}_S \leqslant \lambda \leqslant \{\{\lambda\}_S^{-1}\}_L^{-1} \tag{4.84}$$

Equation (4.84) is the mathematical formulation of the well-known rule that the true effective conductivity of MHM is higher and lower than those estimated from 'adiabatic' and 'isothermal' sectioning, respectively [1]. The physical meaning of the relevant assumptions [eqs. (4.75) and (4.79), respectively], may be assessed from consideration of the flat circuit of two randomly connected resistances (Fig. 4.3(a)). The upper bound of the effective resistance [i.e. eq. (4.74)] would then correspond to the elimination of all transversal connections (Fig. 4.3(b)) since the finite resistances are replaced by infinite ones (circuit breakers).

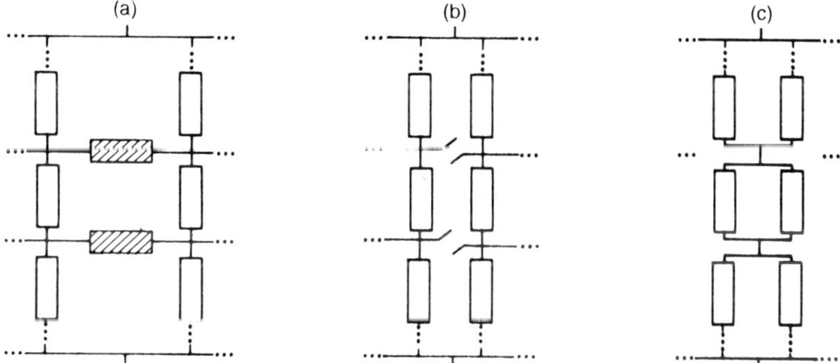

Fig. 4.3. Randomly connected resistance circuits: initial (a); that corresponding to eq. (4.74) (b); that corresponding to eq. (4.78) (c)

In its turn, the lower bound [i.e. eq. (4.79)] would be obtained assuming zero resistance for all transversal connections (Fig. 4.3(c)).

The general eq. (4.84) was used to estimate the upper (λ_+) and the lower (λ_-) bounds of the effective conductivity of MHM with the following special unit cell geometries [1, 43].

(i) Spherical inclusion in a cube:

$$\lambda_- = \lambda_2 + (\lambda^* - \lambda_2)\pi_2 \tag{4.85a}$$

$$\lambda_+ = [(1 - \pi_1)/\lambda_2 + \pi_1/\lambda_c]^{-1}. \tag{4.85b}$$

(ii) Cubical inclusion in a cube:

$$\lambda_- = \lambda_2 \frac{\lambda_1 - (\lambda_1 - \lambda_2)(1 - \varphi^{2/3})\varphi^{1/3}}{\lambda_1 - \varphi^{1/3}(\lambda_1 - \lambda_2)} \tag{4.86a}$$

$$\lambda_+ = \lambda_2 \frac{\lambda_2 + (\lambda_1 - \lambda_2)\varphi^{1/3}}{\lambda_2 + (\lambda_1 - \lambda_2)\varphi^{2/3}(1 - \varphi^{1/3})} \tag{4.86b}$$

(iii) Interpenetrating structures:

$$\lambda_- = \lambda_1[C^2 + a(1 - C)^2 + 2aC(1 - C)/(aC + 1 - C)] \tag{4.87a}$$

$$\lambda_+ = \lambda_1 \left/ \frac{1 - C}{C^2 + a(1 - C^2)} + \frac{C}{C(2 - C) + a(1 - C)^2} \right. \tag{4.87b}$$

Eqs. (4.87) may be compared with eq. (4.88) derived using somewhat modified method of unit cell sectioning [44]:

$$\lambda = \lambda_1 \frac{C^2 + aC(1 - C)}{aC(1 - C) + 1 - C + C^2} + \frac{a[C(1 - C) + a(1 - c)^2]}{C(1 - C) + a(1 - C + C^2)} \tag{4.88}$$

In the above eqs. (4.85)–(4.88), $\lambda^* = [2\lambda_2/(a - 1)\pi_1]\{1 - [1/(a - 1)\pi_1 \ln [(a - 1)\pi_1 + 1]\}$, $\lambda_c = \lambda_2(1 - 1/a)\pi_2/I_p(1)$, $\pi_1 = 2(3\varphi/4\pi)^{1/3}$, $\pi_2 = \pi^{1/3}(3\varphi/4)^{2/3}$, $p = 1 + a/(1 - a)\pi - a = \lambda_2/\lambda_1 I_b(z) = 0.5/b^{1/2} \ln [(z - b^{1/2})/(z + b^{1/2})]$ $(b > 0)$, $I_b(z) = 1/|b|^{1/2} \arctan (z/|b|^{1/2})$ $(b < 0)$, and the geometrical parameter C is obtained from the relationship, $3C^2 - 2C^3 = \varphi$.

4.2 THEORETICAL ASPECTS: PHYSICAL APPROACHES

4.2.1 STEP-BY-STEP AVERAGING APPROACH

The concept of a sharp, 'mathematical' (i.e. infinitely thin) interface between the components implicit in all the 'pragmatic' approaches considered so far had to be modified to allow for the existence of a boundary interphase (BI) in two-component polymer systems (cf. Part I). This 'physical' aspect is, however, explicitly accounted for by the Step-by-Step Averaging (SSA) approach [45–47].

This approach is based essentially on the following postulates [1].

(i) The transport properties of a binary system with a disordered structure are identical to those of an appropriately chosen, 'mutually adequate' (as concerns isotropicity, mechanical stability, geometrical equivalence of components, etc.) one with an ordered structure.

(ii) The effective conductivities of an ordered, macroscopic binary system and of its microscopic unit cell are the same.

(iii) Any multi-component system may be systematically reduced to the binary case.

Having defined the mutually adequate ordered system and its unit cell, we have to proceed with the SSA technique according to which any macroscopic property of MHM may be obtained as the final result of a step-by-step calculation of that property for each next level (or scale) of the whole spectrum of structural microheterogeneities involved. More precisely, we start by defining the lowest level of structural microheterogeneity (that is, the BI thickness, Δr) and calculate its effective property; a similar calculation is repeated at each next structural scale [i.e. the BI of thickness Δr and a single particle of component 2 of size $2r$; the isolated cluster (IsC) formed by several particles coated with BI, each of the latter assumed quasi-homogeneous, an infinite cluster (InC) composed of quasi-homogeneous isolated clusters, etc.] until the macroscopic property sought for is obtained.

The practical application of the SSA technique requires incorporation of two basic models: the percolation model and the model of an equivalent element.

The Percolation Model

This model is intended to apply to a large variety of physical systems in which the geometrical phase transition (i.e. the reversible transition between non-bonded and bonded states) may occur [48–54]. The behavior of percolating systems is frequently studied numerically on lattices which may be regarded as assemblies of bonds and sites. In this case, we would differentiate the problem of bonds from the problem of sites; namely, the former is concerned with monitoring the transition from a non-bonded to a bonded state with increasing concentration φ, whereas in the latter case such a transition is studied on many lattice sites. The critical concentration φ_c corresponding to transition from isolated groups of particles [isolated clusters (IsC)] to an 'infinite' (i.e. spanning the whole lattice) cluster (InC) defines the 'percolation threshold'. In other words, $\varphi \geq \varphi_c$ is the domain of existence of an InC, while only IsC are assumed to exist at $\varphi < \varphi_c$. The percolation thresholds for regular two- and three-dimensional lattices are shown in Table 4.1.

Consider the 'infinite' square lattice in which the probabilities that each pair of nearest sites is either connected by a bond or not, is φ and $1 - \varphi$, respectively. The characteristic dimension L_n of bonded regions (i.e. those with uninterrupted sequences of bonds between different sites) will be a rapidly rising function of

Table 4.1 Site- and bond-percolation thresholds in regular two- and three-dimensional lattices

Lattice type	Coordination number (Z)	Packing density (η)	Percolation threshold		Critical number of bonds* ($Z\varphi_c^b$)	Critical volume fraction ($\eta\varphi_c^s$)
			φ_c^b	φ_c^s		
Triangular	3	0.61	0.6527	0.70	1.96	0.427
Square	4	0.79	0.500	0.59	2.00	0.466
Hexagonal	6	0.91	0.3473	0.50	2.08	0.455
Tetrahedral (diamond)	4	0.34	0.39	0.43	1.56	0.143
Cubic:						
Simple	6	0.52	0.25	0.31	1.50	0.161
Body-centered	8	0.68	0.18	0.24	1.44	0.163
Face-centered	12	0.74	0.12	0.20	1.44	0.148
Dense hexagonal	12	0.74	0.12	0.20	1.44	0.148

*For three-dimensional lattices, $Z\varphi_c^b \cong d/(d-1)$.

φ. The InC spanning the whole lattice is created at φ_c, and a further increase in the number of new bonds above φ_c is accompanied by coalescence of IsC to an InC. The probability of coalescence to an InC of IsC will be higher the larger their size; therefore, in the region $\varphi > \varphi_c$ the IsC of size L_c are assumed to be located in the discontinuities ('holes') of an InC.

The mean-square gyration radius of a finite cluster is

$$\langle L_n^2 \rangle = \sum_{i=1}^{n} |r_i - r_0|^2 / n$$

where $r_0 = \sum_{i=1}^{n} r_i/n$ is the apparent center of gravity, n is the number of sites in a finite cluster, and r_i is the coordinate of the ith bond; the summation is carried out over all n bonds involved in IsC.

The correlation function for a random array of mass points, defined as

$$K(r, r') = \frac{1}{n} \sum_{i=1}^{n} g(r_i')g(r_i + r_i') = \frac{\langle g(r)g(r+r') \rangle}{\langle g(r) \rangle}$$

may be considered as a measure of the mean density of a cluster at distance $|r'|$ from an arbitrary point r.

The correlation length χ (i.e. the measure of cluster connectivity on a percolating lattice) is defined as

$$\chi = \left\langle \left[\frac{1}{n} \sum_{i=1}^{n} (r_i - r_0)^2 \right]^{1/2} \right\rangle$$

Let S_n be the number of n-bond clusters per one bond of an infinite lattice;

then the sum $\sum nS_n$ will be the fraction of the bonds belonging to finite clusters, and the ratio

$$P_n = \frac{nS_n}{\sum_n nS_n},$$

is the probability that an arbitrarily chosen bond benlongs to the n-bonds cluster.

The mean number of bonds (i.e. the mean mass) in a finite cluster is

$$\langle M \rangle = \sum_{i=1} n^2 S_n / \sum nS_n$$

Given these results, the correlation length for the case $\varphi < \varphi_c$ may now be expressed as

$$\chi = \left[\sum_{n=1}^{N} L_n^2 n^2 S_n \bigg/ \sum_{n=1}^{N} n^2 S_n \right]^{1/2}$$

whereas χ goes to infinity on the approach to the percolation threshold (i.e. $|\varphi - \varphi_c| = |\Delta\varphi| \to 0$) as (Fig. 4.4)

$$\chi = |\Delta\varphi|^{-\nu} \tag{4.89}$$

where ν is the corresponding critical index.

In these conditions (i.e. when the system is in the vicinity of φ_c), the density of an InC (i.e. the probability that an arbitrarily chosen bond belongs to an InC)

$$P_\infty(\varphi) = n_k \bigg/ \sum_{i=1}^{N} n_i$$

also scales with $\Delta\varphi$ as

$$P_\infty(\varphi) \sim |\Delta\varphi|^{\beta} \tag{4.90}$$

where β is the relevant critical index. A typical pattern of evolution of $P_\infty(\varphi)$ with φ is shown in Fig. 4.5.

Comparing eqs. (4.89) and (4.90) we obtain

$$P_\infty(\varphi) \sim \chi^{-\beta/\nu} \tag{4.90a}$$

It is pertinent to point out here that in the scale interval between the lattice constant, l_0, and the correlation length χ, all properties of an InC are similar to those at the critical point. More precisely, in the interval $l_0 \ll L \ll \chi$ the InC acquires the property of self-similarity (or scale invariance), which is a characteristic feature of 'fractals'.

The critical indices for percolating systems may be evaluated within the framework of two theoretical models, i.e. the bond-site model (i) [53] and the blob model (ii) [54].

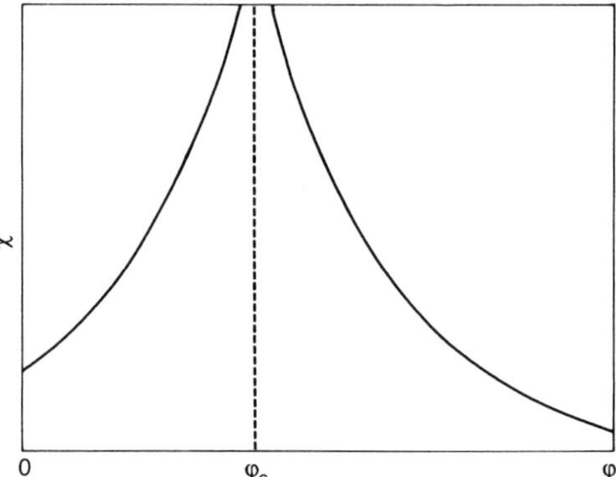

Fig. 4.4. Composition dependence of the correlation length near the percolation threshold (schematic)

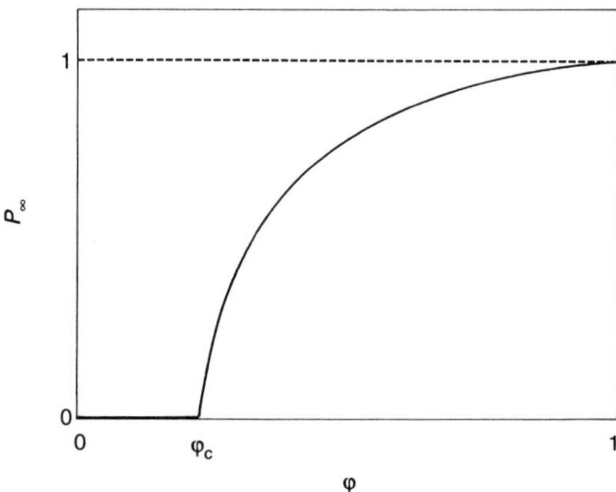

Fig. 4.5. Composition dependence of $P_\infty(\varphi)$ near the percolation threshold (schematic)

(i) The InC is visualized as a network with correlation length χ as a characteristic mesh size (Fig. 4.6(a)). The network junctions (sites) are connected by macrobonds with contour length

$$L_b \sim |\Delta\varphi|^{-\zeta} \tag{4.91}$$

which may exceed χ. According to numerical estimates [48, 49], the critical index $\zeta = 1$ for any space dimensionality d, whereas $v = 1.33$ or 0.8–0.8 for $d = 2$ or 3, respectively. Comparing eqs. (4.89) and (4.91) we may conclude that on the approach to the critical state, χ is predicted to increase faster compared with L_b.

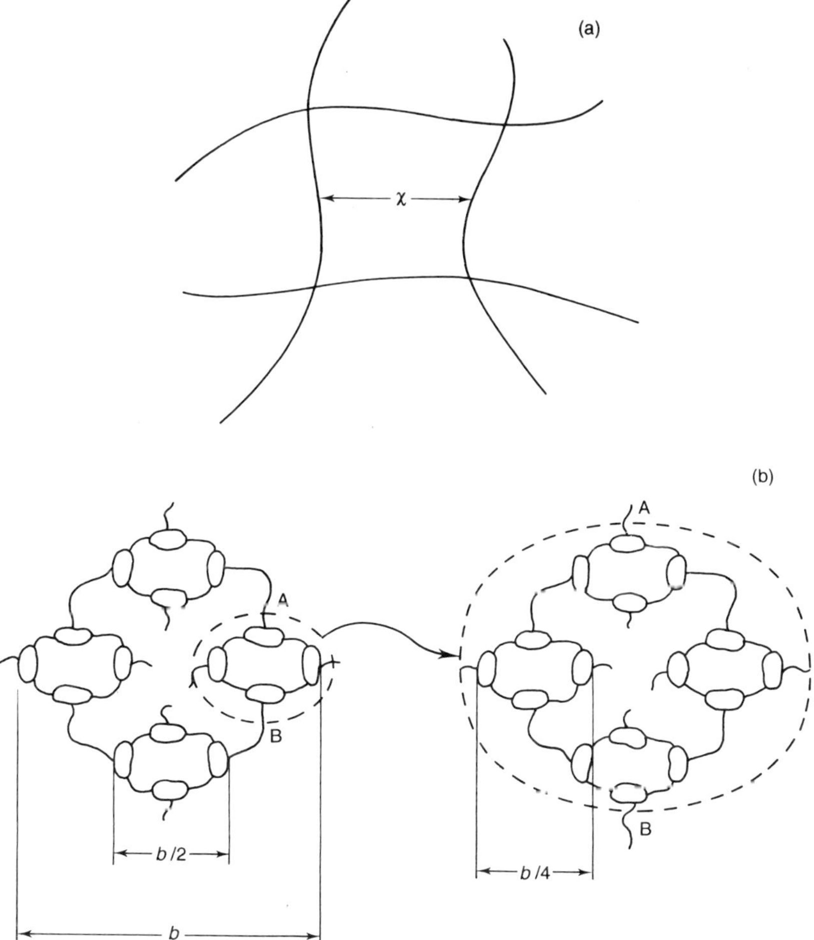

Fig. 4.6. Bond-site model (a) and blob model (b)

(ii) The InC is modelled as an array of blobs connected by macrobonds
(Fig. 4.6(b)). The structure is assumed to duplicate itself at each next length
scale, namely each blob of size b consists of several smaller blobs of size $b/2$
connected by single macrobonds, etc.; thus, at each structural scale the system
is self-similar. It could be shown [54] that the mean cumulative length of non-
overlapping bonds L_b in any portion of InC scales as $L_b \sim b^{1/\nu}$. Therefore, setting
$b = \chi$ we obtain

$$L_b \sim \chi^{1/\nu} \sim |\Delta\psi|^{-1} \tag{4.92a}$$

$$L_b/b \sim b^{1/(\nu-1)} \sim |\Delta\varphi|^{\nu-1} \tag{4.92b}$$

Thus, in contrast to case (i) [cf. eq. (4.91)], the relative bond length L_b/b is predicted to increase with the scale length b.

The above data may be used now to analyze the conductivity Λ of the percolating system. Let the latter be sectioned into d-dimensional cubes ($L > b$) of linear dimensions $b \sim \chi$ and conductivity $\Lambda_b = \Lambda_1/b$ (where Λ_1 is the conductivity of a single bond). The number of cubes per unit length of a row is $1/b$ and the number of parallel rows is b^{1-d}; therefore, the total conductivity will simply be $\Lambda = \Lambda_b b^{2-d} = \Lambda_1 b^{1-d}$. Setting $b = \chi$ we obtain

$$\Lambda = \Lambda_1 |\Delta\varphi|^{v(d-1)} = \Lambda_1 |\Delta\varphi|^t \tag{4.93}$$

where $t = v(d-1)$ is the critical conductivity index.

In one of the first attempts to account explicitly for percolation effects in conductivity of MHM [17, 18], three concentration intervals were considered for the case of highly conductive inclusions in a poorly conductive matrix ($\Lambda_2/\Lambda_1 \leqslant 5 \times 10^{-4}$):

    (i) $\varphi < \varphi_c \Rightarrow \Lambda = \Lambda_1(1 - 5\varphi)^{-1}$;
    (ii) $\varphi_c \leqslant \varphi \leqslant 0.5 \Rightarrow \Lambda = 1.6\Lambda_1(\varphi - \varphi_c)^{1.6}$;
    (iii) $\varphi > 0.5 \Rightarrow$ eq. (4.25).

In the case of not-too-different conductivities of components ($3 \times 10^{-2} \leqslant \Lambda_2/\Lambda_1 < 1$), the percolation effects were considered insignificant, and eq. (4.25) for the effective medium model was recommended throughout.

Another attempt to account explicitly for the percolation effects within the framework of an effective medium model yielded the following result [55]:

$$\varphi\frac{\lambda_1 - \lambda}{\lambda_1 + 2\lambda} + (1 - \varphi)\frac{\lambda_2 - \lambda}{\lambda_2 + 2\lambda} = 0 \tag{4.94}$$

Equation (4.94) was expected to apply in the case of not-too-much-different conductivities of components; moreover, it was derived assuming $\varphi_c = 0.3$, which differs significantly from the best theoretical estimates (cf. Table 4.1). In this respect, much more promising seems to be the approach based on the real space renormalization group (RSRG) theory [56], which provides for reasonable values of the percolation threshold (0.62 and 0.15 for two-dimensional, square tesselation and for three-dimensional, cubic tesselation, respectively) and is intended to apply to MHM whose components differ in their intrinsic conductivities by two and more orders of magnitude.

Apparently, the same conclusion applies to the alternative approach, which is conceptually similar to the RSRG method but is based on a slightly different tesselation technique [51, 52]. At small concentrations ($0 \leqslant \varphi \leqslant \varphi_c$), each IsC represented by a cubic inclusion with linear dimension $l_2 = \varphi^{1/3}$ is separated by distance $2(1 - l_2)$ from its neighbors (Fig. 4.7); the jump-like transition into InC by the mechanism of linking together the individual IsC by single bonds of cross-sectional area $S = [(\varphi - \varphi_c)/(1 - \varphi_c)]^t$ is assumed at higher concentrations

($\varphi > \varphi_c$). In the intermediate interval $\varphi \cong 1 - \varphi \cong 0.5$, each component creates its own InC; however, in the interval $\varphi > 0.5$ the InC of component 2 is assumed, first, to decrease in size, to degenerate, subsequently, into an IsC at $\varphi = 1 - \varphi_c$ and, finally, to disappear at $\varphi \Rightarrow 1$. The relevant geometrical parameters of the percolation model discussed above are collected in Table 4.2.

The effective properties of the percolation model (Fig. 4.7(a)) are derived by sectioning the BRE into four sections with faces parallel and/or perpendicular to the external fields. As an example, consider the case $\varphi < 0.5$ with geometrical parameters specified by Table 4.3. Let section I be filled with component 1; hence, the conductivity of this section is $\lambda_\mathrm{I} = \lambda_1$. In sections II and III the components are connected in series; hence

$$\lambda_\mathrm{II} = \{1/\lambda\}_{l_2}^{-1} = [l_2/\lambda_1 + (1 - l_2)/\lambda_2]^{-1}$$
$$\lambda_\mathrm{III} = \{1/\lambda\}_{l_1}^{-1} = [l_1/\lambda_1 + (1 - l_1)/\lambda_2]^{-1}$$

Finally, $\lambda_\mathrm{IV} = \lambda_2$ for section IV.

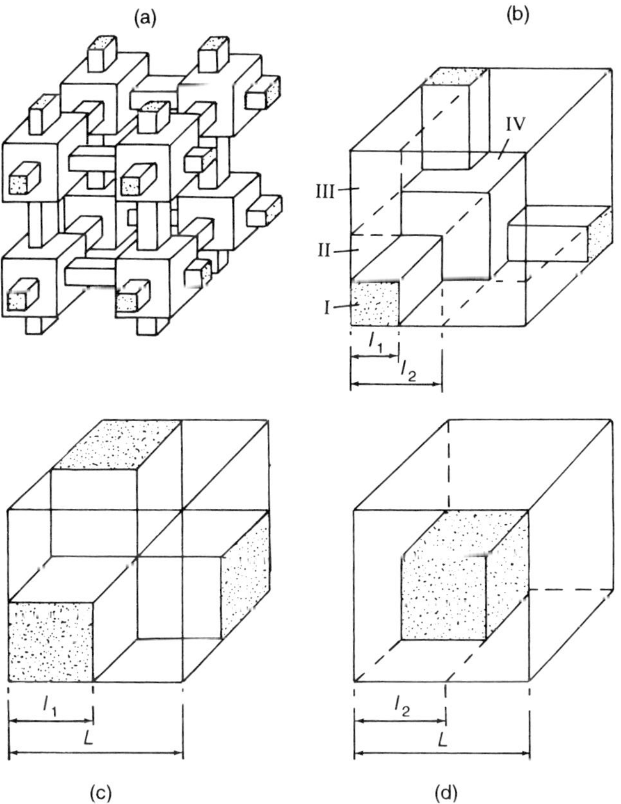

Fig. 4.7. Percolation model (a) and unit cells of BRE at $\varphi > \varphi_c$ (b), $\varphi \cong 0.5$ (c) and $\varphi < \varphi_c$ (d)

Table 4.2 Geometrical parameters of the percolation model*

Range of validity	$\langle S_1 \rangle$	$\langle S_2 \rangle$	$\langle S_3 \rangle$	$\langle S_4 \rangle$	$\langle l_1 \rangle$	$\langle l_2 \rangle$
$0 \leqslant \varphi \| \varphi_c$	0	$\varphi^{2/3}$	0	$1 - \varphi^{2/3}$	0	$\varphi^{1/3}$
$\varphi_c \leqslant \varphi \leqslant 0.5$	$\dfrac{\varphi - \varphi_c}{1 - \varphi_c^{1/3}}$	$\varphi_c^{2/3} - \langle S_1 \rangle$	$2\langle l_1 \rangle \langle l_2 \rangle$	$1 - \varphi_c^{2/3} - \langle S_3 \rangle$	$S_1^{1/2}$	$\varphi_c^{1/3}$

*$S_1 = \langle S \rangle + (\langle S_1 \rangle - \langle S \rangle)g(a)$; $\langle S \rangle = [(\varphi - \varphi_c)/(1 - \varphi_c)]^t$; $g(a) = 5.53a - 8.3a^2 + 3.23a^3 + 0.54a^4$.

Table 4.3 Geometrical parameters and volume fractions of components in different sections of BRE

Section no.	Basal area	Renormalized volume contents	
		Component 1	Component 2
I	$\langle l_1 \rangle^2$	1	0
II	$\langle l_2 \rangle^2 - \langle l_1 \rangle^2$	$\langle l_2 \rangle$	$1 - \langle l_2 \rangle$
III	$2\langle l_1 \rangle(1 - \langle l_2 \rangle)$	$\langle l_1 \rangle$	$1 - \langle l_1 \rangle$
IV	$1 - \langle l_2 \rangle^2 - 2\langle l_1 \rangle(1 - \langle l_2 \rangle)$	0	1

Thus, the effective conductivity of the percolation model will be obtained by averaging out the contributions from all four sections assuming normal orientation of the heat flux to their cross-sectional areas, i.e.

$$\lambda = \lambda_1 S_1 + \lambda_{II} S_2 + \lambda_{III} S_3 + \lambda_2 S_4 \tag{4.95}$$

As already mentioned, eq. (4.95) was derived for the case $\varphi \leqslant 0.5$; in the opposite case (i.e. $\varphi > 0.5$) the indices in eq. (4.94) and in Table 4.2 should be changed, i.e. $\lambda_1 \Leftrightarrow \lambda_2$ and $\varphi \Leftrightarrow 1 - \varphi$. The quality of the fit of eq. (4.95) to the relevant experimental data may be assessed from Fig. 4.8 [47].

Voronoi Polyhedra. The Model of an Equivalent Element

As already emphasized, any structural model of heterogeneous polymers claiming validity had to account explicitly for the smearing out of a sharp, 'mathematical' interface into a 'physical' BI of finite thickness Δr. Apparently, this requirement was met by the model of an equivalent element (EE) [47, 57, 58] which was based on the following physical ideas (Fig. 4.9). The initially isolated particles of component 1 coated with a BI of thickness Δr are assumed to coalesce, first, into IsC with a constant limiting volume content of component 1, φ^*, which may vary from about 0.3 for particles of very irregular shape to 0.6 for the

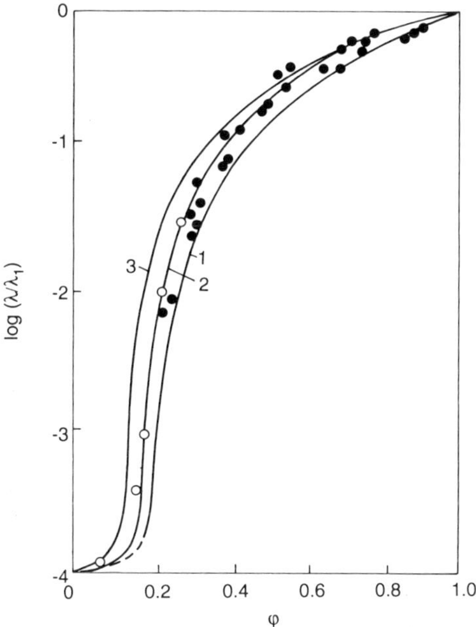

Fig. 4.8. Composition dependence of the relative heat conductivity of tungsten bronzes at 300 K. Solid lines were calculated by eq. (4.95) with $\varphi_c = 0.25$, $t = 2$ (1), $\varphi_c = 0.15$, $t = 1.6$ (2) and $\varphi_c = 0.12$, $t = 1.6$ (3)

dense random packing of spheres [59, 60]. The concentration φ' of such IsC with $\varphi^* = $ const will increase with the nominal concentration φ until an InC is formed at $\varphi' = \varphi_c$. The BRE of the latter is represented by Voronoi polyhedra which are constructed by the intersection of planes drawn normal to the vectors connecting the centers of the neighboring particles at their midpoints. In this fashion, a system of Voronoi polyhedra is formed with their faces tangent to the points of BI contact. In the absence of actual contact points in InC, the faces of the polyhedra should be drawn so as eventually to become tangent to the points of contact after the appropriate smooth shift of particle centers (Fig. 4.9(d)). As a result, the InC is sectioned into different polyhedra with the number of faces dependent on the coordination number N_c of the corresponding particle. In other words, the BRE for this structure would be a polyhedron with the number of faces equal to the mean value of N_c, which is defined as [45–47].

$$N_c = [(1 - \varphi^*) + 3 + \sqrt{(1 + \varphi^*)^2 - 10(1 - \varphi^*) + 9}/2(1 - \varphi^*) \qquad (4.96)$$

It becomes obvious that the limiting volume content of component 1, φ^*, corresponds to a perfect, void-free filling of the total space available by polyhedral BREs.

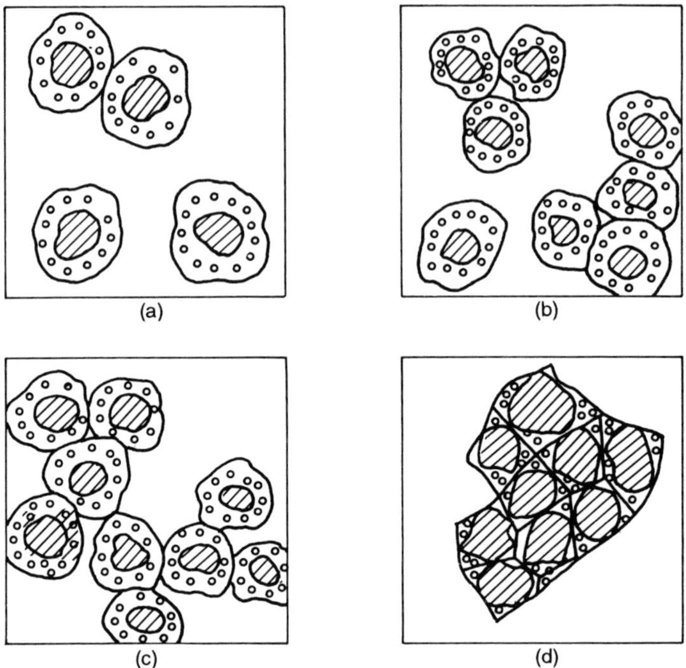

Fig. 4.9. Schematics of the MHM structure at $\varphi < \varphi_c$ (a), $\varphi \cong \varphi_c$ (b) and $\varphi > \varphi_c$ (c), and of the tesselation of InC into Voronoi polyhedra (d)

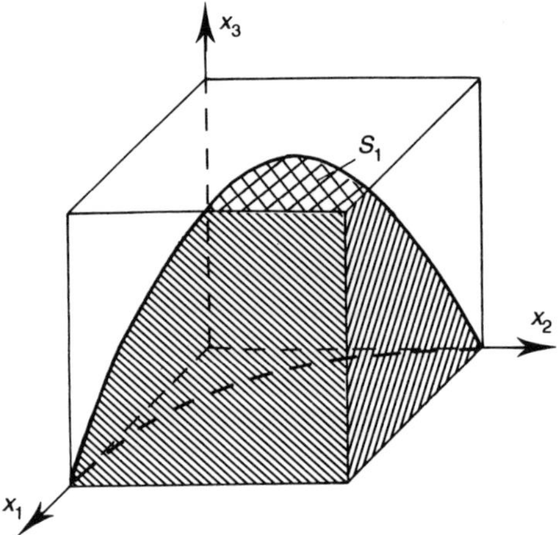

Fig. 4.10. Portion of a sphere embedded into a cube

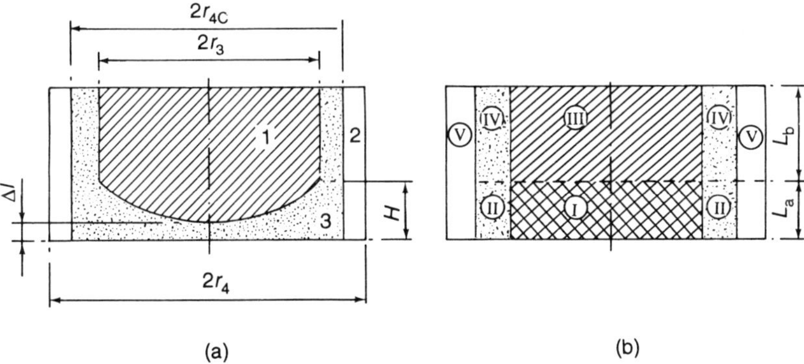

(a)   (b)

Fig. 4.11. Model of an equivalent element (a) and schematic of its partitioning (b).
1—filler particle; 2—polymer phase; 3—boundary interphase

The effective properties of such a BRE will be determined for a simplified
case of a spherical particle embedded into a cube assuming the distance between
the opposite faces of the latter smaller than the particle diameter $2r$ (Fig. 4.10).
In this case, one should differentiate between the interparticle gap, Δl (cf. Chapter
1) and the BI thickness, Δr; in fact, there might be situations when either $\Delta l = \Delta r$
(presumably, in the case of strong interfacial interactions when all continuous
phases available are transformed into a BI), or $\Delta r = 0$ and $\Delta l \neq 0$ (in the other
extreme case of very weak interfacial interactions).

The effective conductivity of the BRE (Fig. 4.11) will be estimated as follows.
Having applied the voltage to the opposite faces of the cube, the resulting
current will comprise three contributions from the currents running through
the particle (contact area S_1), through the matrix and the particle connected
in series (cross-sectional area S_{12}) and through the matrix (cross-sectional
area S_2).

Since the three areas, S_1, S_{12} and S_2, specify the interparticle contacts in a
polyhedron with any number of faces, it is convenient to model the contact
region by a cylindrical body (i.e. the inner current channel) which is, in fact,
the EE [1, 57, 58, 61, 62].

The EE appropriate for polymeric MHM is shown in Fig. 4.11. Section I
comprises a spherical segment of component 1 inscribed into a cylinder and a
gap of thickness Δl; thus, its height and basal area may be defined as $h_1 = r + \Delta l - (r - r_3^2)^{1/2}$ and $S_1 = \pi r_3^2$, respectively. Similar quantities for the remaining
sections, as well as other pertinent geometrical parameters of an EE, are defined
below:

$$h_{II} = h_I, \qquad S_{II} = \pi(r_4^2 - r_3^2);$$
$$h_{III} = (r^2 - r_3^2)^{1/2}, \qquad S_{III} = S_I;$$
$$h_{IV} = h_{III}, \qquad S_{IV} = S_{II};$$

$$h_V = h_1 + h_{II}, \qquad S_V = \pi r_4^2.$$

$$h' = h/r = 1 + \Delta l' - (1 - y_3^2)^{1/2}; \qquad y_3 = r_3/r = 2(N_c - 1)^{1/2}/N_c;$$

$$y_4 = r_4/r = y_3/\varphi^{*1/3}; \qquad y_{4c} = r_{4c}/r = y_3/(1 - v_{BI})^{1/3}; \qquad \Delta l' = \Delta l/r$$

$$v_{BI} = \varphi^*[(1 + \Delta r/r)^3 - 1].$$

The SSA technique will be now applied to evaluate the contributions to the effective conductivity of an EE from the following structures:

(a) section $I \Rightarrow \lambda_a$;
(b) sections $(I + II) \Rightarrow \lambda_b$;
(c) sections $(III + IV) \Rightarrow \lambda_c$;
(d) sections $(I - II) + (I - II) + (III + IV) \Rightarrow \lambda_d$;
(e) total EE $[V + (I + II) + (III + IY)] \Rightarrow \lambda$.

The effective conductivity of section I will be determined by eq. (4.97):

$$\lambda_a = S_1^{-1} \int_{(S_1)} \lambda(x_1, x_2) dx_1\, dx_2 \tag{4.97}$$

which after transformations to polar coordinates and to those normalized by the mean radius of inclusions r will be replaced by eq. (4.97a):

$$\lambda_a = (\pi y_3)^{-1} \int_0^{2\pi} d\theta \int_0^{y_3} \lambda(\rho)\rho\, d\rho \tag{4.97a}$$

where $\lambda(\rho)$ is the conductivity of the differential cylindrical element (DCE) of section I, which may be evaluated using eq. (4.83) for the lower bound, i.e.

$$\lambda(\rho) = \{[\lambda^{-1}]_{L_1(\rho)}^{1,BI}\}^{-1} \tag{4.98}$$

where the superscripts 1, BI refer to components (component 1 and BI, respectively) over which the averaging is performed, and the subscript $L_1(\rho)$ defines the geometry of the averaging procedure.

The expanded form of eq. (4.98) is

$$\lambda(\rho) = \{L_1(\rho)/\lambda_1 + [1 - L_1(\rho)]/\lambda_{BI}\}^{-1} \tag{4.99}$$

where λ_{BI} is the conductivity of the BI, $L_1'(\rho) = [(1 - y_3^2)^{1/2} - z_2]/(z_1 - z_2)$ is the length of the DCE normalized by the total length of section I, $z_1 = 1 + \Delta l'$ and $z_2 = (1 - y_3)^{1/2}$.

Substitution of eq. (4.99) into eq. (4.97a) yields

$$\lambda_a = B[\lambda_{BI}(z_1 - z_2)/y_3^2 \tag{4.100}$$

where

$$B = \frac{2}{(1 - a_{BI})^2} \left\{ [1 + \Delta l' - a_{BI}(1 - y_3^2)^{1/2}] \ln \frac{1 + \Delta l' - (1 - z_1^2)^{1/2} + a_{BI}X}{\Delta l' + a_{BI}[1 - (1 - y_3^2)^{1/2}]} \right.$$

$$\left. - [1 - (1 - z_1^2)^{1/2}](1 - a_{BI}) \right\}$$

$$X = (1 - z_1^2)^{1/2} - (1 - y_3^2)^{1/2}$$

$$a_{BI} = \lambda_{BI}/\lambda_1$$

Averaging the conductivity over the sections (b), (c) and (d) yields, respectively

$$\lambda_b = \{\lambda\}_{\langle S_I \rangle}^{(a,BI)} = \lambda_a \langle S_I \rangle + \lambda_{BI}(1 - \langle S_I \rangle) \tag{4.101}$$

$$\lambda_c = \{\lambda\}_{\langle S_I \rangle}^{(1,BI)} = \lambda_1 \langle S_I \rangle + \lambda_{BI}(1 - \langle S_I \rangle) \tag{4.102}$$

$$\lambda_d = \{\lambda^{-1}\}_{L_a}^{(b,c)} = [L_a/\lambda_b + (1 - L_a)/\lambda_c]^{-1} \tag{4.103}$$

where $\langle S_I \rangle = (r_3/r_{4c})^2$ is the normalized cross-sectional area of section I.

Finally, averaging over sections V and (d) yields the effective conductivity of EE:

$$\lambda_{EE} = \{\lambda\}_{\langle S_{II} \rangle}^{(V,d)} = \lambda_V \langle S_{II} \rangle + \lambda_d(1 - \langle S_{II} \rangle) \tag{4.104}$$

where $\langle S_{II} \rangle = (r_{4c}/r_4)^2$ is the normalized cross-sectional area of the region b + c.

4.3 TESTS OF THEORETICAL PREDICTIONS

4.3.1 ANALYTICAL TEST

The influence of the structural parameters involved in the EE model on the heat conductivity of a MHM will be assessed now from numerical calculations of the following representative cases (see below).

(i) Fixed parameters: $\lambda_2 = 0.2$ W/m.K, $\varphi_c = 0.15$, $\varphi^* = 0.6$, $t = 1.8$, $\Delta l/r = 0.01$; variable parameters: $\lambda_1, \lambda_{BI}/\lambda_2$.

A common feature of all theoretical plots (Fig. 4.12) is the onset of an accelerated rise of λ at φ_c, which becomes more prominent the higher is the ratio λ_{BI}/λ_2, and the occurrence of another relatively weak discontinuity located symmetrically at $1 - \varphi_c$. Apparently, the pattern of the composition dependence of λ is affected more by the ratio λ_{BI}/λ_2 than by λ_1/λ_2 (e.g. a negative slope is invariable observed whatever the latter ratio if $\lambda_{BI}/\lambda_2 < 1$).

(ii) Fixed parameters: $\lambda_2 = 0.2$ W/m.K, $\lambda_1 = 2$ W/m.K, $\varphi^* = 0.6$, $\varphi_c = 0.15$, $t = 1.8$; variable parameters: $\lambda_{BI}/\lambda_2, \Delta l/r$.

The discontinuity at $1 - \varphi_c$ becomes sharper and the slope becomes higher the smaller is the ratio $\Delta l/r$ (Fig. 4.13). As before, the larger is the λ_{BI}/λ_2 ratio, the higher becomes the slope.

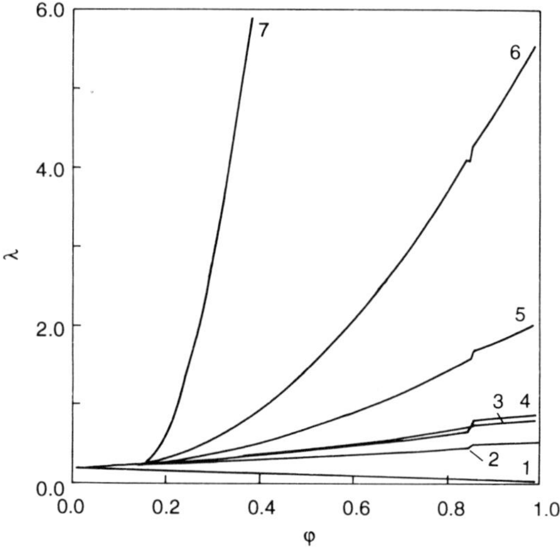

Fig. 4.12. Composition dependence of the heat conductivity of MHM for case (i) calculated assuming $\lambda_1 = 2\text{–}200$ and $\lambda_{\mathrm{BI}}/\lambda_2 = 0.1$ (1); $\lambda_1 = 2, 20, 200$ and $\lambda_{\mathrm{BI}}/\lambda_2 = 1$ (2, 3, 4); $\lambda_1 = 2, 20, 200$ and $\lambda_{\mathrm{BI}}/\lambda_2 = 10$ (5, 6, 7)

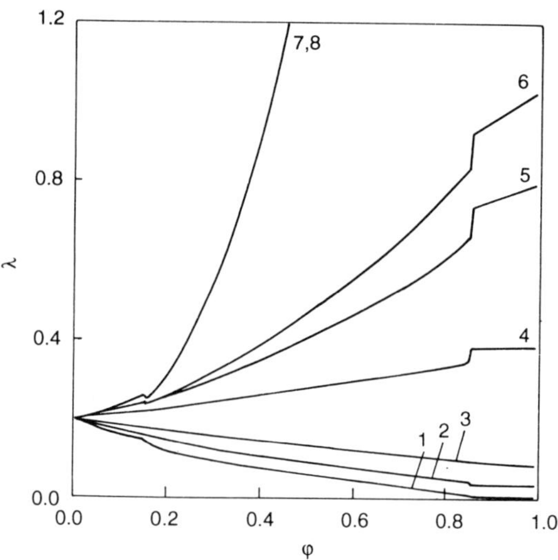

Fig. 4.13. Composition dependence of the heat conductivity of MHM for case (ii) calculated assuming $\lambda_{\mathrm{BI}}/\lambda_2 = 0.1$ and $\Delta l/r = 10^{-1}$ (1), 10^{-2} (2), 10^{-3} (3); $\lambda_{\mathrm{BI}}/\lambda_2 = 1$ and $\Delta l/r = 10^{-1}$ (4), 10^{-2} (5), 10^{-3} (6); $\lambda_{\mathrm{BI}}/\lambda_2 = 10$ and $\Delta l/r = 10^{-1}\text{–}10^{-5}$ (7, 8)

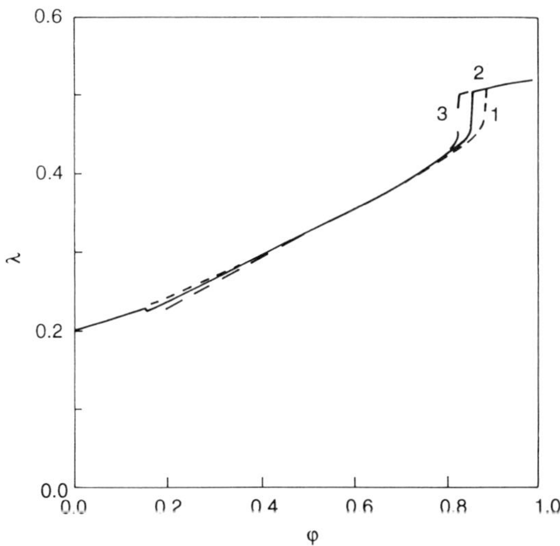

Fig. 4.14. Composition dependence of the heat conductivity of a MHM for case (iii) calculated assuming $\varphi_c - 0.12$ (1), 0.15 (2) and 0.18 (3)

(iii) Fixed parameters: $\lambda_2 = 0.2\,\text{W/m.K}$, $\lambda_1 = 2\,\text{W/m/K}$, $\lambda_{\text{BI}}/\lambda_2 = 1$, $\varphi_c = 0.15$, $\Delta l/r = 0.01$, $t = 1.8$; variable parameter: φ^*.

Both the slope and the jump at $1 - \varphi_c$ tend to increase the lower is φ^* (Fig. 4.14).

(iv) Fixed parameters: $\lambda_2 - 0.2\,\text{W/m/K}$, $\lambda_1 - 2\,\text{W/m.K}$, $\lambda_{\text{BI}}/\lambda_2 - 1$, $\varphi^* - 0.6$, $\Delta l/r = 0.01$, $t = 1.8$; variable parameter: φ_c.

As might have been expected, the general pattern of the theoretical curves remained essentially unchanged, except for the location of discontinuities at φ_c and at $1 - \varphi_c$ (Fig. 4.15).

Summarizing, the most crucial influence on the pattern of λ vs. φ plots appears to be exerted by the ratios $\lambda_{\text{BI}}/\lambda_2$ and $\Delta l/r$. The point of concern is, however, the failure of most of the theoretical curves to extrapolate to λ_1 as φ approaches unity above $1 - \varphi_c$. The origin of this discrepancy, of course, stems from the very nature of the adopted structural model which explicitly assumes the existence of a BI with altered heat conductivity around filler particles in the entire interval from $\varphi \geqslant 0$ to $\varphi \leqslant 1.0$. In other words, the hypothetical, ultimate state of a disperse phase at $\varphi \Rightarrow 1.0$ is the mosaic-like structure of an InC composed of particles coated with a BI at the highest packing density φ^*. This picture does seem physically reasonable since there is no realistic mechanism for a disperse phase above $1 - \varphi_c$ to be converted into a monolithic, void-free body with intrinsic conductivity λ_1. Hence, it is no surprise that the only case when the heat conductivity of MHM, λ, is predicted to increase from the lower theoretical limit $(\lambda_2 = 0.2\,\text{W/m.K}$ at $\varphi = 0)$ to the upper theoretical limit $(\lambda_1 = 2.0\,\text{W/m.K}$ at $\varphi = 1.0)$ corresponds to $\lambda_{\text{BI}}/\lambda_2 = \lambda_1/\lambda_2$ (curve 5 in Fig. 4.12).

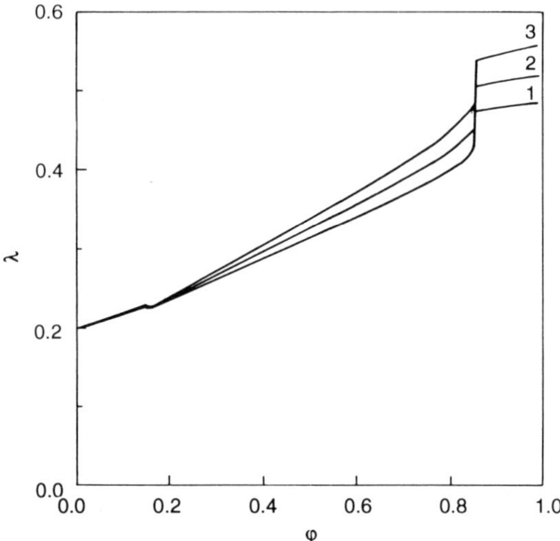

Fig. 4.15. Composition dependence of the heat conductivity of a MHM for case (iv) calculated assuming $\varphi^* = 0.4$ (1), 0.6 (2) and 0.8 (3)

The other important observation is that the ultimate value of λ at $\varphi = 1.0$ becomes closer to λ_1, the lower is the $\Delta l/r$ ratio (cf. curves 4, 5 and 6 in Fig. 4.13). It is thus likely that the fitting parameters Δl and λ_{BI} are, in fact, interrelated rather than mutually independent.

It should be recognized, however, that in view of the obvious experimental limitations the pattern of λ evolution in the composition gap between $\varphi = 1 - \varphi_c$ and $\varphi = 1$ seems to have little practical importance. Therefore, $1 - \varphi_c$ may be regarded as a natural upper composition limit for the quantitative applicability of the SSA technique.

4.3.2   EXPERIMENTAL TEST: FILLED POLYMERS

The heat conductivity of composite materials is notorious for its extreme sensitivity to even the smallest structural defects like binder-free voids between disperse particles which may occur, say, in samples prepared by mechanical mixing and subsequent hot-pressing of granulated polymers and inorganic powders [63, 64]. Thus, theoretical predictions should be tested only on those samples that where prepared in conditions ensuring the elimination of such hazards.

It turned out, however, that in the majority of documented cases (e.g. [65–69]) the discrepancy between the experimental data obtained in routine heat conductivity measurements and the predictions of 'pragmatic' approaches [e.g. eqs. (4.19), (4.20), etc.] increased the higher was the filler content φ, even for apparently defect-free samples. As a first guess, this discrepancy might have originated from the neglect of the 'particle crowding' effect at higher φ values,

which could, however, be explicitly accounted for semi-empirically by the
'physical' equation, eq. (4.105) [70]:

$$\lambda/\lambda_2 = \frac{1 + AB\varphi}{1 - B\psi\varphi} \tag{4.105}$$

where $B = (\lambda_2/\lambda_1 - 1)/(\lambda_2/\lambda_1 + A)$, $\psi = 1 + [(1 - \varphi^*)/\varphi^{*2}]\varphi$ and A is the appropriate geometrical factor. In fact, the quality of the data fit to eq. (4.105)
compared with eqs. (4.19) or (4.20) could be considerably improved by a proper
selection of the 'best' values of the parameters A and φ^* (maximum packing
fraction of the filler) [70–72].

More extensive tests of predictions of different rigorous theories by several
sets of experimental heat conductivity data obtained over broad intervals of
compositions, temperatures and/or pressures [47, 73] will be illustrated below.

Filled Cross-linked Epoxies [74, 75]

The test specimens were prepared by adding the desired amount of filler (powders
of non metals [74] and metals [75] with particles sizes between 20 and 150 μm)

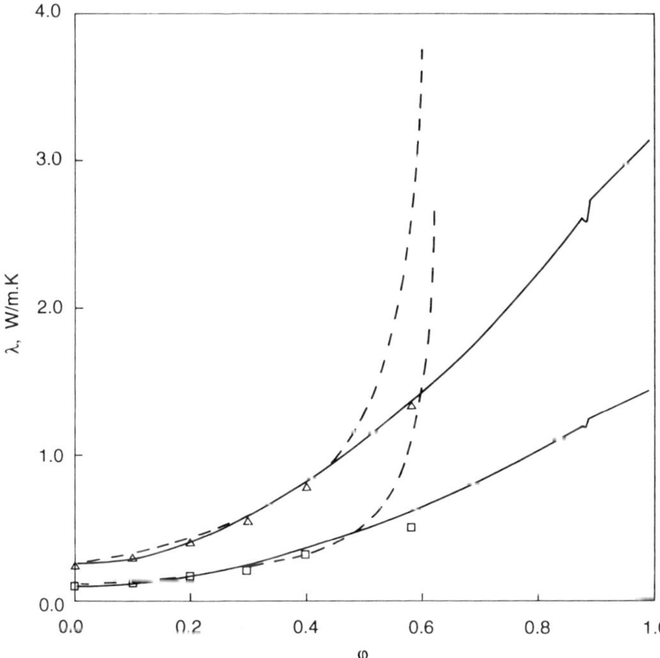

Fig. 4.16. Composition dependence of the heat conductivity of the copper-filled epoxy
at 20 K ($\square$) and 300 K ($\triangle$). Broken curves were calculated by eq. (4.23), solid curves by
the SSA model assuming $\lambda_2 = 0.1$ W/m·K, $\lambda_1 = 5200$ W/m·K, $\varphi_c = 0.12$, $\varphi^* = 0.3$, $\Delta l/r =$
0.017, $t = 1.6$, $\lambda_{BI}/\lambda_2 = 3.5$ (20 K) and $\lambda_2 = 0.25$ W/m·K, $\lambda_1 = 395$ W/m·K, $\varphi_c = 0.12$,
$\varphi^* = 0.55$, $\Delta l/r = 0.01$, $t = 1.5$, $\lambda_{BI}/\lambda_2 = 2.5$ (300 K)

to the base epoxy resin/hardener formulation which was subsequently cured at 373 K (1 h) and post-cured at 453 K (2 h). The heat conductivity of the bar-shaped and disc-shaped specimens was measured in the temperature intervals 2–20 K and 20–300 K, respectively.

In the temperature interval below 10 K the acoustic mismatch of heat-conducting phonons at the polymer–filler interfaces manifested itself as $\lambda < \lambda_2$, whereas at higher temperatures the 'normal' increase in $\lambda > \lambda_2$ with φ was observed. The composition dependence of λ at all temperatures above 10 K was claimed to be adequately represented by eq. (4.23) [74, 75], although the agreement with SSA theory was apparently better, especially at the highest filler loadings (cf. the broken and the solid lines, respectively, in Fig. 4.16).

Polyethylene (PE)/NaCl [76]

Cylindrical specimens with filler volume content from $\varphi = 0$ (pure PE) to $\varphi = 1.0$ (pure NaCl) were prepared by a thorough mechanical mixing of components (both were powders with an average particle size of 200–300 µm) and subsequent compacting of mixtures at room temperature (to ensure good thermal contact between the components and also to minimize the hazards of residual voids, the lower limit of the compacting pressure was far in excess of the indentation hardness of both PE and NaCl). The heat conductivity was measured (estimated experimental inaccuracy about 4%) by the transient hot-wire method [77] in the temperature interval 120–320 K and in the pressure range 0.2–1.75 GPa.

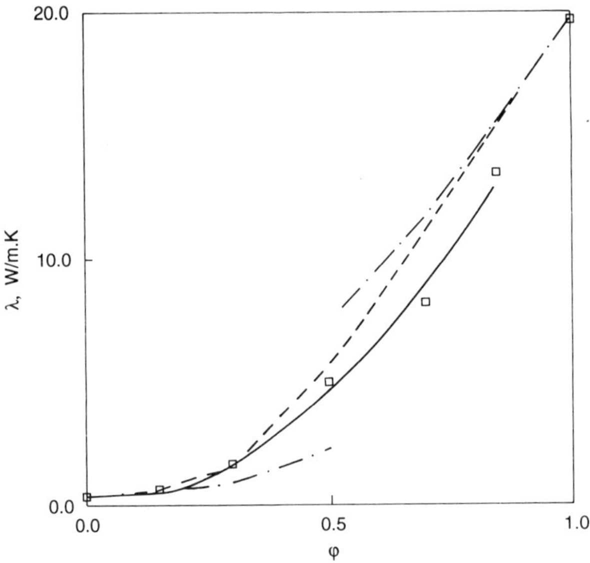

Fig. 4.17. Effective heat conductivity of PE/NaCl at 120 K and 0.2 GPa and predictions of SSA model (solid line), eq. (4.17) (dash- dot line) and eq. (4.18) (broken line)

Table 4.4 Parameters of the SSA model for filled PE (heat conductivity in [W/m K])

Filler	λ_1	λ_2	λ_{BI}/λ^2	t	φ_c	φ^*	Δr
NaCl							
$P = 0.2\,\text{GPa}$							
$\quad T = 120\,\text{K}$	0.371	19.61	39	1.6	0.15	0.49	0.01
$\quad T = 320\,\text{K}$	0.393	5.76	12.5	1.6	0.15	0.43	0.01
$P = 1.73\,\text{GPa}$							
$\quad T = 120\,\text{K}$	0.582	26.54	37	1.6	0.15	0.49	0.01
$\quad T = 320\,\text{K}$	0.725	7.95	10	1.6	0.12	0.70	0.01
AgCl							
$P = 0.11\,\text{GPa}$							
$\quad T = 120\,\text{K}$	0.353	2.26	5.7	1.6	0.12	0.54	0.01
$\quad T = 320\,\text{K}$	0.364	0.84	2.2	2.0	0.12	0.39	0.01
$P = 1.75\,\text{GPa}$							
$\quad T = 120\,\text{K}$	0.582	3.12	5.6	1.6	0.12	0.50	0.10
$\quad T = 320\,\text{K}$	0.725	1.12	1.5	1.9	0.12	0.33	$<10^{-3}$
Quartz (filler size 2r in microns)							
$\quad 2r = 16$	0.300	9.6	4.5	2.0	0.13	0.49	0.01
$\quad 2r = 156$	0.300	5.9	6.0	1.9	0.17	0.53	0.01
Al_2O_3	0.300	8.0	13	2.0	0.17	0.53	0.01
Graphite	0.300	209	10.5	1.6	0.12	0.50	0.01

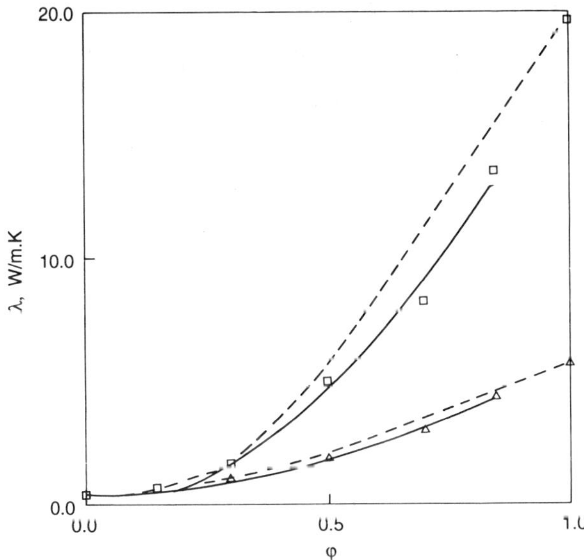

Fig. 4.18. Effective heat conductivity of PE/NaCl at 0.2 GPa and 120 K ($\square$) and 320 K ($\triangle$) and predictions of the SSA model (solid lines) and eq. (4.18) (broken lines)

As can be seen from a typical isothermal–isobaric plots of λ vs. φ (Fig. 4.17), the quality of fit between the experimental data and the theoretical curve constructed according to the SSA technique with the set of parameters listed in the Table 4.4 (solid line) looks better compared with the predictions of eqs. (4.94) and (4.105) (broken and broken-dotted lines, respectively). Essentially similar results were obtained at other pressures and/or temperatures (Fig. 4.18 may serve as a representative example).

In view of the different dependencies of heat conductivities of both polymer and filler on temperature and pressure, a more stringent test of theoretical predictions would provide either a semi-log plot of the ratio λ/λ_2 vs. φ (Fig. 4.19) or, rather, a double-log plot of λ/λ_2 vs. λ_1/λ_2 (Fig. 4.20). Again, the experimental data seem to be in better agreement with the predictions of the SSA approach (solid lines) than with those of either the EMA approach, eq. (4.94), or the RSRG approximation (broken and broken-dotted lines, respectively).

PE/AgCl [78]

The instrumentation for heat conductivity measurements as well as the method of specimen preparation were essentially the same as in [76], except for a slightly broader temperature interval of measurements (100–300 K) and a lower initial pressure (0.1 GPa).

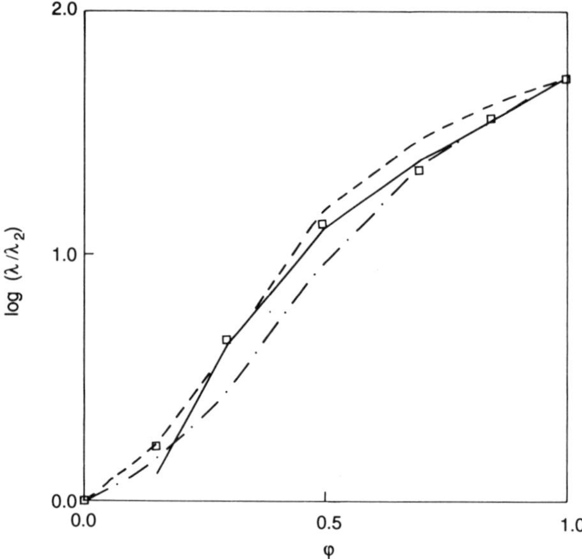

Fig. 4.19. Composition dependence of the ratio λ/λ_2 of PE/NaCl at $\lambda_1/\lambda_2 = 52$ ($\square$) and predictions of SSA (solid line), EMA (broken line) and RSRG (dash-dot line)

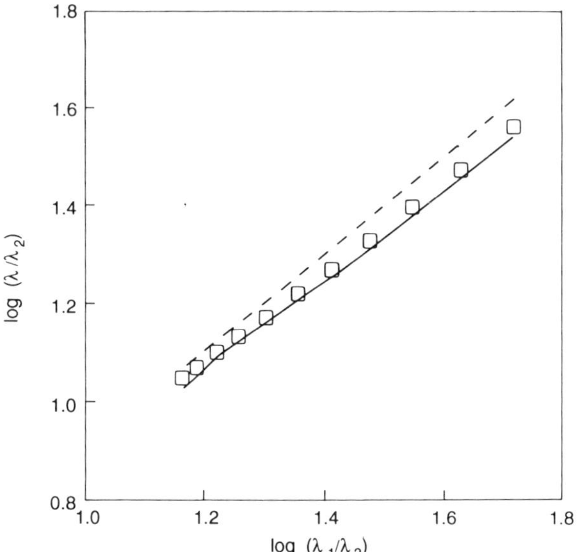

Fig. 4.20. Dependence of λ/λ_2 on λ_1/λ_2 for PE/NaCl at $\varphi = 0.85$ ($\square$), and predictions of the SSA (solid line) and EMA (broken line) approaches

As was the case with PE/NaCl, application of the SSA technique with the set of parameters listed in the Table 4.4 proved superior to the other approaches.

PE/Quartz, PE/Al$_2$O$_3$, PE/Graphite [79, 80]

Disc-shaped specimens with filler volume content up to 0.8 were prepared either by melt-casting (MC) or by compression molding (CM) of binary mixtures of powdery components (PE with an average particle size of $10\,\mu m$, two quartz fractions with average particle sizes of 16 and $156\,\mu m$, two fractions of Al$_2$O$_3$ with particle sizes of 9 and $65\,\mu m$ and graphite with particles of unspecified size). The heat conductivity was measured at 323 K and normal pressure. The erratic behavior of λ for MC specimens above a (particle size-dependent) 'critical' filler content indicated the onset of the formation of macrovoids, while such effects were not observed in the case of CM specimens for which the values of λ increased smoothly with φ.

Judging by the representative data in Figs. 4.21 and 4.22, the theoretical curves constructed according to the SSA approach with fitting parameters shown in Table 4.4 (solid lines) adequately describe the experimental data for CM specimens up to the highest filler content, the quality of the fit being at least comparable with that observed for the following semi-empirical relationship [79] (broken lines):

$$\log \lambda = \varphi C_1 \log \lambda_1 + (1 - \varphi)C_2 \log \lambda_2 \tag{4.106}$$

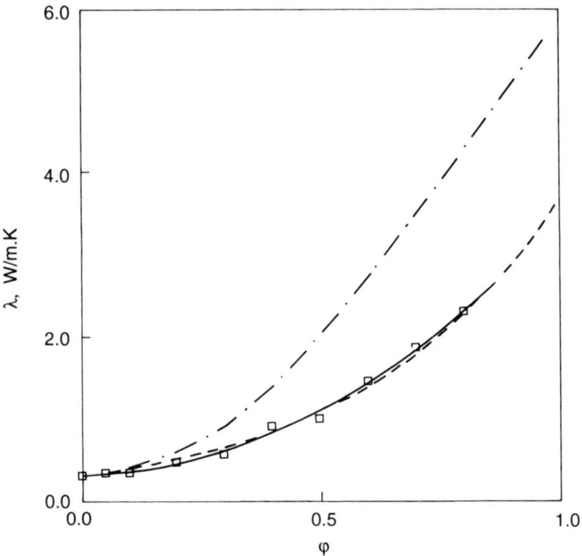

Fig. 4.21. Effective heat conductivity of PE/quartz ($\square$), and predictions of the SSA (solid line), EMA (dash-dot line) approaches and eq. (4.106) (broken line)

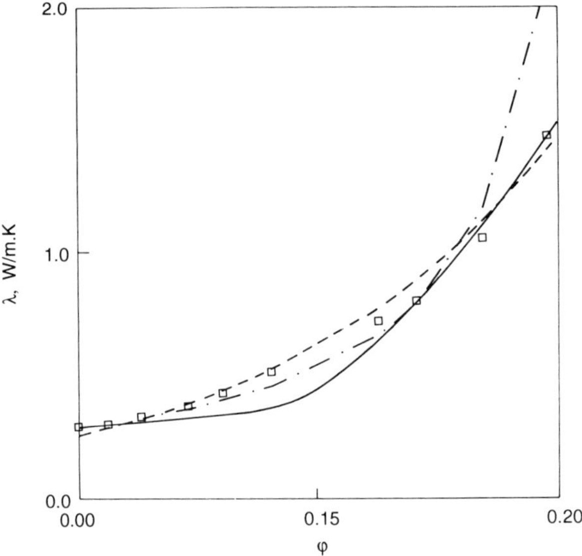

Fig. 4.22. Effective heat conductivity of PE/graphite ($\square$) and predictions of the SSA (solid line), EMA (dash-dot line) and eq. (4.106) (broken line)

where C_1 and C_2 are the fitting factors claimed to account for the probability of either the aggregation of filler particles into conductive chains or morphological changes in the polymer phase in filled samples, respectively (the broken lines in Figs. 4.21 and 4.22 were constructed by eq. (4.106) assuming $1.0 < C_1 < 1.5$ and $C_2 \cong 1.0$ [79]).

4.3.3 EXPERIMENTAL TEST: POLYMER BLENDS

Experimental data are scarce; moreover, the composition dependence is not always readily amenable to a meaningful theoretical analysis since the heat conductivities of the components are of the same order of magnitude (especially when both components are non-crystalline). As a typical example, the heat conductivity in the melt state (i.e. in the temperature interval 400–490 K) of a melt-blended polystyrene (PS)/polycarbonate (PC) pair, estimated by substituting the experimental values of thermal diffusivity a, specific volume v and heat capacity C_p into the standard relationship $\lambda = C_p a/v$, decreased approximately linearly with composition from $\lambda_2 = 0.32\,\mathrm{W/m.K}$ for PC to $\lambda_1 = 0.25\,\mathrm{W/m.K}$ for PS, as might have been expected for incompatible components [81,82]. However, the negative deviations from linear additivity observed at low content

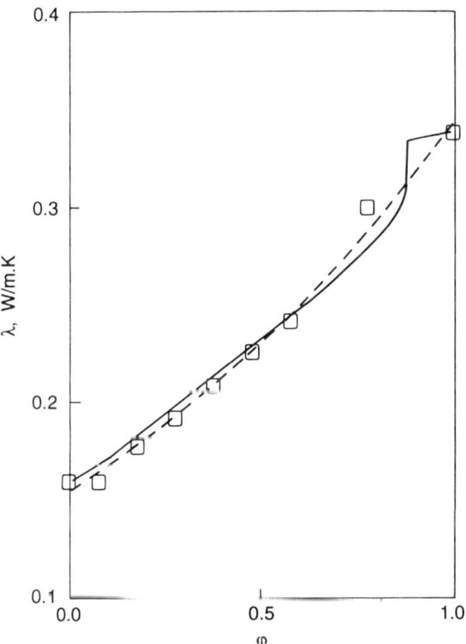

Fig. 4.23. Composition dependence of the heat conductivity of PE/PS blend ($\square$) and predictions of eq. (4.106) (broken line) and of the SSA model (solid line)

of either component were attributed to a limited mutual solubility of PC and PS.

Another example of an incompatible polymer pair is provided by melt-blended specimens of oligomeric PE ($\lambda_2 = 0.34\,\mathrm{W/m.K}$) and PS ($\lambda_1 = 0.16\,\mathrm{W/m.K}$) containing a fixed small amount of a compatibilizing additive SEBS (PS/polybutadiene/PS triblock copolymer) [83]. The room-temperature heat conductivity vs. composition plots (Fig. 4.23) could be empirically represented by eq. (4.106) with two sets of parameters for PE content below and above the apparent 'phase inversion' point (broken line in Fig. 4.23). In contrast, only a single set of parameters ($\Delta l/r = 0.017$, $\varphi^* = 0.15$, $\varphi_c = 0.4$, $t = 1.8$, $\lambda_{\mathrm{BI}}/\lambda_2 = 2.5$) was required to fit the same data to SSA theory (solid line in Fig. 4.23). The same conclusion could be made after a similar treatment of the data on heat conductivity of PE/SEBS and PS/SEBS blends [84] (the best fit parameters from the SSA model were $\varphi_c = 0.4$ and 0.9, $\lambda_{\mathrm{BI}}/\lambda_2 = 0.55$ and 1.3 for the former and for the latter systems, respectively, and $\Delta l/r = 0.01$, $t = 1.8$ and $\varphi^* = 0.15$ for both).

4.4 CONCLUSIONS

As follows from the detailed analysis of the current theories of heat conductivity in MHM, the 'pragmatic' approaches neglecting the structural features of the latter are in approximate agreement with the experimental data at fairly small content of the disperse component, whereas the 'physical' approaches, which account explicitly for the existence of a BI and/or the occurrence of percolation phenomena, are applicable over the entire composition range. Among the latter approaches the SSA technique was shown to describe quantitatively the experimental data on the composition dependence of the heat conductivity of various polymer materials over a much broader composition interval (i.e. from φ_c to $1 - \varphi_c$) compared with the EMA or RSRG methods. Moreover, the critical exponent for conductivity, t, which ensured the optimum fit of the SSA predictions to the experimental data (Table 4.4), was also reasonably close to the corresponding theoretical estimates. Perhaps the quality of the data fit to the theoretical predictions might have been improved even further, provided φ_c and φ^* were treated as fitting parameters rather than fixed, universal constants. In fact, there is experimental evidence suggesting that φ_c changes with polymer–filler interfacial energy and/or polymer melt viscosity [84, 85], whereas φ^* depends on particle shape [72]. Although, as follows from model calculations (Figs. 4.12 and 4.13), the exact value of either of these two parameters appears to be of secondary importance, the product $\varphi_c \varphi^*$ is of value as a theoretical measure of the minimum filler content ($\varphi_{\min}$) for the onset of an InC in real MHM. In this respect, $\varphi_{\min} = \varphi_c \varphi^*$ of the SSA approach is conceptually similar to the critical volume fraction of a conducting polymer, $\varphi_{\min} = \varphi_c \varphi_c = \varphi_c^2$, invoked to account for the 'double percolation' phenomenon in phase-separated, binary polymer blends [86].

It should be remarked at this point that the basic assumption of structural 'symmetricity' of each component near the corresponding percolation threshold (i.e. of a topological equivalence of component 1 at φ_c and of component 2 at $1 - \varphi_c$) implicit both in the SSA technique and in the conceptually similar RSRG approach, may well be an oversimplification of the real situation [56]. The other point of concern is the ratio λ_{BI}/λ_2 derived from curve-fitting procedures for filled polymers, which sometimes may be of the order of λ_1/λ_2 itself. Taking into consideration that the heat conductivity of bulk, isotropic polymers is orders of magnitude smaller than the intrachain contribution $\lambda_\parallel$, but comparable with the interchain contribution $\lambda_\perp$ [87,88], the observed inequality, $\lambda_{BI} \gg \lambda_2$, would imply that the structure of a BI is radically different from the unperturbed mother polymer phase. This hypothetical, higher-conductivity structure of the BI might have involved chain orientation, somehow providing for the increased contribution from $\lambda_\parallel$ [89,90], and/or better interchain packing favoring heat conductivity in exactly the same way as does densification under high pressure [91,92]. Lacking a quantitative, physically realistic model of the BI, it is preferable at the present stage, however, to regard both Δl and λ_{BI} as fitting variables of the SSA approach, rather than real 'material' parameters of each particular two-component polymer system [73].

REFERENCES

1. Dulnev G. N. and Zarichnyak Yu. P. (1974) *Thermal Conductivity of Mixtures and Composites*, Energhia, Leningrad (in Russian).
2. Misnar A. (1968) *Thermal Conductivity of Solids, Liquids, Gases and Their Mixtures*, Mir, Moscow (in Russian).
3. Shermergor T. D. (1977) *Theory of Elasticity of Microheterogeneous Media*, Nauka, Moscow (in Russian).
4. Khoroshun L. P. and Maslov B. P. (1980) *Computer Techniques for Calculation of the Physical Mechanical Properties of Composites*, Naukova Dumka, Kiev (in Russian).
5. Hashin Z. (1983) Analysis of composite materials, *J. Appl. Mech.*, **50**, 481–505.
6. Maxwell J. C. (1873) *Treatise on Electricity and Magnetism*, Oxford University Press, Oxford.
7. Lorenz L. (1880) Ueber die Refraktion Konstante, *Wied. Ann.*, **11**, 70–75.
8. Lorentz H. A. (1880) Ueber die Beziehung zwischen der Fortpflanzungsgeschwindigkeit des Lichtes und der Koerperdichte, *Wied. Ann.*, **9**, 641–648.
9. Malyshev V. A. and Malyshev P. V. (1987) Estimation of the effective dielectric permittivity of a periodic array of solid bodies, *Dokl. Acad. Nauk UkrSSR*, Ser. A, **12**, 48–53.
10. Rayleigh L. W. (1882) On the influence of obstacles arranged in rectangular order upon the properties of a medium, *Phil. Mag.*, **34**, 19–22.
11. Meredith R. E. and Tobias C. W. (1960) Resistance to potential flow through cubical array of spheres, *J. Appl. Phys.*, **31**, 1270–1273.
12. Bruggeman D. A. G. (1935) Berechnung verschiedener physikalischer Konstanten von heterogenen Substanzen, *Ann. Phys.*, **24**, 636–650.

13. Odelevsky V. I. (1951) Calculation of the effective conductivity of heterogeneous systems, *Zhurn. Tekh. Fiz.*, **21**, 667–685.
14. Kondorsky E. K. (1951) On the theory of coercetive force and magnetic susceptibility of ferromagnetic powders, *Dokl. Acad. Nauk USSR*, **80**, 197–200.
15. Weinberg A. K. (1966) Magnetic susceptibility, electric conductivity, dielectric permittivity and heat conductivity of media with spherical and ellipsoidal inclusions, *Dokl. Acad. Nauk USSR*, **169**, 543–547.
16. Fricke H. A. (1924) A mathematical treatment of the electric conductivity of disperse systems, *Phys. Rev.*, **24**, 12–15.
17. Webman I., Jortner Z. and Cohen M. N. (1976) Electronic transport in alkalitungsten bronzes, *Phys. Rev.*, **13**, 213–224.
18. Webman I., Jortner Z. and Cohen M. N. (1975) Numerical simulation of electrical conductivity in microscopically inhomogeneous materials, *Phys. Rev.*, Ser. B, **11**, 2885–2892.
19. Buyevich Yu. A., Korneev Yu. A. and Shelikova I. N. (1975) On the heat and mass transfer in a disperse flux, *Inzh. Fiz. Zhurn.*, **30**, 979–982.
20. Buyevich Yu. A. and Korneev Yu. A. (1976) Effective heat conductivity of a disperse medium at low Pekle numbers, *Inzh. Fiz. Zhurn.*, **31**, 607–612.
21. Esksler B. S. (1979) On the effective heat conductivity and viscosity of disperse media, *Inzh. Fiz. Zhurn.*, **37**, 110–117.
22. Buyevich Yu. A. (1973) On the effective heat conductivity of grainy materials, *Zhurn. Prikl. Matem. Tekhn. Fiz.*, **4**, 57–66.
23. R. H. Davis (1986) The effective thermal conductivity of a composite material with spherical inclusions, *Int. J. Thermophys.*, **7**, 609–620.
24. Lichtenecker K. and Rother K. (1931) Die Herleitung des logarithmischen Mischungsgesetzes des allgemeinen Prinzipien der stationaren Stroemung, *Phys. Z.*, **32**, 3255–3267.
25. Herring C. (1960) Inhomogeneities in electrical and galvanometric measurements, *J. Appl. Phys.*, **31**, 1939–1953.
26. Kudinov V. A. and Moizhes B. Ya. (1972) Effective conductivity of non-homogeneous, isotropic media, *Zhurn. Tekhn. Fiz.*, **42**, 591–599.
27. Hashin Z. and Shtrikman S. A. (1962) A variational approach to the theory of the effective magnetic permeability of multi-phase materials, *J. Appl. Phys.*, **33**, 3125–3131.
28. Yermakov G. A., Fokin A. G. and Shermergor T. D. (1974) Calculation of the bounds for effective dielectric permittivities of non-homogeneous dielectrics, *Zhurn. Tekhn. Fiz.*, **44**, 249–254.
29. Rosen B. W. and Hashin Z. (1970) Effective thermal expansion coefficient and specific heat of composite materials, *Int. J. Eng. Sci.*, **8**, 157–161.
30. Dykhne A. M. (1967) On the calculation of kinetic coefficients of media with random inhomogeneities, *Zhurn. Tekhn. Fiz.*, **52**, 264–267.
31. Kazantsev V. P. (1979) Variational estimates of the effective conductivity of media with macroscopic inclusions, *Izv. VUZ'ov, Fiz.* **5**, 53–59.
32. Stepanov S. V. (1970) On the heat conductivity of two-phase systems, *Inzh. Fiz. Zhurn.*, **18**, 247–252.
33. Beran M. and Molyneux L. (1965) Boundaries of effective properties of binary heterogeneous disordered systems, *Quart. Appl. Math.*, **24**, 107–114.
34. Phan-Thien N. and Milton G. W. (1982) New bounds on the effective thermal conductivity of N-phase materials, *Proc. Roy. Soc. Ser, A*, **380**, 333–348.
35. Levin V. M. (1968) On the estimates of effective conductivity coefficients of multiphase materials, *Prikl. Mekh. Tekhn. Fiz.*, **2**, 52–55.
36. Novikov V. V. (1986) Two-bounds estimates of heat and electrical conductivity of microheterogeneous materials, *Inzh. Fiz. Zhurn.*, **50**, 866–867.

37. Hardi G. G., Littlewood D. E. and Polya G. (1948) *Inequalities*, Gosizdatinlit, Moscow (in Russian).
38. Kolmogorov A. N. and Fomin S. V. (1981) *Elements of the Theory of Functions and of Functional Analysis*, Nauka, Moscow (in Russian).
39. Uitrin A. I. and Belousov V. Ya. (1985) Probabilistic–statistical methods of calculation and optimization of structural parameters of microheterogeneous composite materials, *Poroshkov. Metallurghia*, **3**, 69–73.
40. Jackson J. L. (1968) Transport coefficients of composite materials, *J. Appl. Phys.*, **39**, 1733–1736.
41. Corriell S. R. and Jackson J. L. (1968) Bounds on transport coefficients of two-phase materials, *J. Appl. Phys.*, **39**, 2329–2334.
42. Hill R. (1965) A self-consistent mechanics of composite materials, *J. Mech. Phys. Solids*, **13**, 213–225.
43. Frey S. (1932) Ueber die elektrische Leitfaehigkeit binarer Aggregat, *Z. Elektrochem.*, **38**, 260–274.
44. Dulnev G. N. and Novikov V. V. (1977) Effective conductivity of systems with interpenetrating structures, *Inzh. Fiz. zhurn.*, **33**, 271–274.
45. Novikov V. V. (1983) The effective thermal expansion coefficient of non-homogeneous materials, *Inzh. Fiz. Zhurn.*, **44**, 969–977.
46. Novikov V. V. (1985) On the estimates of effective elasticity moduli of non-homogeneous materials, *Prikl. Mekh. Tekhn. Fiz.*, **5**, 146–153.
47. Privalko V. P., Novikov V. V. and Yanovsky Yu. G. (1991) *Principles of Thermophysics and Rheophysics of Polymer Materials*, Naukova Dumka, Kiev (in Russian).
48. Kirkpatrick S. (1973) Percolation and conduction, *Rev. Mod. Phys.*, **45**, 574–582.
49. Shklovsky V. I. and Efros A. L. (1979) *Electronic Properties of Semiconductor Alloys*, Nauka, Moscow (in Russian).
50. Stauffer d. (1985) *Introduction to Percolation Theory*, Taylor & Francis, London and Philadelphia.
51. Dulnev G. N. and Novikov V. V. (1979) Conductivity of non-homogeneous systems, *Inz. Fiz. Zhurn.*, **36**, 900–909.
52. Dulnev G. N. and Novikov V. V. (1983) Percolation theory and conductivity of non-homogeneous media. The basic model of a non-homogeneous medium, *Inzh. Fiz. Zhurn.*, **45**, 136–141.
53. Skal A. S. and Shklovsky v. I. (1974) Topology of an infinite cluster in the percolation theory and the theory of jump-like conductivity, *Fiz. Tekhn. Poluprovodnikov*, **8**, 1585–1592.
54. Vinogradov A. P. and Sarychev A. K. (1983) The structure of conducting channels and the metal–dielectric transition in composites, *Zhurn. Eksper. Teor. Fiz.*, **85**, 1144–1151.
55. Torquato S. (1987) Thermal conductivity of disordered heterogeneous media from the microstructure, *Rev. Chem. Eng.*, **4**, 151–204.
56. N. Shah and J. M. Ottino (1986) Effective transport properties of disordered, multi phase composites. Application of real-pace renormalization group theory, *Chem. Eng. Sci.*, **41**, 283–296.
57. Dulnev G. N. and Novikov V. V., (1979) On the conductivity of filled heterogeneous systems, *Inzh. Fiz. Zhurn.*, **37**, 657–661.
58. Novikov v. V. (1988) Influence of interphase layer and of structural parameters on the conductivity coefficient of polymer composites, *Komposits. Polim. Mater.*, **38**, 17–23.
59. Guyon E. (1982) Percolation et matiere en grains, *Compt. Rend.*, **294**, 115–130.
60. Ziman J. M. (1979) *Models of Disorder*, Cambridge University Press, London–New York–Melbourne.

61. Novikov V. V. and Klimenko V. S. (1988) Heat conductivity of pseudoalloys, *Tekhn. Vysok. Temper.*, **2**, 10–13.
62. Novikov V. V., Dmitriev M. V., Shapovalov I. P. and Tartakovskaya L. N. (1987) Estimation of the effective dielectric properties of glass-ceramics with account of structural parameters, *Tekhn. Sredstv Svyazi*, **3**, 264–269.
63. Kusy R. and Corneliussen R. (1975) The thermal conductivity of nickel and copper dispersed in poly(vinyl chloride), *Polym. Eng. Sci.*, **15**, 107–112.
64. Kanari K. (1977) Thermal conductivity of composite materials, *Kobunshi*, **26**, 557–561.
65. Kline D. J. (1961) Thermal conductivity studies of polymers, *J. Polymer Sci.*, **50**, 441–450.
66. Sundstrom D. W. and Lee Y.-D. (1972) Thermal conductivity of polymers filled with particulate solids, *J. Appl. Polymer Sci.*, **16**, 3159–3167.
67. Hansen D. and Tomkiewicz R. (1975) Heat conduction in metal-filled polymers: the role of particle size, shape and orientation, *Polymer Eng. Sci.*, **15**, 353–356.
68. Privalko P. V. (1983) Thermophysical properties of filled polymers, *Prom. Teplotekhnika*, **3**, 66–76.
69. Shut N. I., Sichkar T. G. and Vozny P. A. (1985) Influence of the boundary layer structure on heat conduction and molecular mobility of filled epoxy systems, *Komposits. Polim. Mater.*, **24**, 18–21.
70. Nielsen L. E. (1974) The thermal and electrical conductivity of two-phase systems, *Ind. Eng. Chem. Fund.*, **13**, 17–20.
71. Bigg D. M. (1979) Mechanical, thermal, and electrical properties of metal fiber-filled polymer composites, *Polymer Eng. Sci.*, **19**, 1188–1192.
72. Mamunya E. P., Davidenko V. V. and Lebedev E. V. (1991) Properties of functionally filled polymer system in function of properties and content of disperse fillers, *Komposits. Polim. Mater.*, **50**, 37–47.
73. Rymarenko N. L., Voitenko A. I. Novikov V. V. and Privalko V. P. (1992) Composition-dependent properties of microheterogeneous polymeric materials: thermal conductivity of filled polyethylene, *Ukr. Polymer J.*, **1**, 259–266.
74. Garrett K. W. and Rosenberg H. M. (1974) The thermal conductivity of epoxy resin/powder composite materials, *J. Phys.*, Ser. D. (Appl. Phys.), **7**, 1247–1258.
75. de Araujo F. F. T. and Rosenberg H. M. (1976) The thermal conductivity of epoxy resin-metal-powder composite materials from 1.7 to 300 K, *J. Phys.*, Ser. D (Appl. Phys.), **9**, 665–675.
76. Hakansson B. and Ross R. G. (1990) Effective thermal conductivity of binary dispersed composites over wide ranges of volume fraction, temperature and pressure, *J. Appl. Phys.*, **68**, 3285–3292.
77. Kutcherov V., Hakansson B., Ross R.G. and Backstrom G. (1994) Experimental test of theories for the effective thermal conductivity of a dispersed composite, *J. Appl. Phys.*, **71**, submitted.
78. Hakansson B., Andersson P. and Backstrom G. (1988) Improved hot-wire procedure for thermophysical measurements under pressure, *Rev. Sci. Instrum.*, **59**, 2269–2275.
79. Agari Y., Ueda A., Tanaka M. and Nagai S. (1990) Thermal conductivity of a polymer filled with particles in the wide range from low to super-high volume content, *J. Appl. Polymer Sci.*, **40**, 929–941.
80. Agari Y., Ueda A. and Nagai S. (1991) Thermal conductivities of composites in several types of dispersion systems, *J. Appl. Polymer Sci.*, **42**, 1665–1669.
81. Yarema G. E., Besklubenko Yu. D. and Privalko V. P. (1982) Thermophysical properties of polystyrene/polycarbonate blends in the melt state under elevated pressures, *Dokl. Acad. Nauk UkrSSR*, Ser. B, **3**, 54–57.

82. Yarema G. E. (1993) Thesis, Institute of Macromolecular Chemistry, Academy of Sciences of Ukraine, Kiev, Ukraine.

83. Agari Y., Ueda A. and Nagai S. (1992) Thermal conductivity of polyethylene/polystyrene blends containing SEBS block copolymer, *J. Appl. Polymer Sci.*, **45**, 1957–1966.

84. Agari Y., Ueda A. and Nagai S. (1993) Thermal conductivities of blends of polyethylene/SABS block copolymer and polystyrene/SEBS block copolymer, *J. Appl. Polymer Sci.*, **47**, 331–337.

85. Sumita M., Asai S., Miyadera N. *et al.* (1986) Electrical conductivity of carbon black filled ethylene–vinyl acetate copolymer as a function of vinyl acetate content, *Colloid. Polymer Sci.*, **264**, 212–217.

86. Sujita M., Abe H., Kayaki H. and Miyasaka K. (1986) Effect of melt viscosity and surface tension of polymers on the percolation threshold of conductive-particle-filled polymeric composites, *J. Macromol. Sci.—Phys.*, Ser. B, **25**, 171–184.

87. Levon K., Margolina A. and Patashinsky A. Z. (1993) Multiple percolation in conducting polymer blends, *Macromolecules*, **26**, 4061–4063.

88. Galeski A., Milczarek P. and Kryszewski M. (1977) Heat conduction in a two-dimensional spherulite, *J. Polymer Sci.: Polymer Phys. Ed.*, **15**, 1267–1281.

89. Choy C. L. (1977) Thermal conductivity of polymers, *Polymer*, **18**, 984–1004.

90. Choy C. L., Luk W. H. and Chen F. C. (1978) Thermal conductivity of highly oriented polyethylene, *Polymer*, **19**, 155–162.

91. Godovsky Yu. K. (1982) *Thermophysics of Polymers*, Khimia, Moscow (in Russian).

92. Barker R. E., Jr., Chen R. Y. S. and Frost R. S. (1977) Influence of pressure and chemical structure on the thermal conductivity of vitreous poly(alkyl methacrylates). I, *J. Polymer Sci.: Polymer Phys. Ed.*, **15**, 1199–1210.

93. Privalko V. P. and Rekhteta T. A. (1992) Effect of pressure on the thermal conductivity of polymers, *J. Therm. Anal.*, **38**, 1083–1102.

5 Thermoelastic Properties

5.1 THEORETICAL ASPECTS: PRAGMATIC APPROACHES

5.1.1 BASIC DEFINITIONS

The effective moduli of elasticity (EME), C_{ijkl}, the effective moduli of compliance (EMC), S_{ijkl}, are defined by eqs. (5.1) [1]:

$$\langle \sigma_{ij} \rangle = C_{ijkl} \langle \varepsilon_{kl} \rangle \tag{5.1a}$$

$$\langle \varepsilon_{ij} \rangle = S_{ijkl} \langle \sigma_{kl} \rangle \tag{5.1b}$$

where the elastic stress tensor, $\langle \sigma_{ij} \rangle$, and the strain tensor, $\langle \varepsilon_{ij} \rangle$,

$$\langle \sigma_{ij} \rangle = V^{-1} \iiint_{(V)} \sigma_{ij}(\boldsymbol{r}) \, \mathrm{d}V \tag{5.2a}$$

$$\langle \varepsilon_{ij} \rangle = V^{-1} \iiint_{(V)} \varepsilon_{ij}(\boldsymbol{r}) \, \mathrm{d}V \tag{5.2b}$$

are the averages over the BRE volume, V.

In the case of localized domains, we can write, by analogy with eqs. (5.1):

$$\sigma_{ij}(\boldsymbol{r}) = C_{ijkl}(\boldsymbol{r})\varepsilon_{kl}(\boldsymbol{r}) \tag{5.3a}$$

$$\varepsilon_{ij}(\boldsymbol{r}) = S_{ijkl}(\boldsymbol{r})\sigma_{kl}(\boldsymbol{r}) \tag{5.3b}$$

where $\sigma_{ij}(\boldsymbol{r})$, $\varepsilon_{ij}(\boldsymbol{r})$, $C_{ijkl}(\boldsymbol{r})$ and $S_{ijkl}(\boldsymbol{r})$ are the random functions of coordinates.

According to the general postulate of elasticity theory of quasi-homogeneous media [2], under homogeneous boundary conditions the stress–strain fields in a large, statistically homogeneous body are also statistically homogeneous throughout, except for the boundary layers near the external surface.

The homogeneous boundary conditions for an elastic body may be defined as

$$U_i(s) = \varepsilon_{ij}^0 r_j \tag{5.4a}$$

$$f_i(s) = \sigma_{ij}^0 n_j \tag{5.4b}$$

where ε_{ij}^0 and σ_{ij}^0 are the constant strains and stresses, respectively, $U_i(s)$ is the displacement of the surface s bounding the volume V, $f_i(s)$ is the force acting on the unit surface s, r_j is the vector-radius and n_j is the vector-normal to s.

The values of the EME and the EMC may be derived as the solutions to the following series [1]:

$$C_{ijmn} = C^{(1)}_{ijkl}\varphi A^{(1)}_{klmn} + C^{(2)}_{ijkl}(1-\varphi)A^{(2)}_{klmn} \tag{5.5a}$$

$$\varphi A^{(1)}_{klmn} + (1-\varphi)A^{(2)}_{klmn} = I_{klmn} \tag{5.5b}$$

and

$$S_{ijmn} = S^{(1)}_{ijkl}\varphi B^{(1)}_{klmn} + S^{(2)}_{ijkl}(1-\varphi)B^{(2)}_{klmn} \tag{5.6a}$$

$$\varphi B^{(1)}_{klmn} + (1-\varphi)B^{(2)}_{klmn} = I_{klmn} \tag{5.6b}$$

where $A^{(i)}_{klmn}$ and $B^{(i)}_{klmn}$ are the tensors defined as

$$\langle \sigma^{(i)}_{kl}\rangle = B^{(i)}_{klmn}\langle \sigma_{mn}\rangle \tag{5.7a}$$

$$\langle \varepsilon^{(i)}_{kl}\rangle = A^{(i)}_{klmn}\langle \varepsilon_{mn}\rangle \tag{5.7b}$$

and I_{klmn} is the fourth-rank unit tensor.

Each of coupled eqs. (5.5) and (5.6) contain three unknowns ($C, A^{(1)}, A^{(2)}$ and $S, B^{(1)}, B^{(2)}$, respectively); thus, the latter outnumber the former. The closure of eqs. (5.5) and (5.6) is possible with the aid of eqs. (5.7), the solution of which, however, requires additional information about the structure of MHM.

The effective thermal expansion coefficient (TEC), α, may be defined as an average deformation in the BRE of the MHM per unit of temperature change, i.e.

$$\langle \varepsilon_{ij}\rangle = \alpha_{ij}\Delta T \tag{5.8}$$

where α_{ij} is the tensor of TEC coefficients, $\Delta T = T - T_0$ and T_0 and T are the initial and final temperatures, respectively.

Stress, strain and temperature are interrelated through the following constitutive equations [3, 4]:

$$\sigma_{ij} = C_{ijkl}(\varepsilon_{kl} - \alpha_{kl}) \tag{5.9a}$$

$$\varepsilon_{ij} = S_{ijkl}(\sigma_{kl} + \beta_{kl}) \tag{5.9b}$$

where $\beta_{ij} = C_{ijkl}\alpha_{kl}$ is the tensor of thermal stress coefficients, and $\alpha_{ij} = S_{ijkl}\beta_{kl}$.

Sometimes, the tensor of thermal stresses β is replaced by the tensor of Grueneisen coefficients, $\gamma_G = \beta/C_\varepsilon$ (where C_ε is the heat capacity at $\varepsilon = \text{const}$) to obtain

$$\gamma_{G,ij}\alpha_{ij} = (C_\sigma - C_\varepsilon)/C_\varepsilon T_0$$

where C_σ is the heat capacity at $\sigma = \text{const}$.

The TEC of a two-component MHM could be rigorously derived from TECs of each component and the effective compliance tensor, S_{ij}, assuming the following conditions to apply to the two model bodies (denoted by indices 'I' and 'II', respectively) [5]:

$$\langle \sigma_{ij}\rangle_I = \sigma^0_{tj}; \qquad \Delta T = 0; \qquad (\varepsilon_{ij})_I = S_{ijkl}(\sigma_{kl})_I \tag{5.10a}$$

$$\langle \sigma_{ij}\rangle_{II} = 0; \qquad \Delta T = \Delta T^0$$

$$(\varepsilon_{iju})_{II} = S_{ijkl}(\sigma_{kl})_{II} + \alpha_{ij}\Delta T^0 \tag{5.10b}$$

It follows from eq. (5.10a) that

$$\langle \varepsilon_{ij}\rangle_{\mathrm{I}} = S_{ijkl}\sigma_{kl}^0, \qquad \langle \varepsilon_{ij}^{(m)}\rangle_{\mathrm{I}} = S_{ijkl}^{(m)}\langle \sigma_{kl}^{(m)}\rangle_{\mathrm{I}} \tag{5.11}$$

where the index $m = 1, 2$ denotes the corresponding component and does not refer to the averaging procedure.

In a similar fashion, eq. (5.10b) yields

$$\langle \varepsilon_{ij}\rangle_{\mathrm{II}} = \alpha\Delta T^0, \qquad \langle \varepsilon_{ij}^{(m)}\rangle_{\mathrm{II}} = S_{ijkl}^{(m)}\langle \sigma_{kl}\rangle_{\mathrm{II}} + \alpha_{ij}^{(m)}\Delta T^0$$

In the absence of volume forces the following relationships should be valid:

$$\iiint_{(V)} \sigma_{ij}\varepsilon_{ij}\,\mathrm{d}V = \langle\sigma_{ij}\rangle\langle\varepsilon_{ij}\rangle V \tag{5.12a}$$

$$\iiint_{(V)} (\varepsilon_{ij})_{\mathrm{I}}(\sigma_{ij})_{\mathrm{I}}\,\mathrm{d}V = \iiint_{(V)} S_{ijkl}(\sigma_{kl})_{\mathrm{I}}(\sigma_{ij})_{\mathrm{II}}\,\mathrm{d}V = 0 \tag{5.12b}$$

$$\iiint_{(V)} (\sigma_{iju})_{\mathrm{I}}(\varepsilon_{ij})_{\mathrm{III}}\,\mathrm{d}V = \iiint_{(V)} (\sigma_{ij})_{\mathrm{I}}\alpha_{ij}\Delta T^0\,\mathrm{d}V = \sigma_{ij}^0\alpha_{ij}\Delta TV \tag{5.12c}$$

Making use of the standard definition

$$\langle q\rangle_I = \sum_{m=1}^{n} \varphi_m\langle\sigma^{(m)}\rangle_{\mathrm{I}} \tag{5.13}$$

and recalling the result of the linear elasticity theory

$$\langle\sigma_{ij}^{(m)}\rangle = B_{ijkl}^{(m)}\sigma_{kl}^0 \tag{5.14}$$

we obtain

$$\sum_{m=1}^{n} \varphi_m B_{ijkl}^{(m)} = I_{ijkl} \tag{5.15}$$

where I is the corresponding unit tensor.

Now, eq. (5.12c) may be rewritten as a summation of integrals over each component to obtain

$$\alpha_{ij} = \sum_{m=1}^{n} \varphi_m \alpha_{kl}^{(m)} B_{ijkl}^{(m)} \tag{5.16}$$

Thus, the effective TEC will be determined by $B_{ijkl}^{(m)}$, which depends, in its turn, on the mean stresses in the components, $\langle\sigma_{kl}^{(m)}\rangle$, generated in the volume V by the surface loads (σ_{kl}^0).

Substituting eqs (5.11) and (5.14) into

$$\langle\varepsilon_{ij}\rangle_{\mathrm{I}} = \sum_{m=1}^{m} \varphi_m\langle\varepsilon_{ij}^{(m)}\rangle_{\mathrm{I}} \tag{5.17}$$

to obtain

$$S_{ijkl} = \sum_{m=1}^{n} \varphi_m S_{ijrs}^{(m)} B_{rskl}^{(m)} \tag{5.18}$$

and eliminating the unknown tensors $B_{ijkl}^{(m)}$ from eqs. (5.15), (5.17) and (5.18), we obtain

$$\alpha_{ij} = \langle \alpha_{ij} \rangle + C_{klmn}^{0} [S_{mnij} - \langle S_{mnij} \rangle][\alpha_{kl}^{(1)} - \alpha_{kl}^{(2)}] \tag{5.19}$$

where

$$C^{0}[S^{(1)} - S^{(2)}] = I \tag{5.20a}$$

$$\langle \alpha_{ij} \rangle = \alpha_{ij}^{(1)} \varphi - \alpha_{ij}^{(2)}(1 - \varphi) \tag{5.20b}$$

$$\langle S_{mnij} \rangle = S_{mnij}^{(1)} \varphi - S_{mnij}^{(2)}(1 - \varphi) \tag{5.20c}$$

As can be seen from eqs. (5.20), the exact value of the effective TEC, α_{ij}, may be obtained provided the effective compliance tensor is available.

Making use of the standard relationships for a MHM consisting of the isotropic components

$$\alpha_{ij}^{(m)} = \alpha_m \delta_{ij}, \qquad S_{ijkl}^{(m)} = \delta_{ij}/3K \tag{5.21}$$

we obtain from eq. (5.19)

$$\alpha_{ij} = \langle \alpha \rangle \delta_{ij} + \frac{\alpha_1 - \alpha_2}{K_1^{-1} - K_2^{-1}}(3S_{klij} - K_R^{-1}\delta_{ij}) \tag{5.22}$$

where

$$K_R^{-1} = \varphi/K_1 + (1 - \varphi)/K_2 = \langle K^{-1} \rangle \tag{5.23}$$

For the texture-devoid MHM eq. (5.22) reduces to the familiar eq. (5.24) [5–11],

$$\alpha = \langle \alpha \rangle + (\alpha_1 - \alpha_2)\frac{K^{-1} - K_R^{-1}}{K_1^{-1} - K_2^{-1}}, \tag{5.24}$$

where K is the effective bulk elasticity modulus.

It should become clear from the above that the effective TEC is obtained as a solution of the relevant elastic problem, that is, determination of the bulk modulus K. Therefore, in the subsequent discussion the main emphasis will be on the analysis of the elastic properties, whereas only final results will be given for the TEC.

As already pointed out above, the solution of the thermoelasticity problem requires supplementary information on the structure of MHM. Apparently, the simplest structures are the connections of components in parallel (i) and in series (ii) (the models of Voigt [12] and Reuss [13], respectively).

(i) The assumption of equal strain, $\varepsilon(r) = \langle \varepsilon \rangle$, implies $A_{klmn}^{(i)} = I_{klmn}$; therefore, we obtain for the Voigt model

$$C_{ijkl}^{(V)} = \varphi C_{ijkl}^{(1)} + (1 - \varphi)C_{ijkl}^{(2)} \tag{5.25}$$

Equation (5.25) may be used to define the bulk and the shear moduli of the isotropic Voigt body by eqs. (5.26a) and (5.26b), respectively

$$K^{(V)} = \varphi K_1 + (1 - \varphi)K_2 \tag{5.26a}$$

$$G^{(V)} = \varphi G_1 + (1 - \varphi)G_2 \tag{5.26b}$$

(ii) The assumption of equal stress, $\sigma(r) = \langle \sigma \rangle$, implies $B_{klmn}^{(i)} = I_{klmn}$; therefore, we obtain for the Reuss model

$$S_{ijkl}^{(R)} = \varphi S_{ijkl}^{(1)} + (1 - \varphi)S_{ijkl}^{(2)} \tag{5.27}$$

from which we derive for the isotropic Reuss body:

$$K^{(R)} = [\varphi/K_1 + (1 - \varphi)/K_2]^{-1} \tag{5.28a}$$

$$G^{(R)} = [\varphi/G_1 + (1 - \varphi)/G_2]^{-1} \tag{5.28b}$$

Equations (5.26) and (5.28), respectively, define the upper and the lower bounds of the EME of a MHM. It can be shown [14] that this conclusion is consistent with the following conditions:

$$\langle \sigma_{ij}(r)\varepsilon_{ij}(r) \rangle = \langle \sigma_{ij} \rangle \langle \varepsilon_{ij} \rangle \tag{5.29}$$

and

$$\sigma_{kl}^{(A)}\varepsilon_{kl}^{(A)} \leqslant \sigma_{kl}^{(B)}\varepsilon_{kl}^{(B)} + 2\sigma_{kl}^{(A)}(\varepsilon_{kl}^{(A)} - \varepsilon_{kl}^{(B)}) \tag{5.30a}$$

$$\sigma_{kl}^{(A)}\varepsilon_{kl}^{(A)} \leqslant \sigma_{kl}^{(B)}\varepsilon_{kl}^{(B)} + 2\varepsilon_{kl}^{(A)}(\sigma_{kl}^{(A)} - \sigma_{kl}^{(B)}) \tag{5.30b}$$

where $(\sigma_{kl}^{(A)}, \varepsilon_{kl}^{(A)})$ and $(\sigma_{kl}^{(B)}, \varepsilon_{kl}^{(B)})$ are two coupled arbitrary functions which specify the elastic properties of a solid body in two states (A and B, respectively).

Substitution of $(\langle \sigma \rangle, \langle \varepsilon \rangle)$ for $(\sigma^{(A)}, \varepsilon^{(A)})$ and $(\langle \sigma^{(V)} \rangle, \langle \varepsilon \rangle)$ for $(\sigma^{(B)}\varepsilon^{(B)})$ into eq. (5.30a) yields

$$\langle \sigma_{kl} \rangle \langle \varepsilon_{kl} \rangle \leqslant \langle \sigma_{kl}^{(V)} \rangle \langle \varepsilon_{kl} \rangle \tag{5.31a}$$

Making a similar substitution, i.e. $(\langle \sigma \rangle, \langle \varepsilon \rangle)$ and $(\langle \sigma \rangle, \langle \varepsilon^{(R)} \rangle)$ for $(\sigma^{(A)}, \varepsilon^{(A)})$ and $(\sigma^{(B)}, \varepsilon^{(B)})$, respectively, into eq. (5.30b), we obtain

$$\langle \sigma_{kl} \rangle \langle \varepsilon_{kl} \rangle \leqslant \langle \sigma_{kl} \rangle \langle \varepsilon_{kl}^{(R)} \rangle \tag{5.31b}$$

Taking into consideration the validity of the following conditions:

$$\langle \sigma_{kl}^{(V)} \rangle = C_{klij}^{(V)} \langle \varepsilon_{ij} \rangle \tag{5.32a}$$

$$\langle \varepsilon_{kl}^{(R)} \rangle = S_{klij}^{(R)} \langle \sigma_{ij} \rangle \tag{5.32b}$$

we can derive from eqs. (5.31)

$$(C_{klij} - C_{klij}^{(V)})\langle \varepsilon_{kl} \rangle \langle \varepsilon_{ij} \rangle \leqslant 0 \tag{5.33a}$$

$$(S_{klij} - S_{klij}^{(R)})\langle \sigma_{kl} \rangle \langle \sigma_{ij} \rangle \leqslant 0 \tag{5.33b}$$

These results allow as to establish the lower and upper bounds for the elastic moduli of an isotropic MHM, i.e.

$$C_{klij}^{(R)} \leqslant C_{klij} \leqslant C_{klij}^{(V)}, \qquad (C^{(R)}S^{(R)} = I) \tag{5.34}$$

$$K^{(R)} \leqslant K \leqslant K^{(V)} \tag{5.34a}$$

$$G^{(R)} \leqslant G \leqslant G^{(V)} \tag{5.34b}$$

Equations (5.33) and (5.34) were derived from purely phenomenological considerations and thus do not account for the geometrical parameters (i.e. the structure) of a MHM and the elastic interactions between the components. Methods allowing for these features will be reviewed below.

5.1.2 METHOD OF DIRECT CALCULATIONS

The basic structure is an isolated spherical inclusion (component 1) embedded into an infinite, continuous medium (component 2). The relevant stress and strain fields may be found by solving the following elasticity problem [15]:

$$(1 - 2v_1)\boldsymbol{u}_1 + \operatorname{grad} \operatorname{div} \boldsymbol{u}_1 = 0 \tag{5.35a}$$

$$(1 - 2v_2)\boldsymbol{u}_2 + \operatorname{grad} \operatorname{div} \boldsymbol{u}_2 = 0 \tag{5.35b}$$

where $\boldsymbol{u}_1$ and $\boldsymbol{u}_2$ are the displacement vectors for components 1 (i.e. sphere) and 2, respectively, and v_1 and v_2 are the corresponding Poisson's coefficients.

The following boundary conditions are assumed to hold at the surface of the sphere:

$$\boldsymbol{u}_1 = \boldsymbol{u}_2; \qquad \sigma_{ij}^{(1)}n_j = \sigma_{ij}^{(2)}n_j \tag{5.36}$$

and at an infinite distance from the center of the sphere:

$$\langle \sigma_{ij}^{(2)} \rangle = \langle \sigma_{ij} \rangle \tag{5.37}$$

The displacements $\boldsymbol{u}_1$ and $\boldsymbol{u}_2$ [and hence the corresponding strains $\varepsilon_{ij}^{(1)}(\boldsymbol{r})$ and $\varepsilon_{ij}^{(2)}(\boldsymbol{r})$] may be determined by solving eqs. (5.35)–(5.37). Taking the average of $\varepsilon_{ij}^{(1)}(\boldsymbol{r})$ over the volume occupied by the inclusion to obtain $A^{(1)}$, we derive, with the aid of eq. (5.7), the following formulae for the effective bulk and shear moduli and for the TEC, respectively [15, 5, 10]:

$$K = K_2\left[1 + \frac{3(1 - v_2)(K_1 - K_2)\varphi}{3K_2(1 - v_2) + K_1(1 + v_2)}\right] \tag{5.38a}$$

$$G = G_2\left[1 + \frac{15(1 - v_2)(G_1 - G_2)\varphi}{G_2(37 - 5v_2) + 2G_1(4 - 5v_2)}\right] \tag{5.38b}$$

$$\alpha = \frac{\varphi\alpha_1 K_1(K_2 + 4G_2/3) + (1 - \varphi)\alpha_2 K_2(K_1 + 4G_1/3)}{K_1 K_2 + 4G_2[\varphi K_1 + (1 - \varphi)K_2]} \tag{5.38c}$$

The validity of eqs. (5.38) is limited to the cases when the nearest neighbors exert a negligible influence on the pattern of the elastic stress distribution around the test particle (i.e. at $\varphi < 0.2$).

5.1.3   SELF-CONSISTENT FIELD APPROACH

The mean stress, $\langle \sigma_{ij}^{(1)} \rangle$, is approximated by mean stresses in a single, isolated spherical inclusion (component 1) embedded into an infinite medium with the effective elastic properties sought for [14, 16–18]. The fundamental difference with the previous method (cf. section 5.1.2) is that the properties of component 2 in eqs. (5.35)–(5.37) are replaced by those of the effective medium. The solution of the relevant boundary problem is [17]

$$\langle \sigma_{ij}^{(1)} \rangle = \left[\frac{K_1 \delta_{ij} \delta_{kl}}{K_1 + a(K_1 - K)} + \frac{G_1(I_{ijkl} - \delta_{ij}\delta_{kl}/3)}{G + b(G_1 - G)} \right] \langle \sigma_{kl} \rangle \qquad (5.39)$$

where $a = (1 + v)/3(1 - v)$, $b = 2(4 - 5v)/15(1 - v)$ and $v = (3K - 2G)/(6K + 2G)$. Combining eqs. (5.6) and (5.39), we obtain [11, 16, 17]

$$\varphi K_1/(K_1 + 4G/3) + (1 - \varphi)K_2/(K_2 + 4G/3) + 5[\varphi G_2/G - G_2)$$
$$+ (1 - \varphi)G_1/(G - G_1)] + 2 = 0 \qquad (5.40\text{a})$$

$$1/(K + 4G/3) = \varphi/(K_1 + 4G/3) + (1 - \varphi)(K_2 + 4G/3) \qquad (5.40\text{b})$$

$$\alpha = \langle \alpha \rangle - \frac{(\alpha_2 - \alpha_1)\varphi(1 - \varphi)(1/K_2 - 1/K_1)}{\langle 1/K \rangle + 3/4G_2} \qquad (5.40\text{c})$$

As is usual practice, the value of G is derived first from eq. (5.40a) as a positive root of the fourth-power equation, this value is then substituted into eq. (5.40b) to calculate K and, finally, the value of the TEC is calculated by eq. (5.40c).

5.1.4   VARIATIONAL APPROACH

In this method the elastic field of a non-homogeneous medium is determined by minimization of the energy functional [19]

$$U = U_0 - 0.5 \int_V (H_{ijkl}P_{ij}P_{kl} - P_{ij}\varepsilon_{ij} - 2P_{ij}\varepsilon_{ij}^0)\,\mathrm{d}V$$

(where $U_0 = 0.5 \int_V \sigma_{ij}^0 \varepsilon_{ij}^0$ and $H_{ijkl}C_{klpq} = I_{ijpq}$), provided the following conditions are valid:

$$\nabla_j(C_{ijkl}^0 \varepsilon_{kl}' + P_{ij}) = 0, \qquad u_i'(s) = 0$$

where $u_i' = u_i - u_i^0$, $\varepsilon_{ij}' = \varepsilon_{ij} - \varepsilon_{ij}^0$, $C_{ijkl}' = C_{ijkl} - C_{ijkl}^0$ and $P_{ij} = \sigma_{ij} - C_{ijkl}^0\varepsilon_{kl} = C_{ijkl}'\varepsilon_{kl}$, the symbols with and without the superscript zero referring to the standard (isotropic and homogeneous) reference body and to the MHM, respectively.

Assuming identical boundary conditions in the MHM and in the reference body, we can derive eqs. (5.41a) and (5.41b) for the bulk modulus and shear modulus, respectively, defining the upper and the lower bounds as those corresponding to the maximum of U (the case of the stiffer component chosen as a reference body) and to the minimum of U (the reverse case of the softer component as a reference body), respectively [19]:

$$-\frac{a_2 D_K}{1 + a_2(1 - \varphi)(K_1 - K_2)} \leqslant K - \langle K \rangle \leqslant -\frac{a_1 D_K}{1 - a_1 \varphi(K_1 - K_2)} \tag{5.41a}$$

$$-\frac{b_2 D_G}{1 + b_1(1 - \varphi)(G_1 - G_2)} \leqslant G - \langle G \rangle \leqslant -\frac{b_1 D_G}{1 - b_1 \varphi(G_1 - G_2)} \tag{5.41b}$$

where $a_i = 3/(3K_i + 4G_i)$, $b_i = 6(K_i + 2G_i)/5G_i(3K_i + 4G_i)$ and $D_X = \varphi(1 - \varphi)(X_1 - X_2)^2$.

Apparently, eqs. (5.41) predict a narrower gap between the upper and the lower bounds than eqs. (5.34). The former bounds are appropriate for the 'soft inclusions in a rigid matrix'-type morphology, whereas the reverse case of 'rigid inclusions in a soft matrix' refers to the lower bounds.

The lower bound of eqs. (5.41) for the modulus of elasticity M of any kind (i.e. Young's, bulk or shear) may be also rewritten in a generalized, semi-empirical form as [20]

$$M/M_1 = (1 + AB\varphi)/(1 - B\psi\varphi) \tag{5.42}$$

where A is the geometrical constant dependent on the particle shape (e.g. $A = (7 - 5v_1)/(8 - 10v_1)$ for spherical inclusions), $B = (M_2 - M_1)/(M_2 + M_1 A)$, $\psi = (1 - \varphi^*)\varphi/\varphi^{*2}$ or $\psi\varphi = 1 - \exp[-\varphi/(1 - \varphi/\varphi^*)]$.

Making use of eqs. (5.41) and (5.24), we obtain for the lower and upper bounds of the TEC [5, 6]

$$\frac{4\varphi(1 - \varphi)G_1(K_1 - K_2)(\alpha_1 - \alpha_2)}{3K_1 K_2 + 4G_1 K_V} \leqslant \alpha - \langle \alpha \rangle \leqslant \frac{4\varphi(1 - \varphi)G_2(K_1 - K_2)(\alpha_1 - \alpha_2)}{3K_1 K_2 + 4G_2 K_V}$$

$$\tag{5.41c}$$

In the special case of a MHM consisting of isotropic components with identical shear moduli ($G = G_1 = G_2$) but different bulk moduli (hence TEC) the effective TEC is rigorously defined as [14, 18]

$$\alpha = \alpha_2 + \frac{\varphi(\alpha_1 - \alpha_2)K_1(K_2 + 4/3G)}{K_1 K_2 + 4/3G[\varphi K_1 + (1 - \varphi)K_2]} \tag{5.43}$$

In this case the upper and the lower bounds may be estimated as

$$\varphi z_1 \leqslant (\alpha - \alpha_2)/(\alpha_1 - \alpha_2) \leqslant \varphi z_2 \tag{5.44}$$

where $z_1 = [1 + (1 - \varphi)(1/K_1 - 1/K_2)/(1/K_1 + 1/K_2)]^{-1}$, $z_2 = [1 + (1 - \varphi)(1/K_1 - 1/K_2)(1/K_1 + 1/K_2)]^{-1}$ and $K_1^* = 4G_2/3$, $K_2^* = 4G_1/3$ (if $(K_1 - K_2)(G_1 - G_2) \geqslant 0$), or $K_1^* = 4G_1/3$, $K_2^* = 4G_2/3$ (if $(K_1 - K_2)(G_1 - G_2) < 0$).

5.1.5 METHOD OF RANDOM FUNCTIONS

In this method all the local parameters are expressed as [21–25]

$$C(r) = \langle C \rangle + C^0(r)$$
$$\varepsilon(r) = \langle \varepsilon \rangle + \varepsilon^0(r)$$

where the symbols with the superscript zero denote the random deviations of the local property from the corresponding volume averages, and $\langle C^0 \rangle = 0$, $\langle \varepsilon^0 \rangle = 0$ is assumed. In this case the mean stresses may be defined as

$$\langle \sigma \rangle = \langle C\varepsilon \rangle = \langle C \rangle \langle \varepsilon \rangle + \langle C^0 \varepsilon^0 \rangle;$$

therefore the effective tensor C will be

$$C = \langle C \rangle + \langle C^0 \varepsilon^0 \rangle \langle \varepsilon \rangle^{-1}$$

Using these results together with the local equilibrium criterion, $[\sigma_{ij}(r)]_{,j} = 0$, we can write

$$C^0(r)[\varepsilon_{ij}(r)]_{,j} = -[C^0(r)\varepsilon(r)]_{,j} \tag{5.45}$$

The general solution of eq. (5.45) is

$$\varepsilon_{ij}(r) = \varepsilon'_{ij}(r) + \int_V G_{ijkl}(r, r')C'_{klmn}(r)\varepsilon_{mn}(r)\,dV \tag{5.46}$$

where $G_{ijkl}(r,r')$ is the Green's function for the strain problem, eq. (5.45), and $\varepsilon'(r)$ is defined by the condition $\langle C \rangle [\varepsilon'_{ij}(r)]_{,j} = 0$.

Application of the conditional moments technique yielded the following general results for the simplest case of isolated inclusions at large separations [26]:

$$K = \langle K \rangle - \frac{\varphi(1 - \varphi)K^{(3)2}}{K^{(4)} + 4m/3} \tag{5.47a}$$

$$G = \langle G \rangle - \frac{\varphi(1 - \varphi)G^{(3)2}}{G^{(4)} + m(9k + 8m)/6(k + 2m)} \tag{5.47b}$$

where $K^{(4)} = \varphi K_2 + (1 - \varphi)K_1, G^{(4)} = \varphi G_2 + (1 - \varphi)G_1, K^{(3)} = K_1 - K_2, G^{(3)} = G_1 - G_2, k = l + 2m/3$, and l and m are arbitrary elastic parameters. Eqs. (5.9), (5.12), (5.24), (5.26), etc. may be derived from eqs. (5.31) by the proper selection of the parameters l and m; for example, $m = G_1 G_2/G^{(4)}$ and $l = K_1 K_2/K^{(4)} - 2m/3$ are recommended when the inclusions are stiffer than the matrix, while in the reverse case of soft inclusions in a rigid matrix, $m = \langle G \rangle$ and $l = \langle K \rangle$ should be used [26].

5.1.6 METHOD OF INTEGRAL SECTIONING

In this method the upper and lower bounds are estimated, as already outlined in section 4.1.11, from consideration of the conditional sectioning of the BRE into prisms (Fig. 4.2(b)) and into layers (Fig. 4.2(c)), each method of sectioning involving the choice of coupled fictive functions for the stress $\sigma'_{ij}(r)$ and for the strain $\varepsilon'_{ij}(r)$ [27, 28].

In the former method, the average stress in the cross-sectional area, $\{\sigma'_{ij}(r)\}_S$, is assumed to be equal to the volume-average stress, $\langle \sigma'_{ij} \rangle$, i.e.

$$\{\sigma'_{ij}(r)\}_S = \langle \sigma'_{ij} \rangle \tag{5.48a}$$

$$\langle \varepsilon'_{ij} \rangle = \langle \varepsilon_{ij} \rangle \tag{5.48b}$$

In the latter method the average strain over the length L in a selected direction, $\{\varepsilon'_{ij}(r)\}_L$, is assumed to be equal to the volume-average strain, i.e.

$$\{\varepsilon'_{ij}(r)\}_L = \langle \varepsilon'_{ij} \rangle \tag{5.49a}$$

$$\langle \sigma'_{ij} \rangle = \langle \sigma_{ij} \rangle \tag{5.49b}$$

Substitution of the couples $(\langle \sigma_{kl} \rangle, \langle \varepsilon_{kl} \rangle)$ and $(\langle \sigma'_{kl} \rangle, \langle \varepsilon_{kl} \rangle)$ for $(\sigma^{(A)}_{kl}, \varepsilon^{(A)}_{kl})$ and $(\sigma^{(B)}_{kl}, \varepsilon^{(B)}_{kl})$ into eq. (5.30a), and a similar substitution of $(\langle \sigma_{kl} \rangle, \langle \varepsilon_{kl} \rangle)$ and $(\langle \sigma_{kl} \rangle, \langle \varepsilon'_{kl} \rangle)$ into eq. (5.30b), respectively, yields

$$\langle \sigma_{kl} \rangle \langle \varepsilon_{kl} \rangle \leqslant \langle \sigma'_{kl} \rangle \langle \varepsilon_{kl} \rangle \tag{5.50a}$$

$$\langle \sigma_{kl} \rangle \langle \varepsilon_{kl} \rangle \leqslant \langle \sigma_{kl} \rangle \langle \varepsilon'_{kl} \rangle \tag{5.50b}$$

The two bounds for C_{ijkl} will be established from consideration of the relationship between $\langle \sigma'_{kl} \rangle$ and $\langle \varepsilon_{kl} \rangle$ and between $\langle \varepsilon'_{kl} \rangle$ and $\langle \sigma_{kl} \rangle$, respectively. Making use of the following general condition

$$\{\sigma'_{kl}\}_S = \{C_{klij}(r)\varepsilon_{ij}(r)\}_S$$

we can write for the linear elasticity case:

$$\langle \sigma'_{kl} \}_S = H_{klij}(x_3)\{\varepsilon_{ij}(r)\}_S \tag{5.51}$$

In the latter expression, $H_{klij}(x_3)$ is the elasticity tensor of the layer of thickness $\mathrm{d}x_k$ oriented normally to the Ox_3 axis (Fig. 4.2):

$$H_{klij}(x_k) = C^{(2)}_{klmn}I_{mnij} + \langle S_1(x_3) \rangle [C^{(1)}_{klmn} - C^{(2)}_{klmn}]A^{(1)}_{mnij}(x_3) \tag{5.52}$$

where $\langle S_1(x_3) \rangle = S_1(x_3)/S(x_3)$ is the cross-sectional area of the BRE in the plane $x_3 = \text{const}$, $S(x_3) = S_1(x_3) + S_2(x_3)$, $I_{ijmn} = (\sigma_{im}\sigma_{jn} + \sigma_{in}\sigma_{jm})/2$ and the tensor $A^{(1)}_{mnij}(x_3)$ is defined as

$$\{\varepsilon^{(1)}_{kl}(r)\}_{S_1} = A^{(1)}_{klmn}(x_3)\{\varepsilon_{mn}(r)\}_S$$

Now, multiplying eq. (5.51) by $[H_{klij}(x_3)]^{-1}$, averaging the result over the distance L (i.e. the variable x_3) and taking into account eqs. (5.50) we obtain

$$\langle \varepsilon_{ij} \rangle = \{[H_{klij}(x_3)]^{-1}\}_L \langle \sigma'_{kl} \rangle$$

or, equivalently,

$$\langle \sigma'_{kl} \rangle = \langle \varepsilon_{ij} \rangle \{[H_{klij}(x_3)]^{-1}\}_L^{-1} \tag{5.53}$$

In a similar fashion, we can write for the linear elasticity case

$$\{\varepsilon'_{kl}\}_L = M_{klij}(x_1, x_2)\{\sigma_{ij}(r)\}_L \tag{5.54}$$

where

$$M_{klij}(x_1, x_2) = S_{klmn}I_{mnij} + \langle L_1(x_1, x_2)\rangle(S^{(1)}_{klmn} - S^{(2)}_{klmn})B^{(1)}_{mnij}(x_1, x_2) \tag{5.55}$$

is the tensor of compliance of the prism with L for the height and $dx_1\,dx_2$ for the basal area, $\langle L_1(x_1, x_2)\rangle = L_1(x_1, x_2)/L(x_1, x_2)$, $L(x_1, x_2) = L_1(x_1, x_2) + L_2(x_1, x_2)$, $L_1(x_1, x_2)$ is the length of the straight line drawn in the component 1 through the BRE parallel to the axis Ox_3, and the tensor $B^{(1)}_{mnij}(x_1, x_2)$ is defined as

$$\{\sigma^{(1)}_{mn}(r)\}_{L_1} = B^{(1)}_{mnkl}(x_1, x_2)\{\sigma_{kl}(r)\}_L \tag{5.56}$$

Multiplying eq. (5.54) by $[M_{klij}(x_1, x_2)]^{-1}$, averaging the result over x_1, x_2 and taking into consideration eq. (5.49), we obtain

$$\langle\sigma_{ij}\rangle = \{[M_{klij}(x_1, x_2)]^{-1}\}_S\langle\varepsilon'_{kl}\rangle$$

which may be rewritten as

$$\langle\varepsilon'_{kl}\rangle = \{[M_{klij}(x_1, x_2)]^{-1}\}_S^{-1}\langle\sigma_{ij}\rangle \tag{5.57}$$

Finally, substituting eqs. (5.53) and (5.57) into eqs. (5.48a) and (5.48b), respectively we obtain, after rearrangement, the following general definition of the lower and upper bounds of the EME, C_{klij}:

$$\{[M_{klij}(x_1, x_2)]^{-1}\}_S \leqslant C_{klij} \leqslant (\{[H_{klij}(x_3)]^{-1}\}_L)^{-1} \tag{5.58}$$

In the case when both components of the MHM are isotropic, the tensors of elasticity modulus, C_{ijkl}, and of compliance, S_{ijkl}, may be decomposed into bulk and deviatoric contributions, i.e. [1]

$$C_{ijkl} = 3KV_{ijkl} + 2GD_{ijkl}$$
$$S_{ijkl} = V_{ijkl}/3K + D_{ijkl}/2G$$

where $V_{ijkl} = \delta_{ij}\delta_{kl}/3$ and $D_{ijkl} = (\delta_{ik}\delta_{jl} + \delta_{il}\delta_{jk} - 2\delta_{ij}\delta_{kl}/3)/2$.

The potential energy of an elastic body may be represented by the sum of the corresponding contributions from bulk compression and from pure shear; therefore, the inequalities (5.58) will also apply to the bulk and deviatoric components of the elasticity modulus tensor. Hence, eq. (5.58) may be rewritten as [27, 28]

$$\{K(x_1, x_2)\}_S \leqslant K \leqslant (\{[K(x_3)]^{-1}\}_L)^{-1} \tag{5.59a}$$

$$\{G(x_1, x_2)\}_S \leqslant G \leqslant (\{[G(x_3)]^{-1}\}_L)^{-1} \tag{5.59b}$$

where $K(x_1, x_2) = [\{n/K\}_L - 2\{d\}_L\{P\}_L/\{KP\}_L]^{-1}$ and $G(x_1, x_2) = \{1/G\}_L^{-1}$ are the bulk and shear moduli, respectively, of a cylindrical differential element, $K(x_3) = \{KP\}_S/\{P\}_S$ and $G(x_3) = \{G\}_S$ are those for a layer-like differential element, $n_i = 9/(3 + 4m_i)$, $d_i = (3 - 2m_i)/(3 - 4m_i)$, $P_i = 6m_i/(3 + 4m_i)$, $m_i = G_i/K_i$ and $i = 1, 2$.

The effective TEC will be evaluated by the following procedure [27, 28]. The MHM is assumed to be under the action of surface forces at $(t - t_0) = 0$, whereas the surface forces vanish at $(t - t_0) \neq 0$. The stress and the strain fields will be denoted as $\sigma_{ij}^0(r)$, $\varepsilon_{ij}^0(r)$ and $\sigma_{ij}(r)$, $\varepsilon_{ij}(r)$ in the former and in the latter cases, respectively. Making use of the condition $\langle \sigma_{ij}\varepsilon_{ij}\rangle = \langle \sigma_{ij}\rangle \langle \varepsilon_{ij}\rangle$, the usual definition of the volume-average of a product

$$\langle \sigma_{ij}^0(r)\varepsilon_{ij}(r)\rangle = (t - t_0)\langle \sigma_{ij}^0(r)\alpha_{ij}(r)\rangle \tag{5.60}$$

may be rewritten as

$$\langle \sigma_{ij}^0(r)\varepsilon_{ij}(r)\rangle = \langle \sigma_{ij}^0\rangle \alpha_{ij}(t - t_0) \tag{5.61}$$

where α_{ij} is the effective TEC tensor. Comparing eqs. (5.60) and (5.61) we obtain

$$\langle \sigma_{ij}^0\rangle \alpha_{ij}(t - t_0) = \langle \sigma_{ij}^0(r)\alpha_{ij}(r)\rangle (t - t_0) \tag{5.62}$$

Let a fictive body be under the stress $\sigma_{ij}'(r)$ at $(t - t_0) = 0$ and let the corresponding strains be the true strains, $\varepsilon_{ij}(r)$, at $(t - t_0) \neq 0$. Assuming, as before, $\{\sigma_{ij}(r)\}_S = \langle \sigma_{ij}'\rangle$, we obtain from eq. (5.62)

$$\alpha_{rr}^{(U)}\langle \sigma_{rr}'\rangle = \{\{\alpha_{rr}\sigma_{rr}'\}_S\}_L \tag{5.63}$$

which may be rewritten as

$$\alpha_{rr}^{(U)} = \alpha_{nl}^{(1)}\langle S_1(x_k)\rangle B_{rrnl}^{(1)}(x_k) + \alpha_{nl}^{(2)}\langle S_s(x_k)\rangle B_{rrnl}^{(2)} \tag{5.64}$$

Now, let a fictive body be under the stress $\sigma_{ij}^0(r)$ at $(t - t_0) = 0$ and let the strain be ε_{ij}'' at $(t - t_0) \neq 0$. Assuming, further, $\{\varepsilon_{ij}''(r)\}_L = \langle \varepsilon_{ij}''\rangle = \alpha_{ij}^{(L)}(t - t_0)$, we can write

$$\alpha_{ij}^{(L)}\langle \sigma_{ij}^0\rangle = \{\{\sigma_{ij}^0(r)\alpha_{ij}(r)\}_L\}_S \tag{5.65}$$

Recalling that $\{\sigma_{kl}^0(r)\}_L = [M_{klmn}(x_1, x_2)]^{-1}\{\varepsilon_{mn}(r)\}_L$, $\{\sigma_{ij}^0(r)\}_{L_i} = B_{ijkl}^{(i)}(x_i, x_j)$ $\{\sigma_{kl}^0(r)\}_L$, $\langle \sigma_{kl}^0\rangle = \{[M_{klij}(x_i, x_j)]^{-1}\}_S\langle \varepsilon_{ij}\rangle$, $M_{ijmn}(x_i, x_j) = S_{ijkl}^{(2)}I_{klmn} + \langle L_1(x_i, x_j)\rangle$ $[S_{ijkl}^{(1)} - S_{ijkl}^{(2)}]B_{klmn}^{(1)}(x_i, x_j)$, we obtain from eq. (5.65)

$$\alpha_{ij}^{(L)} = \{N_{kl}(x_i, x_j)/M_{klmn}(x_i, x_j)\}_S\{[M_{mnij}(x_i, x_j)]^{-1}\}_S^{-1} \tag{5.66}$$

where $N_{kl}(x_i, x_j) = \langle L_1(x_i, x_j)\rangle \alpha_{rr}^{(1)} B_{rrkl}^{(1)}(x_i, x_j) + \langle L_2(x_i, x_j)\rangle \alpha_{rr}^{(2)} B_{rrkl}^{(2)}(x_i, x_j)$.

The results obtained may be expressed as

$$\alpha_{rr}^{(L)} \leqslant \alpha_{rr} \leqslant \alpha_{rr}^{(U)} \tag{5.67}$$

Thus, the upper bound of the TEC and the lower bounds of the EME were obtained by application of eqs. (5.48), whereas the lower bound of the TEC and the upper bounds of the EME were obtained from eqs. (5.49). In the case when both components of the MHM are isotropic, eq. (5.67) may be rewritten as

$$\alpha^{(L)} < \alpha < \alpha^{(U)} \tag{5.68}$$

where $\alpha^{(L)} = \{N(x_k)\}_L$, $\alpha^{(U)} = \{N(x_i, x_j)M(x_i, x_j)\}_S\{M(x : x_j)_S^{-1}$, $N(x_k) = \{\alpha a\}_S/\{a\}_S$, $N(x_i, x_j) = \{a\}_L + 2(\alpha_1 - \alpha_2)(b_1 - b_2)\langle L_1(x_i, x_j)\rangle \langle L_2(x_i, x_j)\rangle/\{a^{-1}\}_L$, $M(x_i, x_j) =$

$[\{E^{-1}\}_L - 2(b_1 - b_2)^2 \langle L_1(x_i, x_j) \rangle \langle L_2(x_i, x_j) \rangle / \{a^{-1}\}_L]^{-1}$, $a_i = 18K_i G_i/(3K_i + 4G_i)$, $b_i = (3K_i - 2G_i)/18K_i G_i$, $E_i = 9K_i G_i/(3K_i + G_i)$ and $i = 1, 2$.

The effective thermoelastic parameters, K, G and α, will now be evaluated for a model MHM with a 'cube-in-cube'-type of the unit cell (Fig. 5.1).

As a first step, the unit cell will be assumed sectioned into two parts (Fig. 5.1(b)) as appropriate for estimation of the lower and upper bounds of the EME and the TEC, respectively. The effective bulk and shear moduli (κ' and γ', respectively) of the rectangular prism (first section) will be obtained by taking the property average along the Ox_3 axis. Since the relative length of the straight line drawn through component 1 is $\langle L_1 \rangle = \varphi^{1/3}$ (hence, $\langle L_2 \rangle = 1 - \langle L_1 \rangle$), we obtain from eqs. (5.59)

$$\kappa' = \left\{ \frac{n_2}{K_2} + \left(\frac{n_1}{K_1} - \frac{n_2}{K_2} \right) \varphi^{1/3} - \frac{[d_2 + (d_1 - d_2)\varphi^{1/3}][P_2 + (P_1 - P_2)\varphi^{1/3}]}{P_2 K_2 + (P_1 K_1 - P_2 K_2)\varphi^{1/3}} \right\}^{-1}$$

$$\gamma' = [\varphi^{1/3}/G_1 + (1 - \varphi^{1/3})/G_2]^{-1}$$

Recalling that the relative area of the prism cross-section by the plane $x_3 = \text{const}$ is $\langle S_1 \rangle = \varphi^{2/3}$ (hence, $\langle S_2 \rangle = 1 - \varphi^{2/3}$) and averaging the unit cell properties over the cross-section $x_3 = \text{const}$, we can determine the lower bounds of the unit-cell EME as

$$K^{(L)} = \frac{K_2 P_2 + (\kappa' P' - K_2 P_2)\varphi^{2/3}}{P_2 + (P' - P_2)^{2/3}} \tag{5.69a}$$

$$G^{(L)} = G_2 + (\gamma' - G_2)\varphi^{2/3} \tag{5.69b}$$

Using the same approach to the analysis of the TEC, we can derive the upper

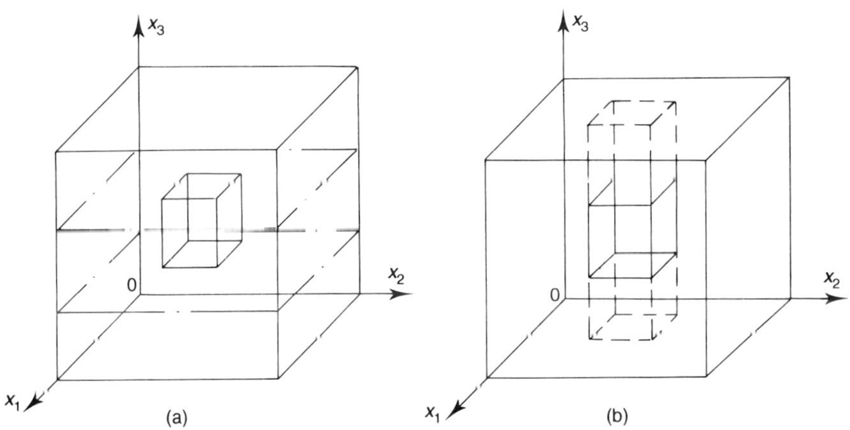

Fig. 5.1. Sectioning of the 'cube-in-cube'-type unit cell to estimate the upper (a) and lower (b) bounds of the moduli of elasticity

bound of the unit cell TEC as

$$\alpha^{(U)} = \frac{\alpha_2(1 - \varphi^{2/3})a_2 + \delta'\varphi^{2/3}a'}{(1 - \varphi^{2/3})a_2 + \varphi^{2/3}a'} \tag{5.69c}$$

where $\delta' = \alpha_2 + (\alpha_1 - \alpha_2)\varphi^{1/3} + 2(\varphi^{1/3} - \varphi^{2/3})(\alpha_1 - \alpha_2)(b_1 - b_2)a_1a_2/[a_2 + (a_1 - a_2\varphi^{1/3}]$ is the TEC of a rectangular prism obtained by averaging along the Ox_3 axis, and $a' = 18\kappa'\gamma'/(3\kappa' + \gamma')$.

The upper bounds of the EME and the lower bound of the TEC will be determined for the unit cell sectioned as shown in Fig. 5.1(a). In this case the property is first averaged over the cross-section $x_3 = \text{const}$, and subsequently along the Ox_3 axis.

Taking into account that the relative area of cross-section per component 1 (i.e. the inclusion) in the coresponding layer is $\langle S_1 \rangle = \varphi^{2/3}$ (hence, $\langle S_2 \rangle = 1 - \varphi^{2/3}$), we obtain from eqs. (5.59)

$$\kappa'' = \frac{K_2P_2 + (K_1P_1 - K_2P_2)\varphi^{2/3}}{P_2 + (P_1 - P_2)\varphi^{2/3}}$$

$$\gamma'' = G_2 + (G_1 - G_2)\varphi^{2/3}$$

Now, the upper bounds of the EME are obtained after averaging along the Ox_3 axis, i.e.

$$K^{(U)} = \left\{ \frac{n_2}{K_2} + \left(\frac{n_1}{\kappa''} - \frac{n_2}{K_2} \right)\varphi^{1/3} - \frac{2[d_2 + (d'' - d_2)\varphi^{1/3}][P_2 + (P'' - P_2)\varphi^{1/3}]}{P_2K_2 + (\kappa''P'' - K_2P_2)\varphi^{1/3}} \right\}^{-1} \tag{5.70a}$$

$$G^{(U)} = [(1 - \varphi^{1/3})/G_2 + \varphi^{1/3}/\gamma'']^{-1} \tag{5.70b}$$

After a similar averaging procedure we estimate the lower bound of the TEC from eq. (5.68):

$$\alpha^{(L)} = \alpha_2 + (\delta'' - \alpha_2)\varphi^{1/3} + \frac{2(\varphi^{1/3} - \varphi^{2/3})(\alpha_2 - \delta'')(b_2 - b_1)a''a_2}{a_2 + (a'' - a_2)\varphi^{1/3}} \tag{5.71}$$

where $\delta'' = [\alpha_1\varphi^{2/3}a_1 + \alpha_2a_2(1 - \varphi^{2/3})]/[a_1\varphi^{2/3} + a_2(1 - \varphi^{2/3})]$ is the TEC of the layer of relative thickness $\langle L_1 \rangle = \varphi^{1/3}$.

It becomes obvious from the above examples that estimation of the lower bounds of the EME and of the upper bound of the TEC involves, essentially, sectioning the appropriate unit cell into prisms and successive averaging of the properties, first, along the selected direction, and finally over the cross-section normal to that direction. On the other hand, the upper bounds of the EME and the lower bound of the TEC are obtained by averaging in the reverse order, namely first over the cross-section normal to the selected direction, and subsequently along that direction.

Apparently, a combination of both methods is also possible. In that case the MHM is conditionally sectioned into two parts, and the effective properties of the first and second parts are estimated by equations appropriate for the upper and lower bounds, respectively. The final averaging over the whole MHM volume yields values that lie between the bounds described by eqs. (5.58) and (5.68).

5.2 THEORETICAL ASPECTS: PHYSICAL APPROACHES

As already stated in Chapter 4, the term 'physical' approach was introduced to differentiate the models which explicitly accounted for the specific structural features of a MHM (i.e. the onset of an 'inifinite' cluster of a disperse component at the percolation threshold φ_c and the transition of a portion of the continuous polymer phase into a 'boundary interphase', BI), from 'pragmatic', phenomenological approaches in which those two fundamental features were neglected. The need to consider the effect of at least one of the latter was recognized in many studies of thermoelastic properties of a MHM (e.g. [29–36]); however, in the discussion to follow the main emphasis will be on the SSA model which explicitly accounts for both the percolation phenomenon and the formation of a BI.

5.2.1 STEP-BY-STEP AVERAGING APPROACH

The Percolation Model

The elastic properties of a percolating system will be completely specified by a Hamiltonian of the selected ensemble of sites and bonds which is characterized by the random distribution of all geometrical parameters (numbers of sites and/or bonds, distances to the most remote elements, etc.). This Hamiltonian should satisfy the following conditions:

(i) Elastic connectivity (i.e. the elasticity moduli of the quasi-lattice must be finite at $\varphi > \varphi_c$ and must vanish at $\varphi \to \varphi_c + 0$).

(ii) Correct representation of the tensor properties of long elastic chains.

(iii) Invariance with respect to rotations in the Hamiltonian free states.

The first condition is apparently met by the Born model; however, this model predicts that both the longitudinal and transversel elasticity constants of a linear chain of N bonds (i.e. the lattice equivalent of a rod of length L) decrease as N^{-1}, whereas we would expect the rods to be more compliant with respect to transversal displacements (i.e. the corresponding elasticity constant should decrease as L^{-3}). In other words, the scalar Born model overestimates the chain stiffness in the vicinity of φ_c and is thus inconsistent with condition (ii).

Apparently, more correct elastic behavior of uninterrupted chains is predicted by the following Hamiltonian which provides for the coincidence of elastic and geometrical percolation thresholds [37–40]:

$$\mathcal{H} = (G/4) \sum_{i,j,k} \alpha_{ij}\alpha_{jk}\delta\Phi_{ijk} + (Q/4a^2)\sum_{i,j}(u_i - u_j)_{\parallel}^2$$

where $\delta\Phi_{ijk}$ is the variance of the angle between bonds $\{(ij),(jk)\}$, a is the lattice constant, α_{ij} is the random variable of $\{0,1\}$ with probabilities $(1 - \varphi)$ and φ, respectively, Q and G are the local elastic constants and $(u_i - u_j)_{\parallel}$ is the variance of displacements of sites i and j parallel to the direction of the bond (i,j).

The stiffness of the percolation network is provided by InC comprising compact, multiple-bond regions connected by reasonably linear single bonds (cf. Fig. 4.5(b)). Thus, the problem of the elastic behavior of a percolating system may be reduced to that of chains of N vectors (i.e. bonds) $\{b_i\}$ each. The Hamiltonian appropriate for such chains may be written as

$$\mathcal{H} = (G/2) \sum_{i=1}^{N} \delta\Phi_i^2 + (Q/2a^2) \sum_{i=1}^{N} \delta b_i^2 \tag{5.72}$$

The elastic constant of the chain of vectors will be determined by the relative angle variance $\delta\Phi_i$ arising from the orientation of the bonds under the action of the stretching force F applied to the chain end. This quantity may be obtained by minimization of the energy, $\mathcal{W} = \mathcal{H} - F\,(R'_N - R_n)$, where $(R'_N - R_N)$ is the chain end displacement from the equilibrium position.

Expressing the work of stretching as

$$F(R'_N - R_N) = (F \times z) \sum_{i=1}^{N} \delta\Phi_i \sum_{j=i}^{N} b_j + (F/a) \sum_{i=1}^{N} b_i\delta b_i$$

(where z is the unit vector normal to the chain plane), and using the exact expression for the angle variance between two bonds after minimization of W, i.e.

$$\delta\Phi_i = [(F \times z)/G] \sum_{j=i}^{N} b_j = [(F \times z)/G](R_N - R_{i-1})$$

(where $\delta b = (a/b)(F \times b)$, R_i is the equilibrium location of the end of vector b_i), we derive from eq. (5.72)

$$\mathcal{H} = F^2 N S_{\perp}^2/2G - F^2 a L_{\parallel}/2Q$$

where

$$S_{\perp}^2 = (1/NF^2) \sum_{i=1}^{N} \{(F \times z)(R_{i-1} - R_N)\}^2$$

is the square of the inertial projection of the site R_i on the direction $F \times z$, and

$$L_{\parallel} = (1/aF^2) \sum_{i=1}^{N} (F \times b)$$

Thus, the force constant (i.e. the stiffness) of the chain may now be expressed as

$$K = G/(NS_\perp^2) \tag{5.73}$$

According to eq. (5.73), the elastic stiffness of the long chain depends both on the chain length, N, and on the shape parameter, $S_\perp$.

It is pertinent to emphasize here that the above results obtained for the two-dimensional case ($d = 2$) may be readily extended to higher dimensionalities. In fact, the chain deformation may be represented, as before, by a succession of rearrangements, where the ith rearrangement involves stretching of the bond b_i and rotation of the bonds $b_i \cdots b_N$ around R_{i-1}. The infinitesimal, d-dimensional rotation may be specified by an array $(d + 1)(d - 2)/2$. This array defines the orientation of a two-dimensional plane in a d-dimensional space, as well as the angle variance $\delta\Phi_i$ characterizing the rotation within that plane. Hence, substitution of the array of $\delta\Phi_i$ into eq. (5.72) yields again eq. (5.73).

The results obtained may now be used to estimate the critical index, τ, of the elastic constant in the vicinity of the percolation threshold, i.e.

$$K \sim (\varphi \quad \varphi_c)^\tau \sim |\Delta\varphi|^\tau \tag{5.74}$$

Let the percolation system be sectioned into cubes of linear dimension $b - \xi$; then the macroscopic elastic constant of the relevant bonded InC may be expressed as

$$K = K_b b^{2-d} \tag{5.75}$$

(where K_b is the elasticity of a single cube). Assuming scale invariance of the elastic behavior (i.e. a self-similar elastic regime), and recalling eq. (4.88), we derive

$$K_b \sim K_1(\varphi - \varphi_c)^n = K_1 b^{n/v} \tag{5.76a}$$

(where K_1 is the elastic constant of a single bond and v is the critical index of the correlation length ξ). Making use of eq. (5.73) we can write, alternatively,

$$K_b \sim 1/LS_b^2 = b^{-2}b^{-1/v} \tag{5.76b}$$

Comparing eqs. (5.76a) and (5.76b), we obtain $n/v = 2 + 1/v$, or $n = 2v + 1$. Combining this result with eqs. (5.74) and (5.75), we obtain

$$K \sim K_1 b^{(1+vd)/v} = K_1|\Delta\psi|^{vd+1}; \tag{5.77}$$

therefore the critical index of elasticity may be expressed as

$$\tau = vd + 1 \tag{5.78}$$

Substitution of $v = 1.33$ ($d = 2$) and $v = 0.8$–0.9 ($d = 3$) into eq. (5.63) yields $\tau = 3.66$ and $\tau = 3.4$–3.7, respectively. It can easily be shown, further, that $\tau - t = 1 + v$ [where t is the critical conductivity index, cf. eq. (4.93)].

The thermoelastic properties of a percolating system were evaluated for the same structural model (Fig. 4.6) [27, 28, 41, 42] which was used before in the discussion of the heat conductivity problem (cf. section 4.2.1). The bulk modulus, κ_i, shear modulus, γ_i, and TEC, δ_i, of every ith section were estimated first by eqs. (5.58) and (5.67) and then averaged over the cross-section of the unit cell to obtain the desired effective properties of the percolating MHM.

Making use of eqs. (5.58) and (5.67), we can write for sections II and III (indices $i = 2$ and 1, respectively)

$$\kappa_i = \left[(n/K)_{\langle l_i \rangle} - \frac{2\{d\}_{\langle l_i \rangle}\{P\}_{\langle l_i \rangle}}{\{KP\}_{\langle l_i \rangle}} \right]^{-1}$$

$$\gamma_i = \{G^{-1}\}_{\langle l_i \rangle}$$

$$\delta_i = \{\alpha\}_{\langle l_i \rangle} + \frac{2\langle l_i \rangle[1 - \langle l_i \rangle](b_1 - b_2)(\alpha_1 - \alpha_2)}{\{a^{-1}\}_{\langle l_i \rangle}}$$

In the next step, the properties are averaged over the sections (II + III) and (I + IV) (indices 3 and 4, respectively), to obtain

$$\kappa_3 = \{\kappa h\}_{F_1}/\{h\}_{F_1}, \qquad \kappa_4 = \{KP\}_{F_2}/\{P\}_{F_2}$$
$$\gamma_3 = \{\gamma\}_{F_1}, \qquad \gamma_4 = \{G\}_{F_2}$$
$$\delta_3 = \{\delta b\}_{F_1}/\{b\}_{F_1}, \qquad \delta_4 = \{\delta b\}_{F_2}/\{b\}_{F_2}$$

where $F_1 = S_3/(S_2 + S_3)$, $F_2 = S_1(S_1 + S_4)$, $h_k = 6\psi/(3 + 4\psi)$, $\psi = \gamma_k/\kappa_k$, $b_k = 18\kappa_k\gamma_k/(3\kappa_k + 4\gamma_k)$ and $k = 1, 2, 3, 4$.

The final step involving averaging over the whole BRE (i.e. sections (II + III) + (I + IV), yields

$$K = \frac{\kappa_3 h_3(\langle S_2 \rangle + \langle S_3 \rangle) + \kappa_4 h_4(\langle S_1 \rangle + \langle S_4 \rangle)}{h_3(\langle S_2 \rangle + \langle S_3 \rangle) + h_4(\langle S_1 \rangle + \langle S_4 \rangle)} \tag{5.79a}$$

$$G = \gamma_3(\langle S_2 \rangle + \langle S_3 \rangle) + \gamma_4(\langle S_1 \rangle + \langle S_4 \rangle) \tag{5.79b}$$

$$\alpha = \frac{\delta_3 b_3(\langle S_2 \rangle + \langle S_3 \rangle) + \delta_4 b_4(\langle S_1 \rangle + \langle S_4 \rangle)}{b_3(\langle S_2 \rangle + \langle S_3 \rangle) + b_4(\langle S_1 \rangle + \langle S_4 \rangle)} \tag{5.79c}$$

Equation (5.79) should apply to a MHM with both large (i.e. EME_2/EME_1, $TEC_1/TEC_2 \leqslant 10^{-4}$) and small (i.e. $1 > EME_2/EME_1$, $TEC_1/TEC_2 > 10^{-2}$) differences between the thermoelastic properties of components.

The Model of an Equivalent Element

The thermoelastic properties of an equivalent element (EE) will be evaluated for each section (Fig. 4.8) as described in detail in the treatment of heat conductivity (cf. section 4.2.1). Estimates of the upper and lower bounds of each property are combined to yields ultimately the values close to the arithmetic means.

The bulk modulus of section I (subscript a) may be expressed as the lower bound in eq. (5.58a), i.e.

$$K_a = S_I^{-1} \iint_{(S_I)} K(x_1, x_2) dx_1 \, dx_2$$

Transformation to polar coordinates normalized by the mean radius of inclusion, r, yields

$$K_a = (1/\pi y_3^2) \int_0^{2\pi} d\theta \int_0^{y_3} K(\rho) \rho \, d\rho \tag{5.80a}$$

where

$$K(\rho) = [\{n/K\}_{L_1(\rho)}^{(1,BI)} - 2\{d\}_{L_1(\rho)}^{(1,BI)} \{P\}_{L_1(\rho)}^{(1,BI)} / \{KP\}_{L_1(\rho)}^{(1,BI)}]^{-1}.$$

In a similar fashion, we derive for the shear modulus of section I

$$G_a = (S_I^{-1}) \iint_{(S_I)} G(x_1, x_2) dx_1 \, dx_2 = (1/\pi y_3^2) \int_0^{2\pi} d\theta \int_0^{y_3} G(\rho) \rho \, d\rho \tag{5.80b}$$

(where $G(\rho) = \{G\}_{L_1(\rho)}^{(1,BI)}$), and for the TEC

$$\alpha_a = \left(\int_0^{2\pi} d\theta \int_0^{y_3} \delta(\rho) \eta(\rho) \rho \, d\rho \right) \Big/ \left(\int_0^{2\pi} d\theta \int_0^{y_3} \eta(\rho) \rho \, d\rho \right) \tag{5.80c}$$

where

$$\delta(\rho) = \{\alpha\}_{L_1(\rho)}^{(1,BI)} + L_1(\rho)[1 - L_1(\rho)](\alpha_{BI} - \alpha_1)(b_{BI} - b_1)/\{a^{-1}\}_{L_1(\rho)}^{(1,BI)}$$

and

$$\eta(\rho) = [\{E^{-1}\}_{L_1(\rho)}^{(1,BI)} - 2L_1(\rho)[1 - L_1(\rho)](b_{BI} - b_1)^2/\{a^{-1}\}_{L_1(\rho)}^{(1,BI)}$$

Section II is connected parallel to I; therefore, the properties of the region $(I + II)$ are the averages over the cross-section $\langle S_1 \rangle$ (subscript b), i.e.

$$K_b = \{KP\}_{\langle S_I \rangle}^{(a,BI)} / \{P\}_{\langle S_I \rangle}^{(a,BI)} \tag{5.81a}$$

$$G_b = \{G\}_{\langle S_I \rangle}^{(a,BI)} \tag{5.81b}$$

$$\alpha_b = \{a\alpha\}_{\langle S_I \rangle}^{(a,BI)} / \{a\}_{\langle S_I \rangle}^{(a,BI)} \tag{5.81c}$$

where $\langle S_1 \rangle = r_3^2 / r_{4c}^2$ is the cross-sectional area of section I normalized to that of the region $(I + II)$, i.e. $S_b = \pi r_{4c}^2$.

In a similar fashion, we obtain for the region $(III + IV)$ (subscript c)

$$K_c = \{KP\}_{\langle S_I \rangle}^{(1,BI)} / \{P\}_{\langle S_I \rangle}^{(1,BI)} \tag{5.82a}$$

$$G_c = \{G\}_{\langle S_I \rangle}^{(1,BI)} \tag{5.82b}$$

$$\alpha_c = \{a\alpha\}_{\langle S_I \rangle}^{(1,BI)} / \{a\}_{\langle S_I \rangle}^{(1,BI)} \tag{5.82c}$$

As soon as the regions $(I + II)$ and $(III + IV)$ are connected in series, the properties of this series, $(I + II) + (III + IV)$ (subscript d), should be the averages along the length $L = r + \Delta l$, i.e.

$$K_d = [\{n/K\}_{L_a}^{(b,BI)} - 2\{d\}_{L_a}^{(b,BI)}\{P\}_{L_a}^{(b,BI)}/\{KP\}_{L_a}^{(b,BI)}]^{-1} \tag{5.83a}$$

$$G_d = \{G^{-1}\}_{L_a}^{(b,BI)} \tag{5.83b}$$

$$\alpha_d = \{\alpha\}_{L_a}^{(b,BI)} + 2L_a(1 - L_a)(b_b - b_{BI})(\alpha_b - \alpha_{BI})/\{a^{-1}\}_{L_a}^{(b,BI)} \tag{5.83c}$$

where $L_a = 1 - z_2/z_1$.

It can easily be verified now that the EE consists of two regions, $(I + II) + (III + IV)$ and V, respectively, connected in parallel. Taking the average over the cross-section $S = \pi r_4^2$, we obtain finally for EE

$$K_{EE} = \{KP\}_{\langle S_{II}\rangle}^{(M,d)}/\{P\}_{\langle S_{II}\rangle}^{(M,d)} \tag{5.84a}$$

$$G_{EE} = \{G\}_{\langle S_{II}\rangle}^{(M,d)} \tag{5.84b}$$

$$\alpha_{EE} = \{a\alpha\}_{\langle S_I\rangle}^{(M,d)}/\{a\}_{\langle S_{II}\rangle}^{(M,d)} \tag{5.84c}$$

where $\langle S_{II}\rangle = (r_{4c}/r_4)^2$ is the normalized cross-sectional area of region d.

5.3 TESTS OF THEORETICAL PREDICTIONS

5.3.1 ANALYTICAL TEST

Predictions of the SSA approach will be now illustrated using the example of the composition dependence of Young's modulus (dimensionless) of a MHM for the following representative cases.

(i) Fixed parameters: $E_1 = 500$, $E_2 = 5$, $v_1 = 0.2$, $v_{BI} = 0.49$, $\varphi_c = 0.15$, $\tau = 3.6$, $\Delta l/r = 0.01$; variable parameters: v_2, E_{BI}, φ^*.

As right expected, the increase in E with φ becomes steeper the higher is E_{BI} and/or the lower is φ^*; v_2 is of secondary importance (Fig. 5.2).

(ii) Fixed parameters: $E_1 = 500$, $E_2 = E_{BI} = 5$, $v_1 = 0.2$, $v_2 = 0.49$, $\varphi_c = 0.15$, $\varphi^* = 0.4$, $\tau = 3.6$, $\Delta l/r = 0.01$; variable parameter: v_{BI}.

The difference between the theoretical curves is not significant over most of the composition interval except near φ^* where the values of E tend to increase the lower is v_{BI} (Fig. 5.3).

(iii) Fixed parameters: $E_1 = E_{BI} = 500$, $E_2 = 5$, $v_1 = 0.2$, $v_2 = 0.49$, $\varphi_c = 0.15$, $\tau = 3.6$; variable parameters: $v_{BI}, \varphi^*, \Delta l/r$.

The values of E tend to increase the higher is $\Delta l/r$ and/or the lower is φ^*; the influence of v_{BI} is less significant (Fig. 5.4).

(iv) Fixed parameters: $E_2 = 5$, $v_1 = 0.2$, $v_2 = 0.49$, $\varphi_c = 0.15$, $\tau = 3.6$, $\Delta l/r = 0.01$; variable parameters: $E_1, E_{BI}, v_{BI}, \varphi^*$.

The theoretical curves become steeper the higher are E_1, E_{BI}, φ^*; the influence of v_{BI} is insignificant (Fig. 5.5).

Summarizing, it is the elastic moduli of the filler and the BI, E_1 and E_2, as

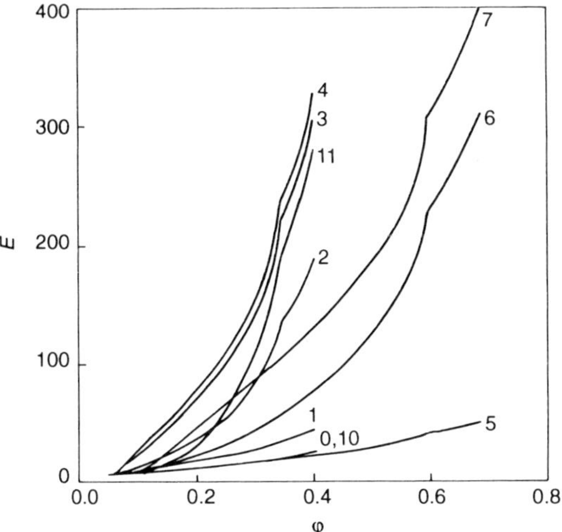

Fig. 5.2. Composition dependence of Young's modulus of a MHM calculated for case (i) assuming $E_{BI} = 5$ (0, 5, 10), 10 (10), 100 (2, 6), 500 (3, 7, 11) and 750 (4, 8); $\varphi^* = 0.4$ (0–4 10, 11) and 0.7 (5–8); $v_2 = 0.49$ (0–8) and 0.3 (10, 11)

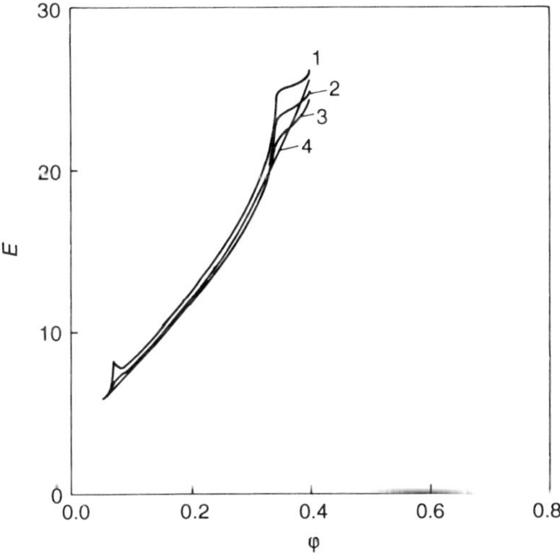

Fig. 5.3. Composition dependence of Young's modulus of a MHM calculated for case (ii) assuming $v_{BI} = 0.2$ (1), 0.3 (2), 0.4 (3) and 0.49 (4)

well as the limiting filler content in InC, φ^*, which determine both the pattern of composition dependence and the absolute values of Young's modulus of a MIIM. Similar model calculations reveal the same trends for bulk and shear moduli; as might have been expected, however, the values of the TEC, α, behave in exactly the opposite manner.

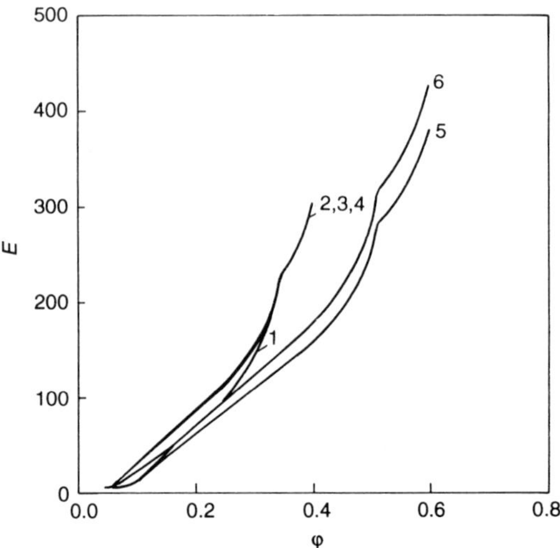

Fig. 5.4. Composition dependence of Young's modulus of a MHM calculated for case (iii) assuming $v_{BI} = 0.49$ (1, 5, 6), 0.1 (2), 0.2 (3) and 0.3 (4); $\varphi^* = 0.4$ (1–4) and 0.6 (5, 6); $\Delta l/r = 0.01$ (1–5) and 0.1 (6)

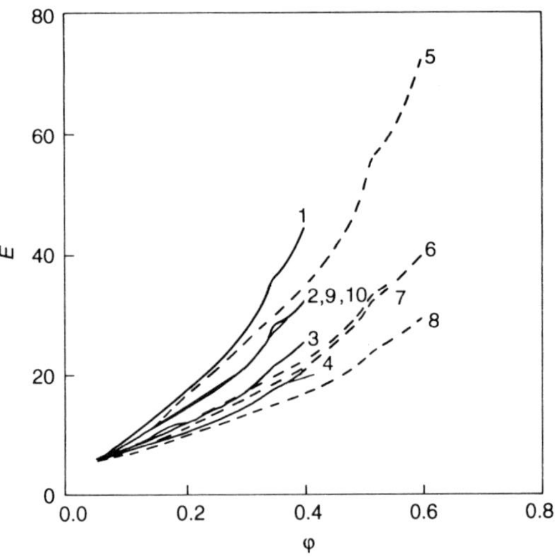

Fig. 5.5. Composition dependence of Young's modulus of a MHM calculated for case (iv) assuming $E_1 = 500$ (1, 3, 5) and 50 (2, 4, 6, 8, 9, 10); $E_{BI} = 5$ (3, 7), 10 (1, 4, 5, 8) and 50 (2, 6, 9, 10); $\varphi^* = 0.4$ (1–4) and 0.6 (5–10); $v_{BI} = 0.49$ (1–8) and 0.2 (9, 10)

5.3.2   EXPERIMENTAL TEST: FILLED POLYMERS

Elasticity Moduli

As can be seen from Figs. 5.6–5.9, the experimental values of Young's modulus (E) of polyethylene (PE) melt-blended with activated calcite (AC, mean particle size $2r = 1$–$10\,\mu m$) [43] and of cross-linked epoxy resin (CE) filled with silica [44] or with glass beads (GB, $2r = 70\,\mu m$) [45], as well as similar data for shear moduli (G) of PE and random (RC) and grafted (GC) copolymers of ethylene and acrylic acid filled with Al ($2r = 100\,\mu m$) [46] are generally in better agreement with the predictions of the SSA model with the fitting parameters from Table 5.1 (solid lines) than with either of the bounds (broken lines) calculated by eqs. (5.41) and (5.42).

Thermal Expansion Coefficient

Values of the TEC in the melt state (i.e. above T_g) for PS and PMMA filled with glass beads [47], styrene/acrylonitrile copolymer filled with crushed glass fibers [48] and metal-filled, cross-linked epoxies [49] changed approximately linearly with composition, whereas negative deviations from linear additivity were observed in the temperature interval below T_g. The latter data could be reasonably approximated by one of the current pragmatic approaches (e.g.

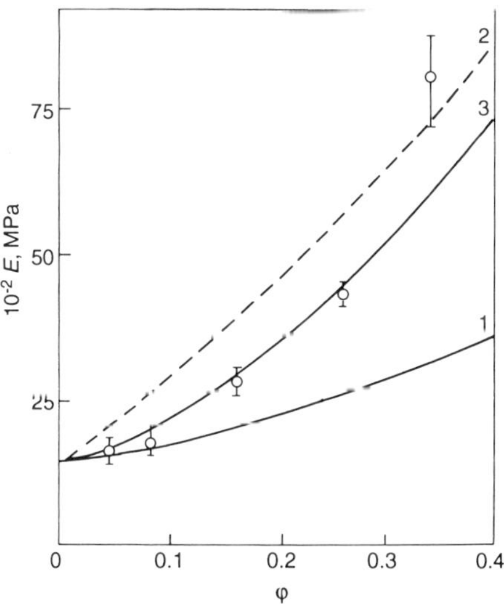

Fig. 5.6. Composition dependence of Young's modulus of PE filled with activated calcite ($\bigcirc$) and predictions of eqs. (5.41) (curves 1 and 2 for the lower and upper bounds, respectively), and the SSA model (curve 3)

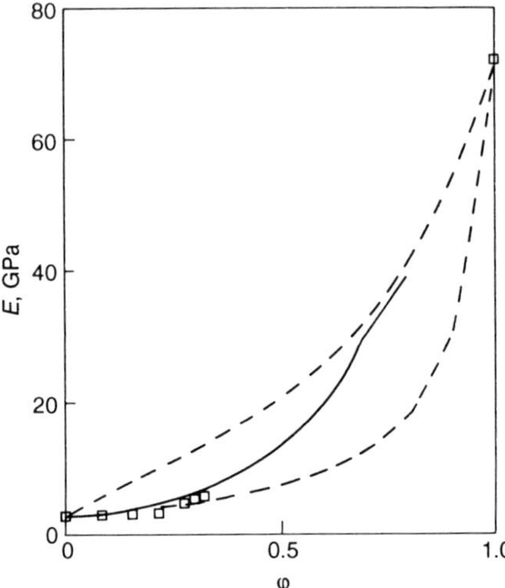

Fig. 5.7. Composition dependence of Young's modulus of an epoxy polymer filled with glass beads ($\square$) and predictions of eq. (5.41) for the upper and lower bounds (broken lines) and the SSA model (solid line)

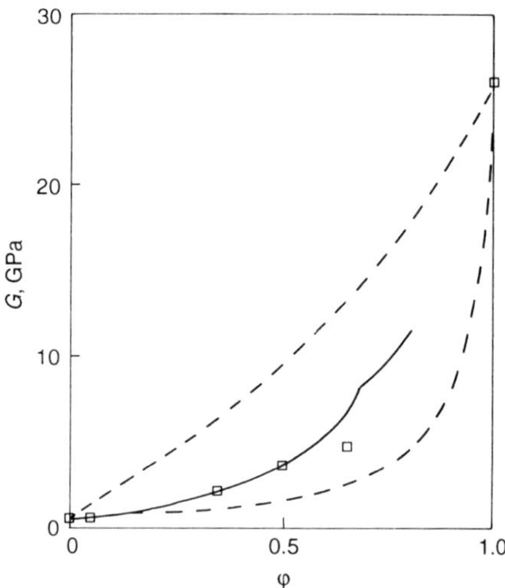

Fig. 5.8. Composition dependence of the shear modulus of PE filled with aluminium ($\square$) and predictions of eq. (5.41) for the upper and lower bounds (broken lines) and the SSA model (solid line)

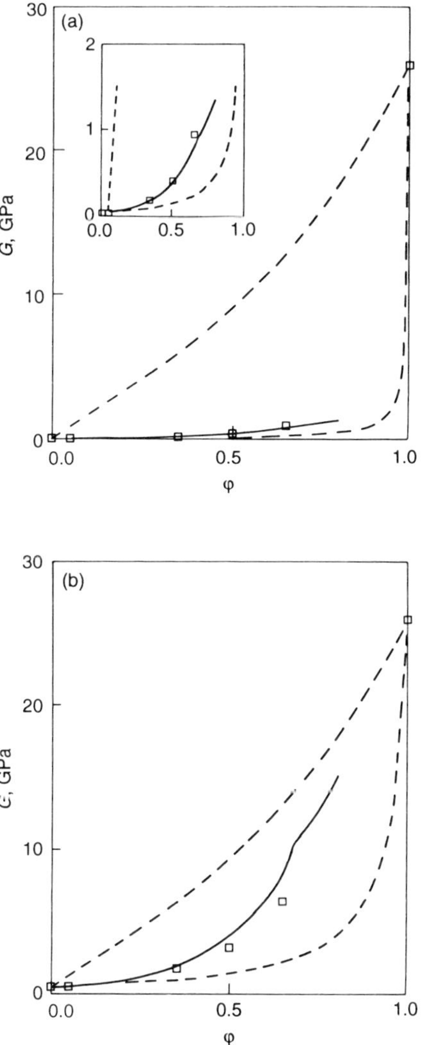

Fig. 5.9. Composition dependence of the shear modulus of random (a) and graft (b) copolymers E/ΛΛ filed with aluminium (□) and predictions of eq. (5.41) for the upper and lower bounds (broken lines) and SSA model (solid line)

⌊11, 50⌋); none of the latter can, however, account explicitly for the the particle size effect (see below).

Monitoring of the temperature dependence of the linear dimensions of cross-linked epoxy resin (ER) filled with unfractionated powders of copper and glass revealed an apparent increase in T_g by some 15–20 K concomitant with

Table 5.1 Parameters of the SSA model for the elastic properties of two-component polymer materials calculated assuming $\tau = 3.6$, $\Delta l/r = 0.05$–0.10 (E, G and K in [GPa])

Material	E	G	K	ν	φ_c	φ^*
			(a) Filled polymers			
AC	2.6	⋯	⋯	0.27		
Silica	73.1	⋯	⋯	0.20		
Al	78	29	87	0.35		
GB	72	30	40	0.20		
PE/AC*					0.15	0.60
matrix	0.15	—	—	0.45		
BI	0.75	—	—	0.45		
PE/Al					0.15	0.35
matrix	1.5	0.53	0.80	0.23		
BI	9.1	4.3	3.4	0.05		
GC/Al					0.15	0.35
matrix	1.45	0.59	0.90	0.23		
BI	6.0	2.9	2.2	0.05		
RC/Al					0.15	0.35
matrix	0.12	0.05	0.07	0.20		
BI	0.55	0.23	0.30	0.20		
CE/GB					0.17	0.67
matrix	26.9	10.1	2.64	0.33		
BI	13.0	4.6	21.7	0.40		
			(b) Polymer blends			
PC/PMMA					0.15	0.60
PC	2.76	1.29	1.89	0.1		
PMMA	2.27	0.87	1.07	0.3		
BI	6.0	2.0	1000	0.49		
PP/EPDM					0.15	0.60
PP	2.09	0.82	1.5	0.27		
EPDM	0.004	0.012	0.60	0.49		
BI	0.005	0.002	0.002	0.1		
			(c) Block copolymers			
PS/PB					0.15	0.60
PS	1.69	0.63	1.66	0.33		
PB	0.0014	0.0005	0.023	0.49		
BI	1.69	0.63	1.66	0.33		

*$\Delta l/r = 0.19$.

a decrease in the TEC (α), the latter effect becoming more pronounced the lower the filler particle size [51].

As can be seen from Fig. 5.10, the experimental data for ER filled with copper (broken lines) may be quantitatively described by SSA theory (solid lines) assuming $\Delta l/r$ to decrease from 0.18 to 10^{-3} as the particle size increases from about $5\,\mu m$ (broken line 1) to $75\,\mu m$ (broken line 3) (in calculations,

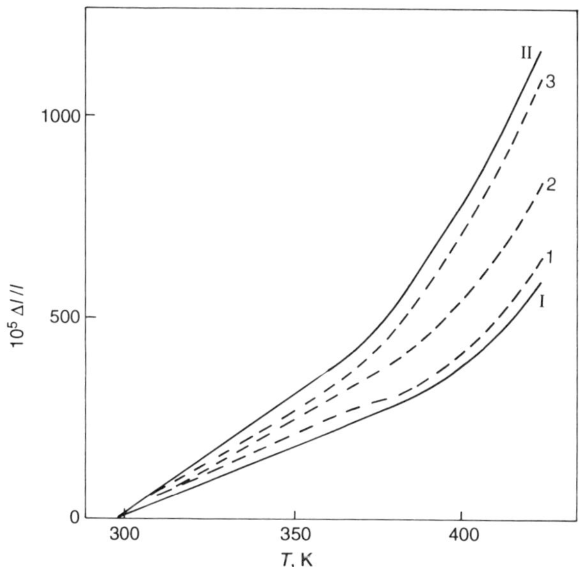

Fig. 5.10. Temperature dependence of the relative length change of epoxy polymer filled with copper particles ($\varphi = 0.3$) of the following sizes (in microns): 2.6–5 (1); 37–45 (2); 63–75 (3). Curves I and II were constructed by the SSA model assuming $\Delta l/r = 0.180$ and 0.001, respectively

the thermoelastic parameters of ER were approximated as [52] E_2 (GPa) $= 34.25 - 24.80 \times 10^{-2}\ T + 77 \times 10^{-5}\ T^2 - 8.76 \times 10^{-7}\ T^3$; $\alpha_2 (10^{-6}\ K^{-1}) = -46.8 + 110.0 \times 10^{-2}\ T - 48.5 \times 10^{-4}\ T^2 + 74.2 \times 10^{-7}\ T^3$; $\nu_2 = 0.28 - 9.17 \times 10^{-6}\ T - 1.2 \times 10^{-6}\ T^2 + 2.5 \times 10^{-9}\ T^3$; those for copper were assumed to be temperature-invariant, i.e. $G_1 = 106.8\ \text{GPa}$, $K_1 = 49.0\ \text{GPa}$, $\alpha_1 = 16.2 \times 10^{-6}\ K^{-1}$; the best-fit values of the thermoelastic parameters of a BI were $K_{BI}/K_2 = G_{BI}/G_2 = \alpha_2/\alpha_{BI} = 5$).

5.3.3 EXPERIMENTAL TEST: POLYMER BLENDS

Elasticity Moduli

As can be seen from Fig. 5.11, the static Young's moduli of melt-blended specimens of isotactic polypropylene (PP) and ethylene–propylene–diene terpolymer (EPDM) [53, 54] may be quantitatively described by both eq. (5.41) for the upper bound (i.e. assuming continuity of the PP phase) and the SSA model with parameters shown in Table 5.1. However, no theoretical model including SSA, apparently, is quantitatively applicable to account for the composition dependence of room-temperature values of E for melt-blended specimens of poly(methyl methacrylate) (PMMA) and polycarbonate (PC) [55] (Fig. 5.12).

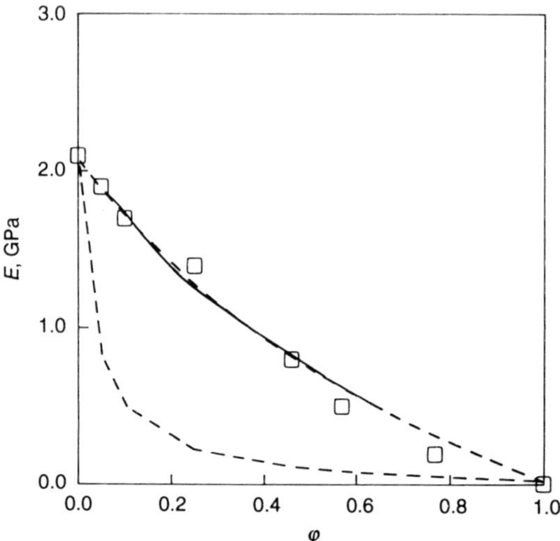

Fig. 5.11. Composition dependence of Young's modulus of PP/EPDM blends ($\square$) and predictions of eq. (5.41) for the upper and lower bounds (broken lines) and the SSA model (solid line)

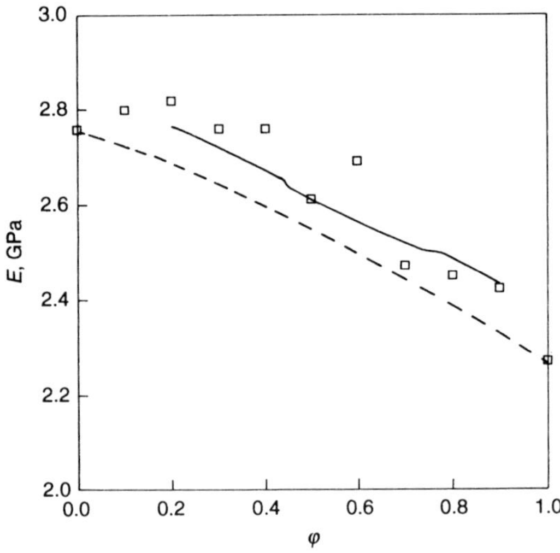

Fig. 5.12. Composition dependence of Young's modulus of PC/PMMA blends ($\square$) and predictions of eq. (5.41) for the upper bound (broken line) and the SSA model (solid line)

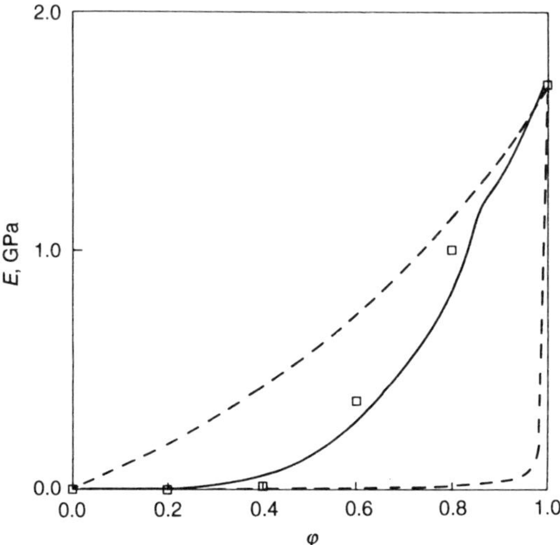

Fig. 5.13. Composition dependence of Young's modulus of S/B block copolymer ($\square$) and predictions of eq. (5.41) for the upper and lower bounds (broken line) and the SSA model (solid line)

5.3.4   EXPERIMENTAL TEST: BLOCK COPOLYMERS

Elasticity Moduli

Both eq. (5.41) for the upper bound and the SSA model with the values of fitting parameters shown in Table 5.1 quantitatively account for the concentration dependence of Young's modulus of styrene/butadiene block copolymer [56, 57] in the entire range of copositions (Fig. 5.13).

5.4   CONCLUSIONS

Evidence presented in this chapter allows us to draw a broad conclusion that the SSA model is superior compared with other current approaches as concerns the quality of fit to experimental data on the composition dependence of the elastic moduli and the thermal expansion coefficient of polymer MHM, provided the corresponding properties of pure components differ by several orders of magnitude (as was the case with filled polymers); otherwise (i.e. in the case of polymer blends and/or block copolymers), the performances of the SSA model and of the best pragmatic approaches are comparable. However, in the case of filled polymers the following considerations are relevant

(i) It was implicit in current theoretical models that the structure (and/or properties) of the components either remains unchanged after formation of a

MHM (as in the case of the 'pragmatic' approaches) or the eventual structural change may occur only in the low-modulus (polymer) component (as in the alternative case of 'physical' approaches). Recent X-ray diffraction experiments, however, revealed a non-negligible change in the lattice parameters of filler particles due to a stress build-up at the matrix–filler interface [58, 59]. As a result, in tensile experiments the local stress acting on filler particles was several-fold larger than the nominal stress in the composite, the observed 'mechanical reinforcement' effect increasing, the smaller the particle size. Thus, the particle size issue finds a different physical explanation; alternatively, these data may also be used to rationalize the similar values for the properties of a BI derived by data treatment within the framework of the SSA model, on the one hand, and of a filler, on the other (cf. Tables 4.4 and 5.1).

(ii) The values of the linear TEC for the melt state (α_l) of two polystyrene samples of different molar masses (PS-3.5 and PS-10) solution-filled with silica smoothly decreased with filler content φ, whereas those for the glassy state (α_g) decreased, passed through a minimum at the apparent 'critical' filler content φ^* (0.43 and 0.27 for PS-3.5 and PS-10, respectively) and then smoothly approached the filler TEC (Fig. 5.14) [60].

The concentration dependence of α_l for both polymers could be reasonably approximated (solid lines in Fig. 5.14) by the simple Turner's equation [11, 50]:

$$\alpha = [(1 - \varphi)\alpha_2 + \varphi\alpha_1 K_1/K_2]/(1 - \varphi + \varphi K_1/K_2) \tag{5.85}$$

assuming $K_1/K_2 = 25$ and 8 for PS-3.5 and PS-10, respectively. The value of $K_1 = 30\,\text{GPa}$ estimated from the experimental value of $K_2 = 1.2\,\text{GPa}$ for molten PS-3.5 [47, 61] is comparable with $K_1 = 45\,\text{GPa}$ for bulk silicate glass [62]: the assumed threefold increase in K_2 for PS-10 compared with PS-3.5 also seems quite reasonable.

Equation (5.85) satisfactorily accounts for the composition dependences of α_g (broken lines in Fig. 5.14 calculated assuming $K_1 = 30\,\text{GPa}$, $K_2 = 5.7\,\text{GPa}$ [30] and $\alpha_2 = 0.75 \times 10^{-4}\,\text{K}^{-1}$) at $\varphi > \varphi^*$. As shown elsewhere [63–65], the critical filler content φ^* corresponds to complete saturation of interactions at the filler–polymer interface when the thickness of the interparticle gap becomes comparable with the gyration radius of an unperturbed macromolecular coil, $\langle R_g \rangle$. Thus, the large negative deviations of the experimental values of α_g from the theoretical curve observed at $\varphi < \varphi^*$ is indicative of the structural hetero-geneity of a continuous polymer phase (i.e. a BI with changed structure and the remaining unperturbed polymer).

(iii) The best fit of experimental data on the composition dependence of the thermoelastic properties of polymer MHM to predictions of either the 'pragmatic' or 'physical' approaches is frequently achieved assuming the limiting filler content φ^* to be a fitting parameter rather than a theoretical constant, $\varphi^* = 0.6$. In fact, treatment of the experimental data on Young's modulus by eq. (5.42) yielded the best-fit value $\varphi^* = 0.5$ for a series of particulate-filled polymers [66]; an even lower value ($\varphi^* = 0.10$) was obtained from a similar analysis of the

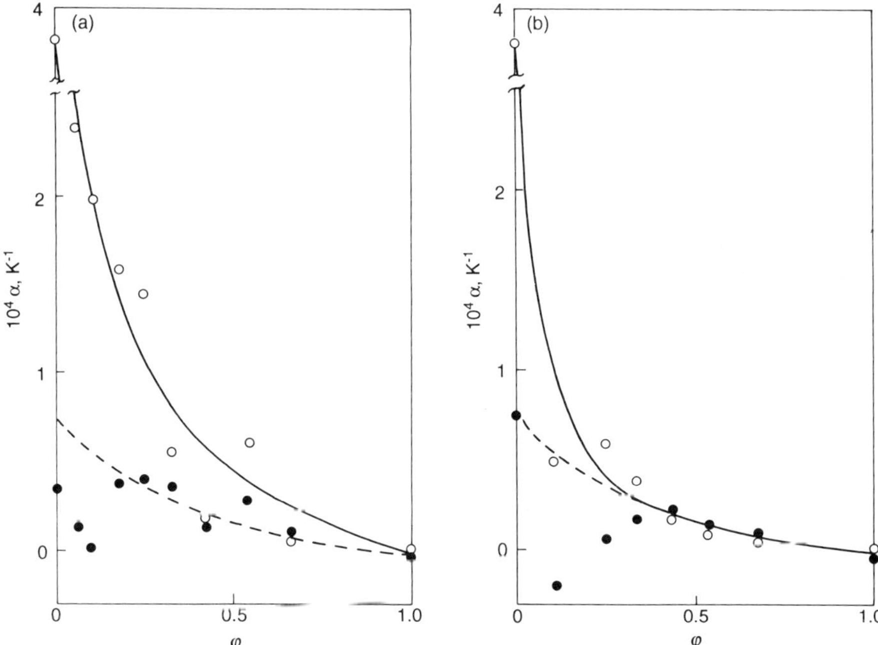

Fig. 5.14. Composition dependence of linear TEC for silica-filled polystyrenes PS-3.5 (a) and PS-10 (b) in the melt ($\bigcirc$) and glassy ($\bullet$) states

data for silica-filled, cross linked furfural-acetone resin [31]. Usually, the observed variation of φ^* is explained by the different abilities of fillers to form spacial structures [31, 66]; alternatively, φ^* may be defined [67] as a theoretical upper limit of filler content above which a MHM would become mechanically and thermodynamically unstable due to a severe loss of conformational entropy of macromolecules if the mean interparticle gap between filler particles $\langle L \rangle$ were to decrease below $\langle R_g \rangle$ [68, 69].

REFERENCES

1. Shermergor T. D. (1977) *Theory of Easticity of Microheterogeneous Media*, Nauka, Moscow (in Russian).
2. Hashin Z. (1966) Theory of mechanical behavior of heterogeneous media, in *Appl. Mech. Surveys*, Spartan Books, pp. 263–279.
3. Boley B. and Weiner J. (1964) *Theory of Temperature Stresses*, Mir, Moscow (Russian translation).
4. Nowacki W. (1975) *Elasticity Theory*, Mir, Moscow (Russian translation).
5. Rosen B. W. and Hashin Z. (1970) Effective thermal expansion coefficients and specific heats of composite materials, *Int. J. Eng. Sci.*, **8**, 157–163.

6. Levin V. M. (1967) On the thermal expansion coefficients of non-homogeneous materials, *Izv. Acad. Nauk USSR*, **1**, 88–93.

7. Cribb J. L. (1968) Shrinkage and thermal expansion of a two-phase material, *Nature (London)*, **220**, 576–581.

8. Shapery R. A. (1968) Thermal expansion coefficients for a mixture of two solids, *J. Comp. Mater.*, **2**, 380–386.

9. Steel T. R. (1968) Determination of the constitutive coefficients for a mixture of two solids, *Int. J. Solids Structure*, **4**, 1149–1153.

10. Kerner E. H. (1956) Elastic and thermoelastic properties of composite media, *Proc. Phys. Soc.*, **69**, 808–813.

11. Holliday L. and Robinson J. (1973) Review: the thermal expansion of composites based on polymers, *J. Mater. Sci.*, **8**, 301–311.

12. Voigt W. (1910) *Lehrbuch der Kristallphysik*, Teubeur, Berlin.

13. Reuss R. (1929) Berechnung der Fleissgrenze von Mischkristallen auf Grund der Plastizitaetsbedingung fuer Einkristall, *Z. Angew. Math. Mech.*, **9**, 41–49.

14. Hill R. (1952) The elastic behavior of a crystalline aggregate, *Proc. Phys. Soc.*, Ser. A, **65**, 349–362.

15. Krivoglaz M. A. and Cherevko A. S. (1959) On the elastic moduli of solid mixtures, *Fiz. Mekh. Mater.*, **8**, 161.

16. Kerner E. H. (1967) Elastic moduli of perfectly disordered composite materials, *Mech. Phys. Solids*, **15**, 319–324.

17. Budiansky B. (1965) On the elastic moduli of some heterogeneous materials, *Mech. Phys. Solids*, **13**, 223–226.

18. Hill R. (1965) A self-consistent mechanics of composite media, *Mech. Phys. Solids*, **13**, 213–225.

19. Hashin Z. and Shtrikman S. (1963) A variational approach to the theory of the elastic behavior of multi-phase materials, *Mech. Phys. Solids*, **11**, 127–134.

20. Nielsen L. E. (1974) *Mechanical Properties of Polymers and Composites*, Marcel Dekker, Inc., New York.

21. Beran M. J. (1965) *Statistical continuum theories*, *Trans. Soc. Rheol.*, **9**, 339–355.

22. Beran M. J. (1971) Application of statistical theories to heterogeneous materials, *Phys. Status Solidi*, Ser. A, **6**, 365–384.

23. Beran M. J. and Melay J. J. (1970) Mean field variations in a statistical sample of heterogeneous linear elastic solids, *Int. J. Solids Structure*, **6**, 1035–1054.

24. Lifshits I. M. and Rosenzweig L. N. (1946) On the theory of elastic properties of polycrystals, *Zhurn. Eksperim. Tekhn. Fiz.*, **16**, 967–973.

25. Fokin A. G. and Shermergor T. D. (1968) On the calculation of elastic moduli of non-homogeneous materials, *Mekh. Polimerov*, **4**, 58–64.

26. Khoroshun L. P. and Maslov B. P. (1980) *Computer Techniques for Calculation of the Physical Mechanical properties of Composites*, Naukova Dumka, Kiev (in Russian).

27. Novikov V. V. (1985) On the determination of effective elasticity moduli of non-homogeneous materials, *Prikl. Mekh. Tekhn. Fiz.*, **5**, 146–153.

28. Privalko V. P., Novikov V. V. and Yanovsky Yu. G. (1991) *Principles of Thermophysics and Rheophysics of Polymer Materials*, Naukova Dumka, Kiev (in Russian).

29. Lipatov Yu. S. (1977) *Physical Chemistry of Filled Polymers*, Khimia, Moscow (in Russian).

30. Privalko V. P., Besklubenko Yu. D. and Lipatov Yu. S. (1979) Low-temperature mechanical behavior of filled glassy polymers, *Komposits. Polim. Mater.*, **2**, 11–17.

31. Kandyrin L. B., Kuleznev V. N., Chernin E. I. *et al.* (1977) On the calculation of elastic moduli of filled polymers at high filler loadings, *Kolloid. Zhurn.*, **39**, 966–969.

32. Papanicolaou G. C. and Theocaris P. S. (1979) Thermal properties and volume

fraction of the boundary interphase in metal-filled epoxies, *Colloid. Polymer Sci.*, 257, 239–246.

33. Lipatov Yu. S. and Babich V. F. (1982) Features of thermomechanical behavior of simple models of composite materials with interfacial layers, *Mekh. Komposit. Mater.*, **2**, 255–232.

34. Babich V. F. and Lipatov Yu. S. (1982) On shift and resolubulity of relaxation maxima with change in properties of the boundary layer in composite materials, *J. Appl. Polymer Sci.*, **27**, 53–62.

35. Theocaris P. S. (1985) The mesophase and its influence on the mechanical properties of composites, *Adv. Polymer Sci.*, **66**, 149–187.

36. Chen X. B. and Wu X. S. (1992) The boundary layer and its influence on the mechanical behavior of composites, *Polymer*, **17**, 3639–3642.

37. Kantor Y. and Witten T. I. (1984) Mechanical stability of tenuous onjects, *J. Phys. Lett.*, **45**, 675–679.

38. Kantor Y. and Webman I. (1984) Elastic properties of random percolating systems, *Phys. Rev. Lett.*, **52**, 1891–1894.

39. Bergman D. J. (1985) Elastic moduli near percolation: universal ratio and critical exponent, *Phys. Rev.*, Ser. B, **31**, 1696–1698.

40. Webman I. and Kantor Y. (1984) The elasticity and vibrational modes of percolating networks and other fractal structures, in *Kinetics of Aggregation and Gelation*, ed. by F. Family and D. F. Landau, North-Holland, Amsterdam, pp. 133–136.

41. Dulnev G. N. and Novikov V. V. (1979) Conductivity of non-homogeneous systems, *Inzh. Fiz. Zhurn.*, **36**, 900–909.

42. Dulnev G. N. and Novikov V. V. (1983) Percolation theory and conductivity of non-homogeneous media. The basic model of a non-homogeneous medium, *Inzh. Fiz. Zhurn.*, **45**, 136–141.

43. Dzenis Yu. A. (1986) Influence of the aggregation of a stiff disperse filler on elastic properties of the polymer composite, *Mekh. Polimerov*, **1**, 14–22.

44. Young R. J. (1977) Failure of brittle polymers by slow crack growth, *J. mater. Sci.*, **12**, 684–692.

45. Hsich H. S. Y., Yanyo L. C. and Ambrose R. J. (1984) A hybrid model for mechanical spectra of filled and unfilled elastomers, *J. Appl. Polymer Sci.*, **29**, 413–425.

46. Boehme R. D. (1968) Aluminum reinforcement of some polyethylenes. *J. Appl. Polym. Sci.*, **12**, 1097–1108.

47. Besklubenko Yu. D. (1980) Thesis, Institute of Macromolecular Chemistry, Academy of Science of Ukraine, Kiev.

48. Kajiyama T., Yoshinaga T. and Takayanagi M. (1977) The effect of thermal stress on the thermal expansion coefficient and glass transition temperature of glass fiber–polymer composites, *J. Polymer Sci.: Polymer Phys. Ed.*, **15**, 1557–1568.

49. Papanicolaou G. C., Paipetis S. A. and Theocaris P. S. (1977) Thermal properties of metal-filled epoxies, *J. Appl. Polymer Sci.*, **21**, 689–701.

50. Raghava R. S. (1988) Thermal expansion of organic and inorganic matrix composites: a review of theoretical and experimental studies, *Polymer Composites*, **9**, 1–11.

51. De Pinhero M. F. F. and Rosenberg H. M. (1980) Thermal expansion of epoxy resin/ particle composites—a size effect, *J. Polymer Sci.: Polymer Phys. Ed.*, **18**, 219–226.

52. Skudra A. N. and Sbitnev O. V. (1982) Temperature dependence of thermal expansion behavior of reinforced plastics, *Mekh. Kompozit. Mater.*, **1**, 12–24.

53. Pukansky B., Tudos F., Kallo A. and Bodor G. (1989) Multiple morphology in polypropylene/ethylene-propylene diene terpolymer blends, *Polymer*, **30**, 1399–1406.

54. Pukansky B., Tudos F., Kallo A. and Bodor G. (1989) Effect of multiple morphology on the properties of polypropylene/ethylene-propylene–diene terpolymer blends, *Polymer*, **30**, 1407–1413.

55. Colarik J., Lednicky F., Pukansky B. and Pegoraro M. (1992) Blends of polycarbonate with poly(methyl methacrylate): miscibility, phase continuity, and interfacial adhesion, *Polymer Eng. Sci.*, **32**, 886–893.
56. Holden G., Bishop E. T. and Legge N. R. (1969) Thermoplastic elastomers, *J. Polymer Sci.*, Part C, **26**, 37–57.
57. Faucher J. A. (1974) The modulus of block copolymers, *J. Polymer Sci.: Polymer Phys. Ed.*, **12**, 2153–2155.
58. Nakamae K., Nishino T. and Xu A. (1992) Studies on mechanical properties of polymer composites by X-ray diffraction: 3. Mechanism of stress transmission in particulate epoxy composite by X-ray diffraction, *Polymer*, **33**, 2720–2724.
59. Xu A., Nishino T. and Nakamae K. (1992) Stress transmission in silica particulate epoxy composite by X-ray diffraction, *Polymer*, **33**, 5167–5172.
60. Privalko V. P., Stanislavsky V. B. and Titov G. V. (1988) Thermal expansion of highly filled polystyrenes, *Vysokomol. Soed.*, Ser. B, **30**, 540–542.
61. Privalko V. P., Besklubenko Yu. D., Lipatov Yu. S., Demchenko S. S. and Khmelenko G. I. (1977) Thermodynamics of filled polystyrene, *Vysokomol. Soed.*, Ser. A, **19**, 1744–1755.
62. Smith J. C. (1976) Experimental values for the elastic constants of a particulate-filled glassy polymer, *J. Res. Natl. Bur. Stand.*, Ser. A, **80**, 45–49.
63. Lipatov Yu. S., Privalko V. P., Demchenko S. S. and Titov G. V. (1985) Energy state of macromolecules in boundary layers of highly filled polystyrene, *Dokl. Acad. Nauk USSR*, **284**, 651–654.
64. Lipatov Yu. S., Titov G. V., Demchenko S. S. and Privalko V. P. (1987) Energetics of interfacial interactions in highly filled polystyrenes, *Vysokomol. Soed.*, Ser. A, **29**, 604–610.
65. Titov G. V. (1993) Thesis, Institute of Macromolecular Chemistry, Academy of Sciences of Ukraine, Kiev.
66. Bigg D. M. (1987) Mechanical properties of particulate filled polymers, *Polymer Composites*, **8**, 115–122.
67. Lipatov Yu. S. and Privalko V. P. (1984) On the criteria of the concept of highly filled polymer, *Vysokomol. Soed.*, Ser. B, **26**, 257–260.
68. Fleer G. J. and Scheutjens J. M. H. M. (1982) Adsorption of interacting oligomers and polymers at an interface, *Adv. Colloid Interface Sci.*, **16**, 341–359.
69. Skvortsov A. M. and Gorbunov A. A. (1986) Conformations of macromolecules in filled polymers, *Vysokomol. Soed.*, Ser. A, **28**, 1941–1946.

Appendix 1

A.1 ELEMENTS OF THE PROBABILITY THEORY

A.1.1 THE PARTITION FUNCTION AND THE PROBABILITY DENSITY

A numerical function, $\chi = \chi(\omega)$, is a random quantity if for any $x(-\infty < x < \infty)$ the ensemble of ω satisfying the condition $\chi(\omega) < x$ is a random event of probability unity, i.e.

$$P(-\infty < \chi(\omega) < \infty) = 1 \tag{A 1.1}$$

The probability of the event $(\chi < x)$ will be expressed as

$$P(\chi < x) = F(x) \tag{A 1.2}$$

which defines $F(x)$ as a partition function of the random quantity χ.

The partition pattern of a random quantity is perfectly continuous provided there is an integrable function $f(x)$ available on the straight line in a real space which satisfies the condition

$$F(x) = \int_{-\infty}^{x} f(z)\,\mathrm{d}z \tag{A.1.3}$$

where $f(z)$ is the partition density (or one-particle partition density).

The common partition function of a series of random quantities, $\chi_1(\omega)$, $\chi_2(\omega),\ldots,\chi_n(\omega)$, is the probability

$$P(\chi_1 < x_1, \chi_2 < x_2,\ldots,\chi_n < x_n) = F(x_1, x_2,\ldots,x_n) \tag{A.1.4}$$

where $F(x_1,x_2,\ldots,x_n)$ is the n-particle partition function. The function $f(z_1, z_2,\ldots,z_n)$, ensuring the validity of the following condition:

$$F(x_1,x_2,\ldots,x_n) = \underbrace{\int_{-\infty}^{x_1} \int_{-\infty}^{x_2} \cdots \int_{-\infty}^{x_n}}_{n} f(z_1,z_2,\ldots,z_n)\,\mathrm{d}z_1\,\mathrm{d}z_2 \cdots \mathrm{d}z_n \tag{A.1.5}$$

at any values of the array $x_1, x_2,\ldots,x_n$, is the n-particle probability density.

When the function $f(x_1, x_2,\ldots,x_n)$ is known, the one-particle partition density, $f(x_i)$, may be defined as

$$f(x_i) = \underbrace{\int_{-\infty}^{\infty} \cdots \int_{-\infty}^{\infty}}_{n-1} f(x_1, x_2,\ldots,x_n)\,\mathrm{d}x_1\,\mathrm{d}x_2 \cdots \mathrm{d}x_{i-1}\,\mathrm{d}x_{i+1} \cdots \mathrm{d}x_n \tag{A.1.6}$$

whereas the two-particle probability density is

$$f(x_i, x_j) = \underbrace{\int_{-\infty}^{\infty} \cdots \int_{-\infty}^{\infty}}_{n-2} f(x_1, x_2, \ldots, x_n)\, dx_1 \cdots dx_{i-1}\, dx_{i+1} \cdots dx_{j-1}\, dx_{j+1} \cdots dx_n \tag{A.1.7}$$

and so on.

For independent events, the following condition holds:

$$f(x_1, x_2, \ldots, x_n) = f(x_1)f(x_2)\cdots f(x_n) \tag{A.1.8}$$

A.1.2 MATHEMATICAL EXPECTATION, MOMENTS, CORRELATION OF RANDOM QUANTITIES

If χ is a random quantity, and $f(\chi)$ is its partition density, then the mathematical expectation of χ is defined as

$$M\chi = \int_{-\infty}^{\infty} \chi f(\chi)\, d\chi \tag{A.1.9}$$

The mathematical expectation of the function $\varphi(\chi)$ of random quantity χ is

$$M\varphi(\chi) = \int_{-\infty}^{\infty} \varphi(\chi)f(\chi)\, d\chi \tag{A.1.10}$$

If χ is a random quantity, its mathematical expectation $M\chi$ is also a random quantity. Given any integer n, the quantity $M\chi^n$ is the nth order moment of the random quantity n, and the mathematical expectation $M(\chi - M\chi)^n$ is the nth-order central moment of the random quantity χ:

$$M(\chi - M\chi)^n = \int_{-\infty}^{\infty} (\chi - M\chi)^n f(x)\, d\chi \tag{A.1.11}$$

An MHM with random distribution of components may be visualized as a medium with the material properties being the random functions of coordinates. In this case, the components of the N-dimensional random vector will be the functions of a material property, $C(r_i)$, at different points, i.e. $\chi_k = C(r_k)$.

The partition function density will be expressed as

$$f[C(r_i)] = f[C(r_1), C(r_2), \ldots, C(r_N)] \tag{A.1.12}$$

The multi-point distribution reduces to a single-point one at the condition $N = 1$.

The geometrical structure of the specimen will be specified by the indicative function $g_\alpha(r_i)$, which will be either unity (at the point r_i chosen in the space occupied by component α) or zero (if the point r_i is outside the space occupied by component α), i.e.

$$g_\alpha(r_i) = \begin{cases} 1 & (r_i \in V_\alpha) \\ 0 & (r_i \bar{\in} V_\alpha) \end{cases} \tag{A.1.13}$$

It is obvious that the function $g_\alpha(r_i)$ cannot be specified for any random MHM; however, it is possible to determine some of its moments. For example, let φ_α be the volume fraction of component α, i.e.

$$\varphi_\alpha = V^{-1} \int_{(V)} g_\alpha(r_i)\, dv = \langle g_\alpha(r_i) \rangle \qquad (A.1.14)$$

Stated otherwise, $\langle g_\alpha(r_i) \rangle$ is the probability that the random point r_i within the volume V belongs to component α, and $g_\alpha(r_i)$ is the corresponding one-point probability density. The angular brackets $\langle \cdots \rangle$ mean the averaging over the volume V; according to the ergodicity theorem, $M\chi = \langle \chi \rangle$.

Statistically, the function $\langle g_\alpha(r_i) \rangle$ provides little information on the structure of the MHM since it cannot specify the location of particles in space. Therefore, for a binary system of two components, α and β, we introduce the two-point moment

$$S_{\alpha\beta}(r_i, r_j) = \langle g_\alpha(r_i) g_\beta(r_i) \rangle \qquad (A.1.15)$$

Henceforth, only statistically homogeneous systems (i.e. those with coordinate-invariant statistical characteristics) will be considered. In this case, the one-point distributions are constant, and the two-point ones will depend on the difference between coordinates, i.e.

$$S_{\alpha\beta}(r_i') = \langle g_\alpha(r_i) g_\beta(r_i + r_i') \rangle \qquad (A.1.16)$$

The physical meaning of the latter function, $S_{\alpha\beta}(r_i')$, may be clarified by consideration of two limiting cases.

(i) If $r_i' \Rightarrow 0$, $g_\alpha(r_i)$ coincides with $g_\beta(r_i + r_i')$; therefore

$$\lim_{r_i' \Rightarrow 0} S(r_i') = \varphi_\alpha^2 \qquad (A.1.17)$$

(ii) If $r_i' \Rightarrow \infty$, we can predict the value of the function $g(r_i)$ at any point provided its value at one point is available, i.e.

$$\lim_{r_i' \Rightarrow \infty} S_{\alpha\beta}(r_i') = \langle g_\alpha(r_i) \rangle \langle g_\beta(r_i + r_i') \rangle = \varphi_\alpha \varphi_\beta \qquad (A.1.18)$$

It follows therefrom that $S_{\alpha\beta}(r_i')$ is, in fact, the measure of probability that the point r_i belongs to component α, and the other point, $r_i + r_i'$, belongs to component β.

The statistically homogeneous material may also be characterized by a two-point correlation function:

$$K_{\alpha\beta}(r_1, r_2) \rangle = \langle \psi_\alpha^0(r_1) \psi_\beta^0(r_2) \rangle = \langle \psi_\alpha(r_1) \psi_\beta(r_2) \rangle - \langle \psi_\alpha(r_1) \rangle \langle \psi_\beta(r_2) \rangle \qquad (A.1.19)$$

where $\psi^0(r_i) = \psi(r) - \langle \psi(r) \rangle$, and $\psi(r)$ is the random function of coordinates.

The correlation function of the indicative function, $g(r_i)$, will be expressed as

$$K_{\alpha\beta}(r_i') = \langle g_\alpha^0(r_i) g_\beta^0(r_i + r_i') \rangle = S_{\alpha\beta}(r_i') - \varphi_a \varphi_\beta \qquad (A.1.20)$$

The distance $l_c = |r'_i|$ over which $K_{\alpha\beta}(r'_i)$ decreases to about zero is called the characteristic correlation length (scale). It serves as one of the most important structural parameters of a MHM. For example, the dimensions of the BRE, l, are specified by the following condition:

$$l \gg l_c \qquad \text{(A.1.21)}$$

The above definitions may now be used to define the random space of the material tensor $C(r_i)$ as

$$C(r) = C_1 g_1(r) + C_2 g_2(r) \qquad \text{(A.1.22)}$$

and the one-point partition density of $C(r)$ as

$$f[C(r)] = \varphi_1 \delta[C(r) - C_1] + \varphi_2 \delta[C(r) - C_2] \qquad \text{(A.1.23)}$$

where $\delta(x)$ is the delta-function.

The corresponding mathematical expectation is

$$M[C(r)] = \int_{-\infty}^{\infty} C(r) f[C(r)] \, dC = \varphi_1 C_1 + \varphi_2 C_2 = \langle C \rangle \qquad \text{(A.1.24)}$$

The nth-order central moment (i.e. dispersion) of the material tensor $C(r)$ will be, finally

$$M[C(r) - \langle C \rangle]^2 = \int_{-\infty}^{\infty} [C(r) - \langle C \rangle]^2 f[C(r)] \, dC = \varphi_1 \varphi_2 (C_1 - C_2)^2 \qquad \text{(A.1.25)}$$

It follows from eq. (A.1.25) that the deviation of the local property $C(r)$ from its mean value will be larger, the larger is the difference between the intrinsic properties of components, while this deviation will be lower, the lower is the volume fraction of either components.

Appendix 2

A.2 MODELING OF THE EFFECTIVE PROPERTIES OF A MHM WITH ANISOTROPIC INCLUSIONS

A.2.1 HEAT CONDUCTIVITY

The simple model of a small, isotropic cube (or rather, parallelepiped) embedded in a large, isotropic parallelepiped which was used in Chapter 4 to analyze the effective conductivity of a MHM will now be extended to the more general case of anisotropic components. Consider a MHM with isolated, anisometric (i.e. with one dimension much longer than the two others) and orthotropic (i.e. with different properties at each of three dimensions) inclusions. The inclusions, of irregular shape, arc replaced by rectangular papallelepipeds of equal volume keeping the ratio of the three principal dimensions unchanged. The symmetry axes of both inclusions and the matrix are assumed coincident with the co-ordinate axes (otherwise, this condition may be achieved by transformation of the coordinate system).

Let each inclusion be surrounded by the surface, S_n, which belongs to the matrix and encompasses the volume V_n such that $V_n/V = \varphi$ (where V is the total volume). The volume element V_n is assumed to be of rectangular shape and to be any size from macroscopic to vanishingly small. Special analysis has shown that replacement of elongated inclusions of irregular shape by parallelepipeds of the same volume does not significantly change the effective property of a MHM.

The unit cell of the relevant structural model and its conditonal sectioning are shown in Fig. A.2.1. The effective heat conductivity of the model MHM will be evaluated as the final result of SSA calculations, i.e. the conductivities of portions I, (I + II) and (I + II + III) will be determined in steps, the relevant structure at each step being considered as quasi-homogeneous.

The effective heat conductivity, λ, is a second-rank tensor, i.e.

$$\lambda = \begin{vmatrix} \lambda_{11} & 0 & 0 \\ 0 & \lambda_{22} & 0 \\ 0 & 0 & \lambda_{33} \end{vmatrix}$$

where λ_{kk} is the effective heat conductivity along the Ox_k axis.

Let us define the lower and upper bounds of the effective heat conductivity (i.e., $\lambda_{kk}^{(L)}$ and $\lambda_{kk}^{(U)}$, respectively) of the unit cell (Fig. A.2.1(a)). The lower bound

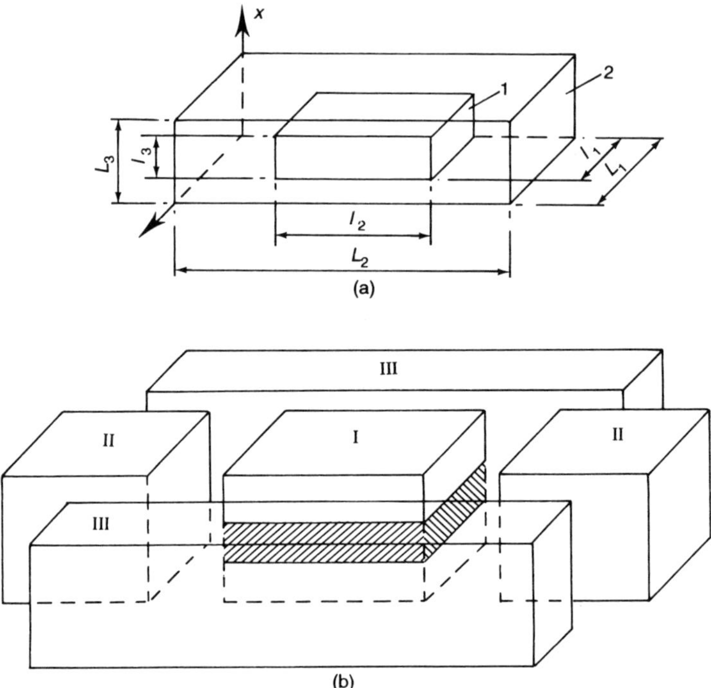

Fig. A.2.1. Model of a MHN with isolated anisotropic inclusions; (a) unit cell (1—inclusion, 2—matrix). (b) conditional sectioning of the unit cell

may be defined as

$$\lambda_{kk}^{(L)} = \lambda_{kk}^{(1,L)}\langle l_i \rangle \langle l_j \rangle + \lambda_{kk}^{(2,L)}[(\langle l_{cj} \rangle + \langle l_j \rangle)(\langle l_{ci} \rangle + \langle l_i \rangle) - \langle l_i \rangle \langle l_j \rangle]$$

$$+ \lambda_{kk}^{(2)}[1 - (\langle l_{ci} \rangle + \langle l_i \rangle)(\langle l_{cj} \rangle + \langle l_j \rangle)] \tag{A.2.1}$$

where

$$\lambda_{kk}^{(1,L)} = [\langle l_{ck} \rangle / \lambda_{kk}^{(c)} + \langle l_k \rangle / \lambda_{kk}^{(1)} + (1 - \langle l_k \rangle - \langle l_{ck} \rangle)/\lambda_{kk}^{(2)}]^{-1}$$

$$\lambda_{kk}^{(2,L)} = [(\langle l_{ck} \rangle + \langle l_k \rangle)/\lambda_{kk}^{(c)} + (1 - \langle l_{ck} \rangle - \langle l_k \rangle)/\lambda_{kk}^{(2)}]^{-1}$$

$\langle l_i \rangle, \langle l_j \rangle$ and $\langle l_k \rangle$ are the relative dimensions of the inclusions along the axes Ox_i, Ox_j and Ox_k, respectively, $\langle l_{ci} \rangle, \langle l_{cj} \rangle$ and $\langle l_{ck} \rangle$ are the corresponding relative thicknesses of the BI, and $\lambda_{kk}^{(c)}, \lambda_{kk}^{(1)}$ and $\lambda_{kk}^{(2)}$ are the heat conductivities of the BI, inclusion and the matrix material along the Ox_k axis.

The upper bound of the heat conductivity along the Ox_k axis is determined by conditional sectioning of the unit cell by equipotential planes drawn normal to the heat flux, i.e.

$$\lambda_{kk}^{(U)} = \{\langle l_k \rangle / \lambda_{kk}^{(1,U)} + \langle l_{ck} \rangle / \lambda_{kk}^{(2,U)} + [1 - (\langle l_k \rangle + \langle l_{ck} \rangle)]/\lambda_{kk}^{(2)}\}^{-1} \tag{A.2.2}$$

where

$$\lambda_{kk}^{(1,\text{U})} = \lambda_{kk}^{(1)}\langle l_j\rangle\langle l_j\rangle + \lambda_{kk}^{(c)}[(\langle l_{cj}\rangle + \langle l_j\rangle)(\langle l_{ci}\rangle + \langle l_i\rangle) - \langle l_i\rangle\langle l_j\rangle]$$
$$+ \lambda_{kk}^{(2)}[1 - (\langle l_{cj}\rangle + \langle l_j\rangle)(\langle l_{ci}\rangle + \langle l_i\rangle)]$$

$$\lambda_{kk}^{(2,\text{U})} = \lambda_{kk}^{(c)}(\langle l_{ci}\rangle + \langle l_i\rangle)(\langle l_{cj}\rangle + \langle l_j\rangle) + \lambda_{kk}^{(2)}[1 - (\langle l_{cj}\rangle + \langle l_j\rangle)(\langle l_{ci}\rangle + \langle l_i\rangle)]$$

If the main axes of the tensor λ do not coincide with the axes of the laboratory system of coordinates, we can use the tensors of rotation by angle, φ, and determine λ' in the laboratory system of coordinates.

As an example, consider the transforation of the heat conductivity tensor λ by rotation of the coordinate system around the Ox_2 axis by the angle φ. Expressing the tensor of rotation as

$$\Psi = \begin{vmatrix} \cos\varphi & 0 & \sin\varphi \\ 0 & 1 & 0 \\ -\sin\varphi & 0 & \cos\varphi \end{vmatrix}$$

we obtain

$$\lambda' = \Psi^{-1}\lambda\Psi = \begin{vmatrix} \lambda_{11}\cos^2\varphi + \lambda_{33}\sin^2\varphi & 0 & (\lambda_{33}-\lambda_{11})\sin\varphi\cos\varphi \\ 0 & \lambda & 0 \\ (\lambda_{33}-\lambda_{11})\sin\varphi\cos\varphi & 0 & \lambda_{11}\sin^2\varphi + \lambda_{33}\cos^2\varphi \end{vmatrix}$$

Thus, the heat conductivity along the Ox_1 axis will be

$$\lambda'_{11} = \lambda_{11}\cos^2\varphi + \lambda_{33}\sin^2\varphi$$

The above structural model may be generalized to account for the unit cells of the 'cube-in-cube' type (i.e. $\langle l_1\rangle = \langle l_2\rangle = \langle l_3\rangle = \varphi^{1/3}$), and/or the 'parallelepiped-in-parallelepiped' type (i.e. $\langle l_1\rangle = 1$, $\langle l_2\rangle = \langle l_3\rangle = \varphi^{1/2}$).

A.2.2 THERMOELASTIC PROPERTIES

Elasticity Moduli and Poisson's Coefficients

Consider the MHM of a layered structure with orthotropic properties of each layer, the symmetry axes of the layers coincident with the coordinate axes (Fig. A.2.2). For each component (i.e. layer) the following relationships are valid:

$$\langle\varepsilon_{11}^{(i)}\rangle = \langle\sigma_{11}^{(i)}\rangle/E_{11}^{(i)} - v_{12}^{(i)}\langle\sigma_{22}^{(i)}\rangle/E_{22}^{(i)} - v_{13}^{(i)}\langle\sigma_{33}^{(i)}\rangle/E_{33}^{(i)} + \alpha_{11}^{(i)}\Delta t \tag{A.2.3a}$$

$$\langle\varepsilon_{22}^{(i)}\rangle = -v_{21}^{(i)}\langle\sigma_{11}^{(i)}\rangle/E_{11}^{(i)} + \langle\sigma_{22}^{(i)}\rangle/E_{22}^{(i)} - v_{23}^{(i)}\langle\sigma_{33}^{(i)}\rangle/E_{33}^{(i)} + \alpha_{22}^{(i)}\Delta t \tag{A.2.3b}$$

$$\langle\varepsilon_{33}^{(i)}\rangle = -v_{31}^{(i)}\langle\sigma_{11}^{(i)}\rangle/E_{11}^{(i)} - v_{32}^{(i)}\langle\sigma_{22}^{(i)}\rangle/E_{22}^{(i)} + \langle\sigma_{33}^{(i)}\rangle/E_{33}^{(i)} + \alpha_{33}^{(i)}\Delta t \tag{A.2.3c}$$

$$\langle\varepsilon_{31}^{(i)}\rangle = \langle\sigma_{31}^{(i)}\rangle/G_{31}^{(i)}, \qquad \langle\varepsilon_{23}^{(i)}\rangle = \langle\sigma_{23}^{(i)}\rangle/G_{32}^{(i)}$$

$$\langle\varepsilon_{12}^{(i)}\rangle = \langle\sigma_{11}^{(i)}\rangle/G_{12}^{(i)} \tag{A.2.3d}$$

where the subscripts $i = 1, 2$ refer to the corresponding component.

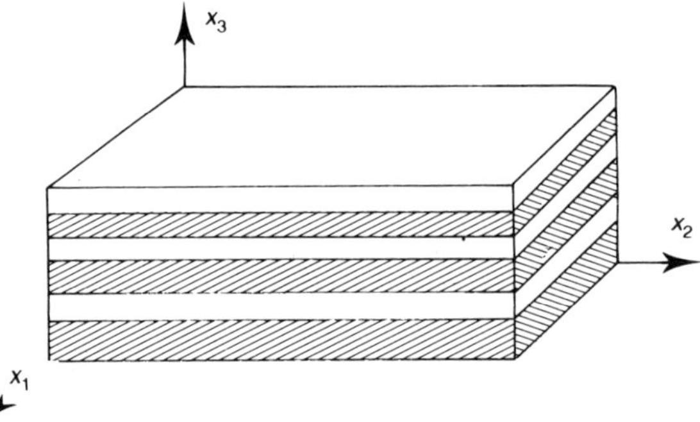

Fig. A.2.2. Model of a MHM of a layered structure

Equations (A.2.3) are complemented by the compatibility conditions

$$\langle \sigma_{33} \rangle = \langle \sigma_{33}^{(i)} \rangle, \qquad \langle \sigma_{kl} \rangle = \varphi \langle \sigma_{kl}^{(1)} \rangle + (1 - \varphi)\langle \sigma_{kl}^{(2)} \rangle \qquad k, l \neq 3 \qquad \text{(A.2.4a)}$$

$$\langle \varepsilon_{33} \rangle = \varphi \langle \varepsilon_{33}^{(1)} \rangle + (1 - \varphi)\langle \varepsilon_{33}^{(2)} \rangle, \qquad \langle \varepsilon_{ij} \rangle = \langle \varepsilon_{ij}^{(1)} \rangle = \langle \varepsilon_{ij}^{(2)} \rangle \qquad i, j = 1, 2$$
$$\text{(A.2.4b)}$$

where φ, as usual, is the volume fraction of component 1.

The longitudinal and transverse values of EME will now be evaluated assuming $\Delta t = 0$.

(i) Longitudinal direction (i.e. along the Ox_1 axis).

The effective Young's modulus along the Ox_1 axis (Fig. A.2.2) is determined assuming $\langle \sigma_{11} \rangle \neq 0$, $\langle \sigma_{kl} \rangle = 0$, $k, l \neq 1$, and $\langle \sigma_{33} \rangle = 0$, making use of the following relationships derived from eqs. (A.2.3) and (A.2.4):

$$M_{11}\langle \sigma_{11}^{(1)} \rangle - S_{12}\langle \sigma_{22}^{(1)} \rangle = \langle \sigma_{11} \rangle / E_{11}^{(2)} \qquad \text{(A.2.5a)}$$

$$-S_{21}\langle \sigma_{11}^{(1)} \rangle + M_{22}\langle \sigma_{22}^{(1)} \rangle = -v_{12}^{(2)}\langle \sigma_{11} \rangle / E_{11}^{(2)} \qquad \text{(A.2.5b)}$$

from which we derive

$$\langle \sigma_{11}^{(1)} \rangle = \frac{\langle \sigma_{11} \rangle [M_{22} - v_{21}^{(2)} S_{12}]}{E_{11}^{(2)}[M_{11}M_{22} - S_{12}S_{21}]} \qquad \text{(A.2.6a)}$$

$$\langle \sigma_{22}^{(1)} \rangle = \frac{\langle \sigma_{11} \rangle [S_{21} - v_{12}^{(2)} M_{11}]}{E_{11}^{(2)}[M_{11}M_{22} - S_{12}S_{21}]} \qquad \text{(A.2.6b)}$$

where

$$M_{11} = \varphi / E_{11}^{(2)} + (1 - \varphi) / E_{11}^{(1)}, \qquad M_{22} = \varphi / E_{22}^{(2)} + (1 - \varphi) / E_{22}^{(1)}$$

$$S_{12} = \varphi v_{12}^{(2)} / E_{22}^{(2)} + (1 - \varphi) v_{12}^{(1)} / E_{22}^{(1)}$$

$$S_{21} = \varphi v_{21}^{(2)} / E_{11}^{(2)} + (1 - \varphi) v_{21}^{(1)} / E_{11}^{(1)}$$

Combining $\langle \varepsilon_{11} \rangle = \langle \varepsilon_{11}^{(1)} \rangle$ from eq. (A.2.4) with eq. (A.2.3) to obtain

$$\langle \varepsilon_{11} \rangle = \langle \sigma_{11}^{(1)} \rangle / E_{11}^{(1)} - v_{12}^{(1)}\langle \sigma_{22}^{(1)} \rangle / E_{22}^{(1)} \qquad \text{(A.2.7)}$$

we derive from eq. (A.2.6):

$$\langle \varepsilon_{11} \rangle = \langle \sigma_{11} \rangle \left\{ \left[\frac{M_{22}}{E_{11}^{(2)}} - \frac{v_{21}S_{12}}{E_{11}^{(2)}} \right] \bigg/ E_{11}^{(1)} \right.$$

$$\left. - \frac{v_{21}^{(1)}}{E_{22}^{(1)}} \left[\frac{S_{21}}{E_{11}^{(2)}} - \frac{M_{11}v_{21}^{(2)}}{E_{11}^{(2)}} \right] \right\} \bigg/ [M_{11}M_{22} - S_{12}S_{21}] \quad (A.2.8)$$

Substituting these results into the standard definition of the efffective Young's modulus of a layered structure along the strate (i.e. the Ox_1 axis), $E_{11} = \langle \sigma_{11} \rangle / \langle \varepsilon_{11} \rangle$, we derive

$$E_{11} = \frac{M_{11}M_{22} - S_{12}^2}{\left[\dfrac{M_{22}}{E_{11}^{(2)}} - \dfrac{S_{12}v_{21}^{(2)}}{E_{11}^{(2)}} \right] \bigg/ E_{11}^{(1)} - \dfrac{v_{12}^{(1)}}{E_{22}^{(1)}} \left[\dfrac{S_{21}}{E_{11}^{(2)}} - \dfrac{M_{11}v_{21}^{(2)}}{E_{11}^{(2)}} \right]} \quad (A.2.9)$$

Making use of the standard definitions of the Poisson's coefficients, v_{21} and v_{31}:

$$\langle \varepsilon_{22} \rangle = -v_{21} \langle \sigma_{11} \rangle / E_{11}, \qquad \langle \varepsilon_{33} \rangle = -v_{31} \langle \sigma_{11} \rangle / E_{11}$$

we derive from eqs. (A.2.3) and (A.2.6):

$$\frac{v_{21}}{E_{11}} = \left[\frac{v_{21}^{(1)}}{E_{11}^{(1)}} \left(\frac{M_{22}}{E_{11}^{(2)}} - \frac{v_{21}^{(2)}S_{12}}{E_{11}^{(2)}} \right) - \left(\frac{S_{21}}{E_{11}^{(1)}} - \frac{M_{11}v_{21}^{(2)}}{E_{11}^{(2)}} \right) \right] \bigg/ E_{22}^{(1)}(M_{11}M_{22} - S_{12}^2)$$

$$(A.2.10a)$$

$$\frac{v_{21}}{R_{11}} = \frac{v_{31}^{(2)}}{E_{11}^{(2)}} + \varphi \left[P_{32} \left(\frac{S_{12}}{E_{11}^{(2)}} - \frac{v_{21}^{(2)}M_{11}}{E_{11}^{(2)}} \right) + P_{31} \left(\frac{M_{22}}{E_{11}^{(2)}} - \frac{v_{21}^{(2)}S_{12}}{E_{11}^{(2)}} \right) \right] \bigg/ (M_{11}M_{22} - S_{12}^2)$$

$$(A.2.10b)$$

(ii) Transversal direction (i.e. along the Ox_3 axis).

The Young's modulus will be derived assuming $\langle \sigma_{33} \rangle \neq 0$, $\langle \sigma_{11} \rangle = 0$ and $\langle \sigma_{22} \rangle = 0$. Making use of eqs. (A.2.3) and (A.2.4), we obtain

$$M_{11}\langle \sigma_{11}^{(1)} \rangle - S_{12}\langle \sigma_{22}^{(1)} \rangle = (1 - \varphi)P_{13}\langle \sigma_{33} \rangle \quad (A.2.11a)$$

$$-S_{12}\langle \sigma_{11}^{(1)} \rangle + M_{22}\langle \sigma_{22}^{(1)} \rangle = (1 - \varphi)P_{23}\langle \sigma_{33} \rangle \quad (A.2.11b)$$

where

$$P_{13} = v_{13}^{(1)}/E_{33}^{(1)} - v_{13}^{(2)}/E_{33}^{(2)}$$

$$P_{23} = v_{23}^{(1)}/E_{33}^{(1)} - v_{23}^{(2)}/E_{33}^{(2)}$$

Equations (A.2.11) yield

$$\langle \sigma_{11}^{(1)} \rangle = (1 - \varphi)\langle \sigma_{33} \rangle (P_{13}M_{22} + P_{23}S_{12})/(M_{11}M_{22} - S_{12}S_{21}) \quad (A.2.12a)$$

$$\langle \sigma_{22}^{(1)} \rangle = (1 - \varphi)\langle \sigma_{33} \rangle (P_{23}M_{11} + P_{13}S_{21})/(M_{11}M_{22} - S_{12}S_{21}) \quad (A.2.12b)$$

Combining $\langle \varepsilon_{33} \rangle = \varphi \langle \varepsilon_{33}^{(1)} \rangle + (1 - \varphi) \langle \varepsilon_{33}^{(2)} \rangle$ from eq. (A.2.4) with eqs. (A.2.3) and (A.2.12) we derive

$$\langle \varepsilon_{33} \rangle = \langle \sigma_{33} \rangle \left[\frac{1}{\langle E_{33} \rangle} - \left(\frac{P_{31}(P_{13}M_{22} + P_{23}S_{12})}{M_{11}M_{22} - S_{12}S_{21}} \right. \right. \\ \left. \left. - \frac{P_{31}(P_{23}M_{11} + P_{13}S_{21})}{M_{11}M_{22} - S_{12}S_{21}} \right) \varphi(1 - \varphi) \right] \quad \text{(A.2.13a)}$$

where

$$P_{31} = v_{31}^{(1)}/E_{11}^{(1)} - v_{31}^{(2)}/E_{11}^{(2)}$$
$$P_{32} = v_{32}^{(1)}/E_{22}^{(1)} - v_{32}^{(2)}/E_{22}^{(2)}$$

and the validity of the equality $v_{ij}/E_{jj} = v_{ji}/E_{ii}$ was assumed.

Having rewritten eq. (A.2.13) as

$$\langle \varepsilon_{33} \rangle = \langle \sigma_{33} \rangle \left[\frac{1}{\langle E_{33} \rangle} - \frac{P_{13}^2 M_{22} + P_{32}^2 M_{11} + 2P_{13}P_{23}S_{12}}{M_{11}M_{22} - S_{12}^2} \varphi(1 - \varphi) \right] \quad \text{(A.2.14)}$$

we can define the transversal Young's modulus as

$$\langle E_{33} \rangle = \left[\frac{1}{\langle E_{33} \rangle} - \frac{P_{13}^2 M_{22} + P_{32}^2 M_{11} + 2P_{13}P_{23}S_{12}}{M_{11}M_{22} - S_{12}^2} \varphi(1 - \varphi) \right]^{-1} \quad \text{(A.2.15)}$$

The corresponding Poisson's coefficients, v_{23} and v_{13}, can now be determined from eq. (A.2.12) making use of the standard relationships $\langle \varepsilon_{11} \rangle = - v_{13} \langle \sigma_{33} \rangle / E_{33}$ and $\langle \varepsilon_{22} \rangle = - v_{23} \langle \sigma_{33} \rangle / E_{33}$, i.e.

$$\frac{v_{23}}{E_{33}} = \frac{v_{23}^{(1)}}{E_{33}^{(1)}} + \frac{(1 - \varphi)}{M_{11}M_{22} - S_{12}^2} \left[\frac{v_{21}^{(1)}(P_{13}M_{22} + P_{23}S_{12})}{E_{11}^{(1)}} - \frac{(P_{23}M_{11} + P_{13}S_{12})}{E_{22}^{(1)}} \right]$$

$$\text{(A.2.16a)}$$

$$\frac{v_{13}}{E_{33}} = \frac{v_{13}^{(1)}}{E_{33}^{(1)}} + \frac{(1 - \varphi)}{M_{11}M_{22} - S_{12}^2} \left[\frac{v_{12}^{(1)}(P_{23}M_{11} + P_{13}S_{21})}{E_{22}^{(1)}} - \frac{(P_{13}M_{22} + P_{23}S_{12})}{E_{11}^{(1)}} \right]$$

$$\text{(A.2.16b)}$$

The shear moduli G_{ik} will be determined assuming $\langle \sigma_{ij} \rangle \neq 0$, $i \neq j$, $\langle \sigma_{ll} \rangle = 0$. Making use of eqs. (A.2.3) and (A.2.4), we obtain for the longitudinal shear modulus:

$$G_{kl} = \varphi G_{kl}^{(1)} + (1 - \varphi)G_{kl}^{(2)} \quad \text{(i.e. } G_{kl} = \langle G_{kl} \rangle) \quad \text{(A.2.17a)}$$

and for the transversal shear modulus:

$$G_{kl} = [\varphi/G_{kl}^{(1)} + (1 - \varphi)/G_{kl}^{(2)}]^{-1} \quad \text{(i.e. } G_{kl} = \langle 1/G_{kl} \rangle^{-1} \quad \text{(A.2.17b)}$$

Thermal Expansion Coefficient

Application of the condition $\langle\sigma_{kl}\rangle = 0$, to the standard definition $\varepsilon_{ll} = \alpha_{ll}\Delta t$, taking account of eqs. (A.2.3) and (A.2.4), yields

$$M_{11}\langle\sigma_{11}^{(1)}\rangle - S_{12}\langle\sigma_{22}^{(1)}\rangle = (\alpha_{11}^{(2)} - \alpha_{11}^{(1)})(1 - \varphi)\Delta t \qquad (A.2.18a)$$

$$-S_{21}\langle\sigma_{11}^{(1)}\rangle + M_{22}\langle\sigma_{22}^{(1)}\rangle = (\alpha_{22}^{(2)} - \alpha_{11}^{(1)})(1 - \varphi)\Delta t \qquad (A.2.18b)$$

Solving eqs. (A.2.18) for $\langle\sigma_{ii}^{(1)}\rangle$ we obtain

$$\langle\sigma_{11}^{(1)}\rangle = (1 - \varphi)\Delta t \frac{(\alpha_{11}^{(2)} - \alpha_{11}^{(1)})M_{22} + (\alpha_{22}^{(2)} - \alpha_{22}^{(1)})S_{12}}{M_{11}M_{22} - S_{12}^2} \qquad (A.2.19a)$$

$$\langle\sigma_{22}^{(1)}\rangle = (1 - \varphi)\Delta t \frac{(\alpha_{22}^{(2)} - \alpha_{22}^{(1)})M_{11} + (\alpha_{22}^{(2)} - \alpha_{22}^{(1)})S_{12}}{M_{11}M_{22} - S_{12}^2} \qquad (A.2.19b)$$

Using eqs. (A.2.19) we can derive for the longitudinal TEC (i.e. along the Ox_1 axis)

$$\alpha_{11} = \alpha_{11}^{(1)} + \frac{1 - \varphi}{M_{11}M_{22} - S_{12}^2}\left(\frac{D_{11}}{E_{11}^{(1)}} - \frac{\nu_{12}^{(1)}D_{22}}{E_{22}^{(2)}}\right) \qquad (A.2.20)$$

where

$$D_{11} = (\alpha_{11}^{(2)} - \alpha_{11}^{(1)})M_{22} + (\alpha_{22}^{(2)} - \alpha_{22}^{(1)})S_{12}$$

$$D_{22} = (\alpha_{22}^{(2)} - \alpha_{22}^{(1)})M_{11} + (\alpha_{11}^{(2)} - \alpha_{11}^{(1)})S_{12}$$

In a similar fashion, the transversal TEC (i.e. along the Ox_3 axis) may be calculated as

$$\alpha_{33} = \varphi\alpha_{33}^{(1)} + (1 - \varphi)\alpha_{33}^{(2)} - \varphi(1 - \varphi)\frac{P_{31}D_{11} + P_{32}D_{22}}{M_{11}M_{22} + S_{12}^2} \qquad (A.2.21)$$

Index